NATO ASI Series

Advanced Science Institutes Series

A series presenting the results of activities sponsored by the NATO Science Committee, which aims at the dissemination of advanced scientific and technological knowledge, with a view to strengthening links between scientific communities.

The Series is published by an international board of publishers in conjunction with the NATO Scientific Affairs Division

A Life Sciences B Physics	Plenum Publishing Corporation London and New York
C Mathematical and Physical Sciences D Behavioural and Social Sciences E Applied Sciences	Kluwer Academic Publishers Dordrecht, Boston and London
F Computer and Systems Sciences G Ecological Sciences H Cell Biology	Springer-Verlag Berlin Heidelberg New York London Paris Tokyo Hong Kong

Series H: Cell Biology Vol. 39

The ASI Series Books Published as a Result of Activities of the Special Programme on CELL TO CELL SIGNALS IN PLANTS AND ANIMALS

This book contains the proceedings of a NATO Advanched Research Workshop held within the activities of the NATO Special Programme on Cell to Cell Signals in Plants and Animals, running from 1984 to 1989 under the auspices of the NATO Science Committee.

The books published as a result of the activities of the Special Programme are:

Vol. 1: **Biology and Molecular Biology of Plant-Pathogen Interactions.** Edited by J.A. Baily. 1986.
Vol. 2: **Glial-Neuronal Communication in Development and Regeneration.** Edited by H.H. Althaus and W. Seifert. 1987.
Vol. 3: **Nicotinic Acetylcholine Receptor: Structure and Function.** Edited by A. Maelicke. 1986.
Vol. 4: **Recognition in Microbe-Plant Symbiotic and Pathogenic Interactions.** Edited by B. Lugtenberg. 1986.
Vol. 5: **Mesenchymal-Epithelial Interactions in Neural Development.** Edited by J.R. Wolff, J. Sievers, and M. Berry. 1987.
Vol. 6: **Molecular Mechanisms of Desensitization to Signal Molecules.** Edited by T.M. Konjin, P.J.M. Van Haastert, H. Van der Starre, H. Van der Wel, and M.D. Houslay. 1987.
Vol. 7: **Gangliosides and Modulation of Neuronal Functions.** Edited by H. Rahmann. 1987.
Vol. 9: **Modification of Cell to Cell Signals During Normal and Pathological Aging.** Edited By S. Govoni and F. Battaini. 1987.
Vol. 10: **Plant Hormone Receptors.** Edited by D. Klämbt. 1987.
Vol. 11: **Host-Parasite Cellular and Molecular Interactions in Protozoal Infections.** Edited by K.-P. Chang and D. Snary. 1987.
Vol. 12: **The Cell Surface in Signal Transduction.** Edited by E. Wagner, H. Greppin, and B. Millet. 1987.
Vol. 19: **Modulation of Synaptic Transmission and Plasticity in Nervous Systems.** Edited by G. Hertting and H.-C. Spatz. 1988.
Vol. 20: **Amino Acid Availability and Brain Function in Health and Disease.** Edited by G. Huether. 1988.
Vol. 21: **Cellular and Molecular Basis of Synaptic Transmission.** Edited by H. Zimmermann. 1988.
Vol. 23: **The Semiotics of Cellular Communication in the Immune System.** Edited by E.E. Sercarz, F. Celada, N.A. Mitchison, and T. Tada. 1988.
Vol. 24: **Bacteria, Complement and the Phagocytic Cell.** Edited by F.C. Cabello and C. Pruzzo. 1988.
Vol. 25: **Nicotinic Acetylcholine Receptors in the Nervous System.** Edited by F. Celementi, C. Gotti, and E. Sher. 1988.
Vol. 26: **Cell to Cell Signals in Mammalian Development.** Edited by S.W. de Laat, J.G. Bluemink, and C.L. Mummery. 1989.
Vol. 27: **Phytotoxins and Plant Pathogenesis.** Edited by A. Graniti, R.D. Durbin, and A. Ballio. 1989.
Vol. 31: **Neurobiology of the Inner Retina.** Edited by R. Weiler and N.N. Osborne. 1989.
Vol. 32: **Molecular Biology of Neuroreceptors and Ion Channels.** Edited by A. Maelicke. 1989.
Vol. 33: **Regulatory Mechanisms of Neuron to Vessel Communication in the Brain.** Edited by F. Battaini, S. Govoni, M.S. Magnoni, and M. Trabucchi. 1989.
Vol. 35: **Cell Separation in Plants: Physiology, Biochemistry and Molecular Biology.** Edited by D.J. Osborne and M.B. Jackson. 1989.
Vol. 36: **Signal Molecules in Plants and Plant-Microbe Interactions.** Edited by B.J.J. Lugtenberg. 1989.
Vol. 39: **Chemosensory Information Processing.** Edited by D. Schild. 1990.

Chemosensory Information Processing

Edited by

Detlev Schild

Physiologisches Institut, Universität Göttingen,
Humboldtallee 23, 3400 Göttingen, FRG

Springer-Verlag Berlin Heidelberg New York
London Paris Tokyo Hong Kong
Published in cooperation with NATO Scientific Affairs Division

Proceedings of the NATO Advanced Research Workshop on Information Processing of Chemical Sensory Stimuli in Biological and Artificial Systems held in Göttingen, FRG, July 23–26, 1989

ISBN-13: 978-3-642-75129-5 e-ISBN-13: 978-3-642-75127-1
DOI: 10.1007/978-3-642-75127-1

Softcover reprint of the hardcover 1st edition 1990

2131/3140-543210 – Printed on acid-free-paper

Preface

In July 1989 a symposium was held at the Physiology Department of the Georg August University, Göttingen, on the physiological, biophysical, biochemical, and technical principles of the coding of chemical substances both in nervous systems and artificial devices. This book is the collection of the papers presented at that meeting. Biological and artificial systems for odor coding both have in common that the stimulus selectivity of the receptor cells (sensors) is usually very poor, and the mechanisms which determine selectivity and sensitivity are largely unknown. However, a poor selectivity allows the coding of an enormous number of stimuli by combinations of receptor activities. In the field of chemosensory information coding there are thus two major problems: the function of the receptors and the network that processes and evaluates the primary information of the sensors. Accordingly, this volume has three parts: sensors, the network following the sensors, and the coding in this network.

The expert secretarial assistance of M. Holtmann in preparing the camera-ready manuscript is gratefully acknowledged.

Göttingen, August 1989 D. Schild

CONTENTS

RESPONSE OF OLFACTORY RECEPTOR CELLS, ISOLATED AND *IN SITU*, TO LOW CONCENTRATIONS OF ODORANTS

Stephan Frings and Bernd Lindemann

Department of Physiology, Universität des Saarlandes
D - 6650 Homburg/Saar
West Germany

To study vertebrate olfactory receptor neurons by patch clamp methods, the cells are often dissociated from the mucosa and kept in isolation during stimulation with odorants dissolved in the bathing solution (1,3,4,5,9,10,11,13,15). However, disappointingly large concentrations of odorants are required to stimulate such cells, if they can be stimulated at all. Using a similar method with frog olfactory cells, we noted that somewhat better odorant responses were obtained if proteolytic enzymes were not used during cell isolation (while the Ca concentrations was kept at 1-5 μM), and if the cells patched were not attached to the bottom of the chamber but freely moving - by means of their sensory cilia - in the bathing solution. The bath concentration of cineole had to be in the order of 500 nanomolar to elicit a response (5). This seemed a reasonable concentration because most published responses, recorded from olfactory nerve fibers, required even larger (micromolar) concentrations of odorants in the mucosal compartment (see Table 3 of ref. 5).

Using olfactory cells which were freely swimming except for attachement to the patch pipette, we found that a reasonable estimate of the cellular membrane potential could be made by recording from the membrane area of the soma in the on-cell current clamp mode. Cineole (500 nM) induced a depolarizing receptor potential ranging from -80 to -40 mV. During maintained odorant exposure the receptor potential vanished spontaneously within 10 seconds. This is not unexpected, "adaptation" being well known from recordings from the olfactory nerve fibers (6,7).

More surprising was the observation that, when odorant exposure was maintained, transient depolarizing receptor potentials of 40 to 50 mV continued to appear at regular

NATO ASI Series, Vol. H 39
Chemosensory Information Processing. Edited by D. Schild

intervals. These "slow waves" repeated every 8 - 10 seconds for up to 30 minutes under on-cell current clamp conditions (Fig. 1). Adaptation to higher odorant concentrations caused cessation of slow waves by arrest of membrane voltage at a depolarized level (Fig. 1, upper right panel). Slow waves could be interrupted reversibly by either removing the odorant or by adding 30-50 μM amiloride to the odorant solution. Membrane voltage then assumed resting values near -80 mV.

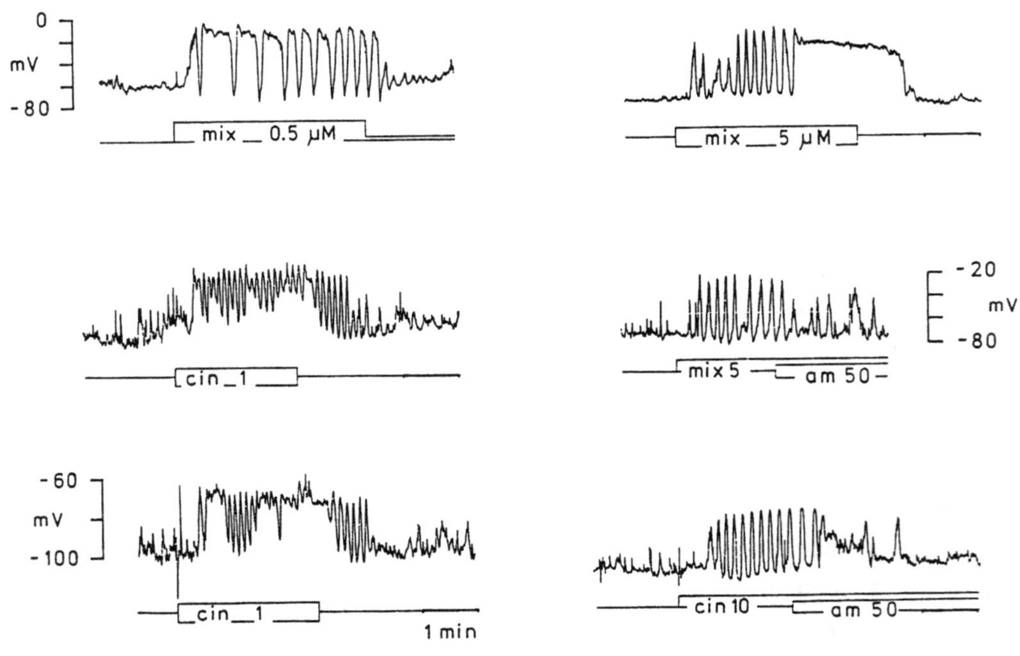

Fig. 1. Response of membrane potential of isolated, free-floating olfactory receptor cells (*Rana esculenta*) to superfusion with a mixture of 15 odorants: acetophenone, anisol, benzaldehyd, butylacetat, camphor, cineole, citral, cyclohexanon, ethylbutyrat, geraniol, 2-heptanon, 4-heptanon, isoamylacetat, linalool, pinacolon (each at 500 nM) or with cineole at 1 or 10 μM. Amiloride (am) was used at 50 μM where indicated. Voltage was recorded from somal membrane patches in the on-cell current clamp mode. Thereby amplitudes were somewhat underestimated, as shown by continued recordings following breakage of the patch. The patch pipette contained a high-K medium (5).

When the patch was broken to obtain whole-cell recording conditions, the odorant-induced slow waves were also seen, but faded within 5 to 10 minutes. This may be due to diffusional "wash-out" of an essential intracellular component by the patch pipette (the inward diffusion of Ca-buffer from the pipette was not responsible). During each depolarizing phase of the slow waves, a burst of action potentials was generated. When recording in the on-cell voltage clamp mode, the bursts appeared as groups of current transients passing through the patched membrane area (Fig. 2A). In this mode, washout being avoided, odorant dependent bursting was observed as long as the seal was stable (occasionally up to 60 min).

Responses similar to those obtained to odorants were seen on exposure with 5 μM 8-bromo-cAMP. The nucleotide induced slow waves and bursts of action potentials. In response to higher concentrations, the slow waves were arrested after a few cycles, the membrane potential becoming constant at about -40 mV, and the generation of action potentials stopped. This behaviour may be related to adaptation. Micromolar concentrations of odorants also caused an adaptation response of this kind (Fig. 1). Following washout of nucleotide or odorant, the membrane potential returned to the resting level.

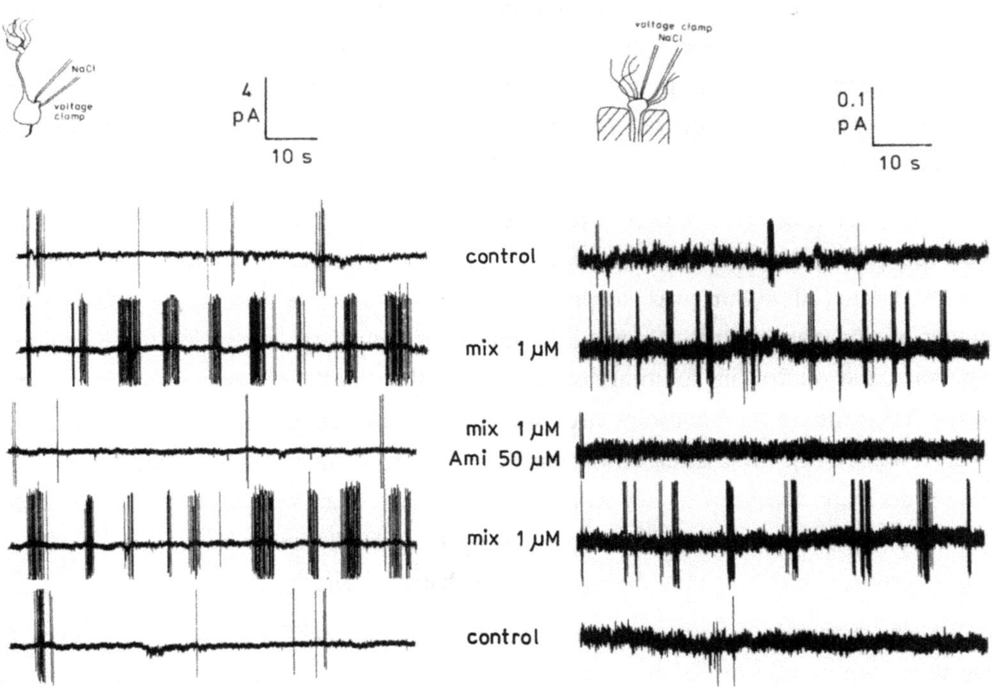

Fig. 2. Left panel: Response of an isolated, free-floating receptor cell to odorants and amiloride. The firing of action potentials in bursts is evident from the strong grouping of current transients, which were recorded in the on-cell voltage clamp mode. The pipette contained a high-Na medium, the pipette potential was clamped to zero (5). Right panel: Similar record obtained from the ciliary knob of a receptor cell *in situ*. Note the smaller amplitude of current transients derived from the ciliary knob.

In these experiments with isolated cells we never saw a phasic or phasic/tonic response to odorants, i.e. at the onset of stimulus an increased spike rate, followed by an adaptive decrease to a lower spike rate, or to the background rate in case of a pure phasic response (6). It is noteworthy, that the difference between the expected phasic or phasic/tonic response and the observed slow-wave induced bursting response may be small in terms of theoretical concepts. Oscillators will change over from the periodic mode to the critically damped mode when just one parameter (the damping constant) is changed.

For olfactory receptor cells, periodic bursting was not explicitly described by other investigators, and few published records indicate this phenomenon (for a summary, see ref. 5). The question arises, therefore, whether an artifact in our experiments, developing during cell isolation or during patch recording from the cell soma, decreased the "damping parameter" and thus induced the bursting response. It is known that damaged cells release diffusible factors which induce potential oscillations in fibroblasts (14). Also, changes in ion concentration in the narrow extracellular space surrounding neurons *in situ* can influence the bursting behaviour of such units (8). In isolated cells the extracellular concentrations would be more nearly constant. To deal with these and similar possibilities, we now recorded from olfactory receptor cells *in situ*, avoiding cell isolation altogether.

Olfactory mucosa, removed from the nose of *Rana esculenta*, was placed in a recording chamber and folded once, such that the mucosal surface faced outward. A patch pipette (20 MOhm) was advanced from the mucosal side towards the folding edge under microscopic control. It passed the mucus of the ciliary space with inside positive pressure, the outflowing solution gently clearing the pipette pathway from mucus and cilia. When the apical membrane area was touched and the positive pressure released, the resistance often increased and a giga-seal developed on applying negative pressure. Odorants were delivered focally with a micropipette whose tip, placed near the patch pipette, emitted a stream of solution flowing tangentially over the mucosal surface. Solutions were fed to the stimulating pipette with electrically actuated valves.

Using high-gain current recording, fast current transients resembling those caused by action potentials were often seen before the patch pipette touched a cellular border, especially when the positive pressure was released "too early". Recording of these transients (sometimes exceeding 10 pA) is probably due to aspiration of free ciliary endings into the pipette, which permit "extracellular" recording of action potential-driven currents through the large ciliary capacitance. This provides an easy method of recording the activity of olfactory receptor cells.

In about half of the cases sealing occured on *sustentacular cells* (epithelial cells (2) with putative supporting function (12)) which had lower input resistances (0.1-0.5 GOhm) and a mean membrane potential of -79 mV (n=33). These cells, as expected from sustentacular cells (12), did not fire action potentials. They often responded to odorants with a depolarization-timecourse reminiscent of the electro-olfactogram (6). This response may be indirect, caused by release of K ions from the receptor cells.

On-cell contact with the apical protrusion, the "ciliary knob", of an *olfactory receptor cell* (2) was evident from the recording of occasional action potentials (background activity). During subsequent whole-cell recording, the input resistance was at least 4 GOhm, as previously found with isolated receptor cells. When odorants (a solution of 15 compounds, each at 0.5 μM) were flushed onto the mucosal surface, 31% of the receptor cells tested responded by firing action potentials at an increased rate. About 15% of the cells tested responded to cineole, others did not respond. Comparing cells, responses to cineole were heterogeneous, comprising a nonperiodic increase in spike rate, firing of one or two bursts or more sustained firing in bursts, as discussed below. Responses of *in situ* cells to odorants occured at lower concentrations than responses of isolated cells. This may be due to a larger number of undamaged sensory cilia being present *in situ.* Background spike rates resumed when the odorants were washed out. The response to odorants also disappeared when 50 μM amiloride was added to the odorant solution.

Burst responses to odorants were occasionally seen during on-cell recording from a ciliary knob (voltage clamp or current clamp) with a tight seal (>10 GOhm). These responses resembled those obtained from isolated cells (Fig. 2A,B). However, the amplitude of the current transients recorded from the patch, which reflect firing of action potentials, was up to tenfold smaller than in records from the soma of isolated cells. Burst responses to odorants were also seen under loose-patch conditions, where damage of the cell membrane by the pipette is less likely. Furthermore, bursting occured when the on-cell pipette contained K or Na as the principal cation. Under on-cell tight-seal conditions (pipette containing high K) bursting was seen but slow waves were not. This may indicate that apical membrane patches are less leaky than the somal membrane patches of isolated cells, and therefore attenuate the potentials of low frequencies from membrane areas outside of the patch too strongly to make them measurable through the patched membrane area.

In some cells, cineole threshold concentrations of 200-500 nanomolar were found, in others concentrations of 1 nanomolar or even as low as 1 picomolar were sufficient to increase the spike rate above background (Fig. 3). Lower concentrations were ineffective in these cases, indicating that a systematic mechanical artifact, arising from the change of solution, was not responsible for the increase in spike rate. Threshold concentrations down to 1 nanomolar were reported in the literature (7). Interestingly, the spike rate at the threshold concentration was rather constant at 0.1-0.13 per second, the repetition frequency of slow waves. It is possible, but not certain, that the threshold spike rate is based on regular slow waves of the membrane potential. At higher concentrations (10 times threshold) the response was a single burst of spikes following the onset of stimulus. This gave the impression of a strong adaptation response. Increasing the cineole concentration from zero to 1000 times threshold did not cause an increase in spike rate at the onset of stimulus, but a burst was fired during washout of the odorant. Similar "off-responses" were also reported by other investigators (6,9). In other experiments an increase of the cineole concentration from 10 times to 100 times threshold did not give rise to a second burst, indicating that strong adaptation had already taken place.

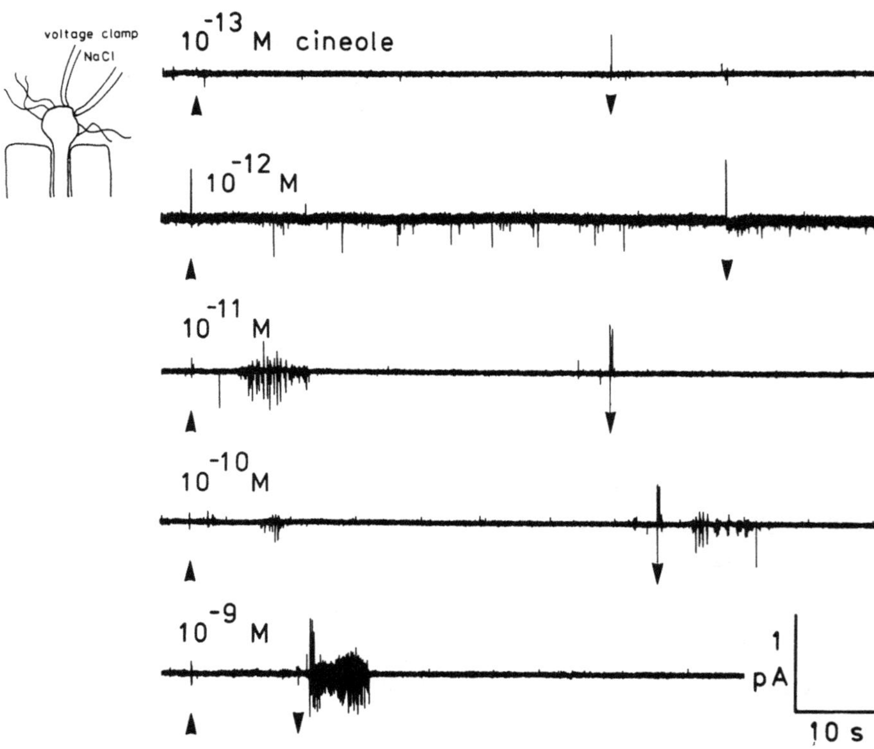

Fig. 3. Recording of current transients, induced by action potentials as described for Fig. 2, in response to increasing concentrations of cineole on the mucosal side. Arrows indicate onset and termination of the chemical stimulus.

"On-responses" as described above were also obtained when 5 μM 8-bromo-cAMP was used instead of cineole. At 10 μM the on-response diminished and an off-response appeared. In other cells, on-responses were induced by cineole in the pico-nanomolar range and by 5-10 μM 8Br-cAMP, but off-responses were not evoked. In several cases the response to the cAMP was blocked by 50 μM amiloride.

It appears, then, that the repetitive-bursting response to odorants, which we described for isolated olfactory receptor cells, is not strictly dependent on cell isolation but also seen with *some* of the cells *in situ.* While slow waves could not be recorded through the unbroken patch at the apical pole of the neurons, the typical bursting pattern of the action potentials suggests that slow waves are, in fact, generated *in situ.* There still remains a possibility that the bursting pattern arises from an artifact, even though not strictly dependent on cell isolation. This will be subject to further investigation.

With *in situ* cells the sustained bursting response was not the only response observed. Several receptor cells of high odorant sensitivity fired one burst at suprathreshold concentrations, thus displaying a more classical response (7) which appeared to include strong adaptation. The dose-response relationship of these cells is presently under investigation. Meanwhile it is rewarding to note that patch clamping of the ciliary knob of *in situ* sensory neurons is possible, and that in the olfactory mucosa the threshold concentration of some neurons to cineole is as low as 1 nanomolar or even 1 picomolar.

ACKNOWLEDGEMENTS - We thank Drs H.Hatt and F.Zufall for pointing out the possibility of sealing patch pipettes to the apical surface of the olfactory mucosa. Support was received from theDeutsche Forschungsgemeinschaft through SFB 246, project C1.

REFERENCES

1. Anderson, P.A.V., Hamilton, K.A. 1987. Intracellular recordings from isolated salamander olfactory receptor neurons. Neuroscience 21: 167-173
2. Andres, K.H. 1969. Der olfaktorische Saum der Katze. Zeitschrift für Zellforschung 96: 250-257
3. Firestein, S., Werblin, F.S. 1987. Gated currents in isolated olfactory receptor neurons of the larval tiger salamander. Proc. Natl. Acad. Sci. USA 84: 6292-6296
4. Firestein, S., Werblin, F.S. 1989. Odor-induced membrane currents in vertebrate-olfactory receptor neurons. Science 244: 79-82
5. Frings, S., Lindemann, B. 1988. Odorant response of isolated olfactory receptor cells is blocked by amiloride. J. Membrane Biol. 105: 233-243

6. Gesteland, R.C. 1971. Neural coding in olfactory receptor cells. Chapter 6 in Handbook of Sensory Physiology Volume IV, pp 132-150
7. Getchell, T.V. 1974. Unitary responses in frog olfactory epithelium to sterically related molecules at low concentrations. J. Gen. Physiol. 64: 241-261
8. Gulrajani, R.M., Roberge, A. 1977. Possible mechanisms underlying bursting pacemaker discharges in invertebrate neurons. Fed. Proc. 37: 2146-2152
9. Kurahashi, T., Shibuya, T. 1989. Membrane responses and permeability changes to odorants in the solitary olfactory receptor cells of newt. Zoological Science 6: 19-30
10. Lynch, J.W., Barry, P.H., 1989. Action potentials initiated by single channels opening in a small neuron (rat olfactory receptor). Biophysical J. 55: 755-768
11. Maue, R.A., Dionne, V.E. 1987. Patch-clamp studies of isolated mouse olfactory receptor neurons. J. Gen. Physiol. 90: 95-125
12. Masukawa, L.M., Hedlund,B., Shepherd, G.M. 1985. Electrophysiogical properties of identified cells in the *in vitro* olfactory epithelium of the tiger salamander. J. Neurosciences 5: 128-135
13. Nakamura, T., Gold, G.H. 1987. A cyclic nucleotide-gated conductance in olfactory receptor cilia. Nature (London) 325: 442-444
14. Oiki, S., Okada, Y. 1988. Factors responsible foroscillations of membrane potential recorded with tight-seal patch electrodes in mouse fibroblasts. J. Membrane Biol. 105: 23-32
15. Trotier, D. 1986. A patch-clamp analysis of membrane currents in salamander olfactory receptor cells. Pflügers Arch. 407: 589-595

Excitation and adaptation of frog olfactory receptor neurones upon stimulation with second messengers and natural odorants[1]

D. Schild, J.A. DeSimone[2], and S. Hellwig

Physiology Dept. University of Göttingen
FRG

INTRODUCTION

Over the last five years it has become feasible to study the olfactory transduction process by the patch-clamp technique. Valuable contributions have come from Maue and Dionne (1987), Firestein and Werblin (1987,1989), Frings and Lindemann (1988), Zufall et al.(1989), Schild (1989), Schmiedel-Jacob et al.(1989), and Trotier (1986). From these studies we now know more about the ionic channels and the whole-cell voltage-gated currents present in these cells, though their physiological role has not been clarified in detail.

There appear to be (i) a TTX-sensitive sodium current (except in the larval tiger salamander where it is TTX-insensitive), (ii) a Ca^{2+}-current, (iii) a Ca-dependent potassium current, (iv) a very slowly inactivating potassium current, and (v) a fast inactivating potassium current (Schild 1989). Odorants seem to depolarize cells by at least two mechanisms. In the tiger salamander (Firestein and Werblin 1989) there seems to be a cation selective channel which is opened upon odor application while in the moth, depolarization is achieved by the closure of potassium channels (Zufall et al. 1989).

1) Supported by the Deutsche Forschungsgemeinschaft, SFB 236/A19
2) JAD was on sabbatical leave from the Dept. of Physiology, Virginia Commonwealth University

NATO ASI Series, Vol. H 39
Chemosensory Information Processing. Edited by D. Schild

One of the current problems seems to be that isolated receptor cells often do not respond to odorants. Investigating the reasons for this might not only reveal shortcomings of the methods applied, it could also provide novel information concerning the transduction process itself.

A second problem, which has recently begun to receive more attention, concerns the second messenger pathways involved in the olfactory transduction process. Though there are biochemical indications (for review see Lancet 1986) for the involvement of cGMP, cAMP, and inositolphosphates these questions have been only rarely addressed from an electrophysiological point of view (Persaud et al. 1988).

In this paper we approach these two points by using two methods, the patch-clamp technique (Hamill et al. 1981) and a whole mucosa preparation (Persaud et al. 1988) which was studied under current clamp conditions. By comaring different types of mucosae, we focus on the actions of cAMP and their relationship to odour stimulation.

METHODS

A. Recording from Isolated Cells

Electrophysiological recordings from isolated receptor cells were made using the patch-clamp method is described by Hamill et al. (1981). The detailed procedure has been given by Schild (1989). A brief summary is as follows. Frogs (Xenopus laevis) were anesthetized with Tricain and decapitated. For most studies the larger air mucosa (located in the main olfactory cavity) was used. The mucosa was removed and placed into a dissociation solution composed of 104 mM NaCl and 2.5 mM KCl. The solution was buffered to pH 7.8 with 10 mM Hepes. No enzymes were used. The tissue was then mechanically disrupted using forceps and scalpel under a dissecting microscope. The tissue pieces were then triturated in the dissociation solution.

A cell suspension was placed in a chamber (100 μl volume). The chamber floor was a glass coverslip coated with Concanavalin A. Cells adhering to the coverslip having the characteristic shape of receptor neurons were chosen for recording. Cells were washed regularly with Ringer's solution (105 mM NaCl, 2 mM KCl, 1 mM $CaCl_2$, 1 mM $MgSO_4$, 10 mM Hepes, pH=7.8, 220 mOsm). All olfactory stimuli were dissolved in Ringer's solution. Solutions were added to the cells from reservoirs suspended above the chamber. The solutions were conducted to the chambers by gravity feed. Recording pipets were pulled and fire-polished using standard methods. The pipet resistance was 5-7 MΩ. Whole cell currents, cell-attached currents, and command voltages were recorded on video tape after pulse code modulation using a Sony 501 unit. For analysis the single cell data were in some cases digitized off-line using a 8-pole Bessel filter, an A/D converter and a PDP 11/73 computer. In other cases the analog data were plotted directly on a strip chart recorder.

B. Recording Whole Mucosa Responses

For transmucosal recording the tissue was placed between two plexiglas Ussing-type chambers. The technique has been described in detail by Persaud et al. (1988). The essential points are as follows: The chambers were connected by a 0.5 mm diameter hole. The mucosa was positioned over a hole in one of the chambers, ciliated side up. The second chamber was then placed over the first and held in the correct position by means of pins from one chamber and matching holes at the opposing corners of the second chamber. The two chambers were pressed together on the mucosa sample with a C-clamp, so the fluid did not leak out around the holes. Each chamber was fitted with fluid inlet and outlet ports which provided means for changing the chamber contents. To insure rapid mixing stainless steel wires were placed near the mucosa to create turbulence in the region of the tissue as fluid passed through the chamber. Ringer's solution and solutions containing various stimuli (see below) were administrated by gravity feed from reservoirs suspended above the chamber. The solution feed was arranged so that either side of the tissue could be stimulated.

Transmucosal potentials were measured in current-clamp using Ag/AgCl electrodes which were connected to the head stage of a patch clamp amplifier. A 10 megaohm resistor was placed in series with the Ag/AgCl electrodes. Potentials were displayed on an oscilloscope and recorded on video tape. Records were subsequently displayed on a strip chart recorder. In all cases a standing transmucosal potential of about 5 mV was measured (basolateral side positive). This result is similar to that reported by Persaud et al. (1988) for Rana catesbeiana.

C. Solution Compositions

In addition to the dissociation solution and Ringer's solution (given above), the following special solutions were used:

1. Pipet Solution (concentrations in mM): NaCl 5; KCl 64; $CaCl_2$ 1; $MgCl_2$ 5; Hepes 10; EGTA 11; pH=7.8; 180 mosm.
2. Amino Acid Cocktail. A mixture of non-essential amino acids (Gibco Limited, number 043-01140 H, 50*) made up in Ringer's solution to a mean concentration of 0.5 mM and titrated to pH = 7.8.
3. Frog Food Extract
   A stimulus solution was prepared from the food pellets the Xenopus were fed while kept in their holding tank. Ten grams of pellets were placed in a few milliliters of water and mixed to the consistency of a soft paste. The paste was diluted with Ringer's solution and stirred for an hour. The mixture was allowed to settle and the upper fluid phase decanted. The fluid extract was then filtered to produce a clear yellow fluid. For use as a stimulus the extract was diluted 1:1 with Ringer's solution. The pH was adjusted to pH 7.8, and the osmolarity adjusted to 220 mosm by the addition of NaCl.
4. Stimulus Cocktail.
   This consisted of a mixture of the following substances: amyl-acetate, cineole, citral, phenylpropanol, terpinol, phenylethyl-amine, triethylamine. The mixture was diluted to 100 μM in each substance with Ringer's solution.
5. The Pharmacological Agents:
   8-bromo-3',5'cAMP (BrcAMP), and phorbol 12,13-dibutyrate (PDB) were obtained from Sigma Chemical Co. All salts and chemicals were reagent grade. Water was double distilled.

RESULTS

The finding that a cocktail of odorants widely used in experiments with amphibia did usually not cause a marked inward current in isolated Xenopus olfactory receptor cells led to questions as to reasons for this apparent failure. It had to be checked whether these cells can generate action potentials, whether they show other odorant responses than a whole-cell inward current, and whether they can be stimulated by second messengers. Fig. 1 shows a single cell's response to the current injection of 2 and 4 pA, respectively.

Fig. 1. Olfactory receptor neuron voltage response to current pulses. Injection of 2 pA drives the cell to threshold. A 4 pA pulse results in repetitive firing. The spikes' maxima were constant in the analog trace and fluctuate here due to the limited plotter resolution.

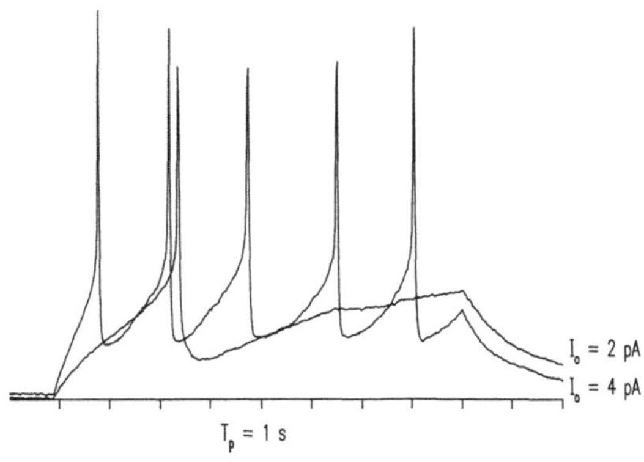

These results demonstrate the remarkable sensitivity of these cells. The fact that these cells are driven to threshold by extremely low currents raises the interesting question about what adaptation mechanisms they utilize at higher generator currents. We will address this point later. Action potentials could be observed not only in the whole-cell configuration but also in the on-cell mode. In some instances a clear increase in spiking rate could be observed following the application of an odorant cocktail (fig. 2) which indicates that Xenopus

Fig. 2 Action potentials of an olfactory neuron. The recording was made in the cell-attached mode and the current shown is mainly the capacitative current resulting from the intracellular voltage changes during action potentials. Obviously, the spike rate increases upon application of the stimulus cocktail.

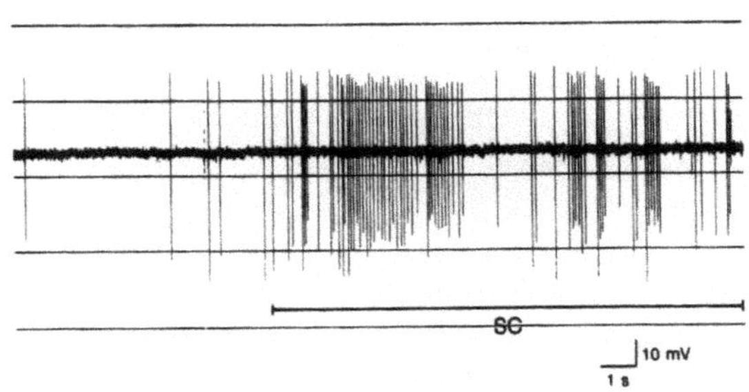

olfactory receptor cells are well capable of reacting to odorants in the on-cell mode of the patch-clamp technique. In this mode the cells are not dialyzed, suggesting that the lack of some key components is responsible for the non-responsiveness in the whole-cell mode. One obvious possibility is the diffusion of second messengers into the patch pipet. As a first attempt, we therefore tried the bath application of BrcAMP to isolated cells (fig.3):

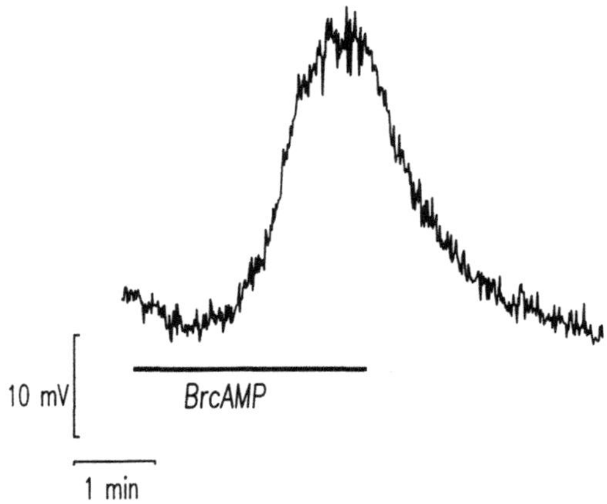

Fig. 3 Receptor neuron response in the current clamp mode upon bath application of 100 μm BrcAMP. A short hyperpolarization is followed by a depolarization.

after a relatively short hyperpolarization the membrane potential depolarizes. The repolarization of the membrane potential is probably mainly due to the diffusion of BrcAMP into the patch pipet. The shape of this response to BrcAMP looks almost identical to the current response of the whole mucosa of Rana catesbeiana when exposed to high concentrations of BrcAMP (fig. 4).

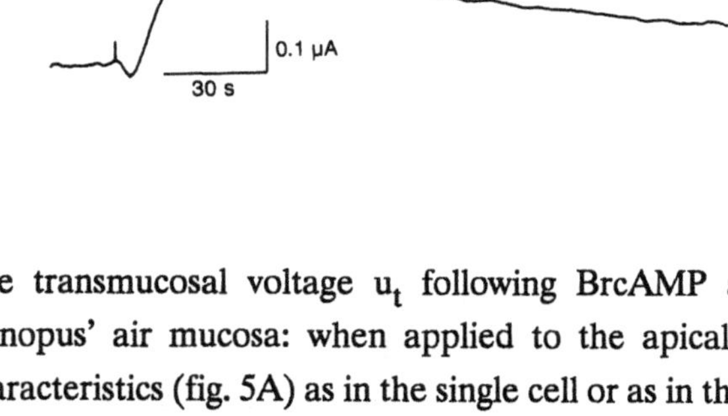

Fig. 4 Whole-mucosa experiment in Rana catesbeiana. The current through the mucosa is shown, the transmucosal voltage was clamped to 0 mV and BrcAMP was applied. Low concentrations of BrcAMP resulted in a single phase response while higher concentrations showed a two-phase response. BrcAMP concentrations were 10 μM (upper trace) and 100 μM (lower trace).

The transmucosal voltage u_t following BrcAMP application was also measured in Xenopus' air mucosa: when applied to the apical side, it clearly showed the same characteristics (fig. 5A) as in the single cell or as in the voltage-clamped current in Rana catesbeiana. In order to localize the effect of BrcAMP we also applied this drug at the basolateral side (fig. 5B) without observing comparable effects. From these recordings it seems clear that cAMP acts predominantly at the ciliar side of the cells and, further, that the action might consist in two antagonistic effects:

Fig. 5 Effects of BrcAMP on the whole mucosa of Xenopus. Recordings were made in the current clamp mode; the transmucosal voltage is shown. (A) Typical two-phase response upon BrcAMP application on the acical side of the mucosa: a negative wave followed by a positive one. (B) Almost no voltage change could be observed if BrcAMP was applied to the basolateral side. (C) When BrcAMP was delivered together with amiloride on the apical side, the drug blocked the positive wave of the response.

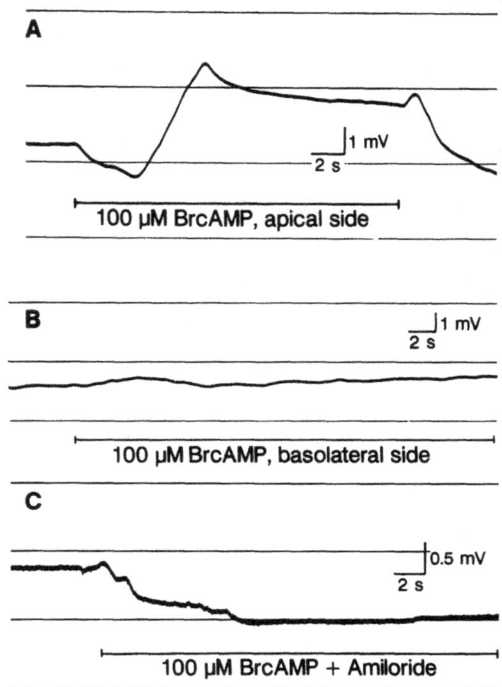

the early wave W_1 could be brought about either by hyperpolarization of single cells on the apical side or by depolarization on the basolateral side, while the following wave W_2 could be caused by a depolarization on the apical side or a hyperpolarization on the basolateral side. Both waves could be separated in two ways: lower concentrations of BrcAMP were followed only by the positive wave W_2 (see fig. 4, upper half). The negative wave (negative on the basolateral side as compared to the apical side) could be isolated by applying amiloride (400 μmolar) to the ciliar side (fig. 5C).
The DAG/IP_3 system was also reported to be involved in olfactory transduction (Restrepo and Bruch 1989).

Fig. 6 Effects of the phorbolester PDB on the whole mucosa of Xenopus. PDB resulted in a positive single phase response.

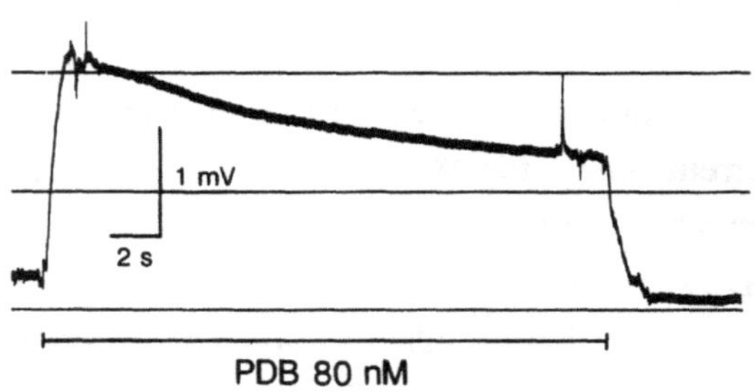

The simplest way to confirm this suggestion from an electrophysiological point of view was to apply the phorbolester PDB to the ciliar side of the whole mucosa and record the change of transmucosal voltage u_t. Fig. 6 shows the result of this experiment: PDB (80 nmolar) seems to depolarize cells on the ciliar side. The latency of the response as compared to the latency of the BrcAMP response is remarkably shorter.

Now it was interesting to see whether odorants when applied to the mucosa would result in similar response patterns of the transmembrane potential u_t. We first tried the amino acid cocktail (ASC) as described in "Methods": the response u_t was surprisingly similar to the response to BrcAMP (fig. 7).

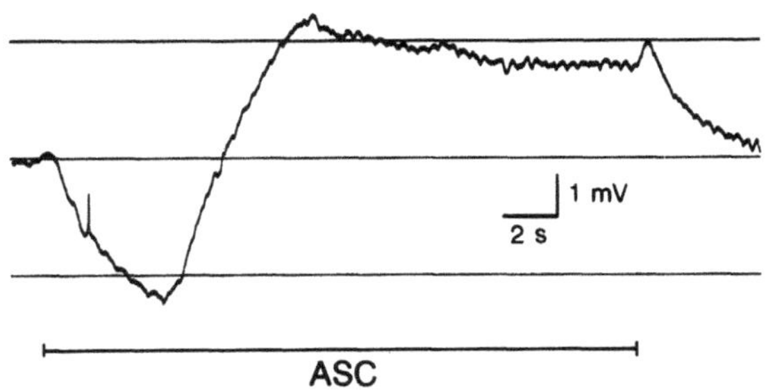

Fig. 7 Transmucosal voltage change following application of a cocktail of amino acids (ASC). Clear negative voltage transient followed by a positive one. Superimposed on this response, there were small amplitude oscillations.

The response to the odorant cocktail looked very similar (not shown). However, the negative early wave (W_1) was not always present in stimulus presentations. Fig. 8A shows an example of a recording in which food extract was applied to the apical side to Xenopus' mucosa. After a short latency there is the positive response wave. The filling of the chamber stopped at the maximum of the response curve, i.e., until that point there was a stimulatory flow while the stimulus solution was stagnant during the rest of the stimulus application. At the decreasing shoulder of the response there was an oscillation of increasing amplitude and decreasing rate. Fig. 8B shows some of these oscillations on a larger time scale.

Oscillations of this type did not occur regularly with stimulus application. Because of their large amplitude these oscillations are probably caused by mass discharges of spiking receptor neurons. Synchronization might be due to the compensatory current in the current clamp. Fig. 8C examplifies the more commonly observed response to food extract; when amiloride was added to the food extract, the response was suppressed (fig. 8D). Another type of oscillatory behaviour of u_t occurred in almost every case after increasing the intracellular cAMP concentration either by BrcAMP, forskoline, theophylline, or by membrane impermeant stimuli such as amino acids (see fig. 6).

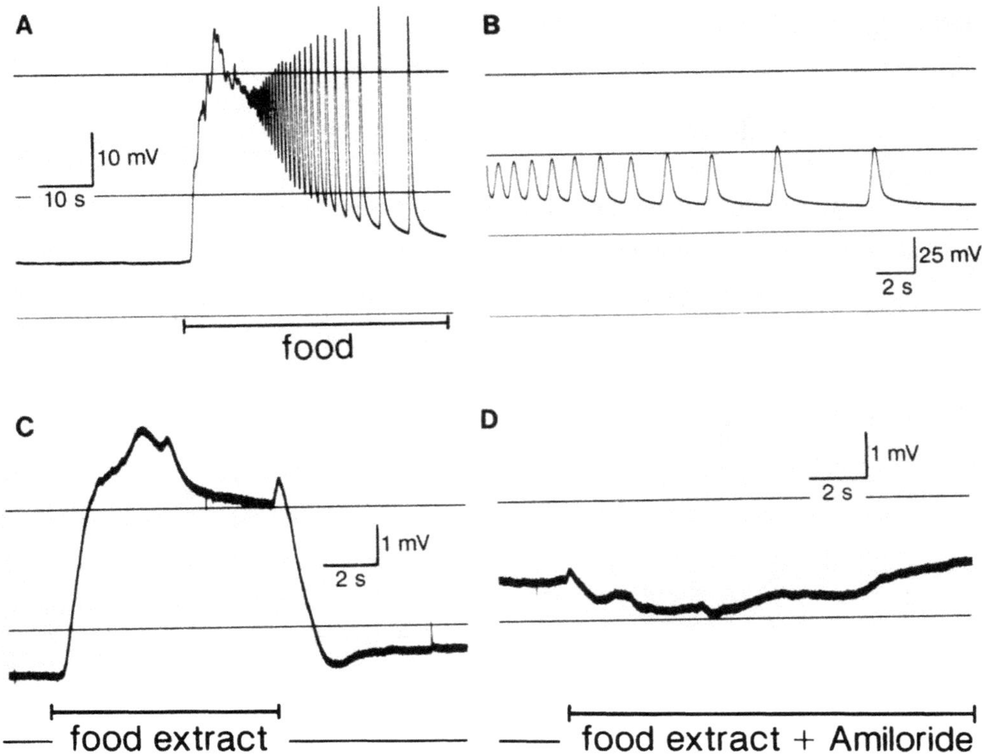

Fig. 8 Voltage responses of the whole mucosa to food extract. Recording in the current clamp mode (I=0). (A) Large depolarizing effect of food extract. The oscillations during the response were observed only rarely; they are shown on a different time scale in part (B) of the figure. (C) Typical whole mucosa response to food extract. (D) The response is blocked by amiloride; in addition a slight hyperpolarization can be observed.

These oscillations resembled relaxation oscillations rather than harmonic oscillations. Frequencies were between 0.3 and 4 Hz and changed as a function of injected current amplitude. A more pronounced oscillation of this type (after BrcAMP application) is shown in fig. 9.

Fig. 9 Low frequency oscillations of membrane potential of the whole mucosa in the current clamp mode (I=0). These oscillations could be observed regularly after increasing the intracellular cAMP concentration.

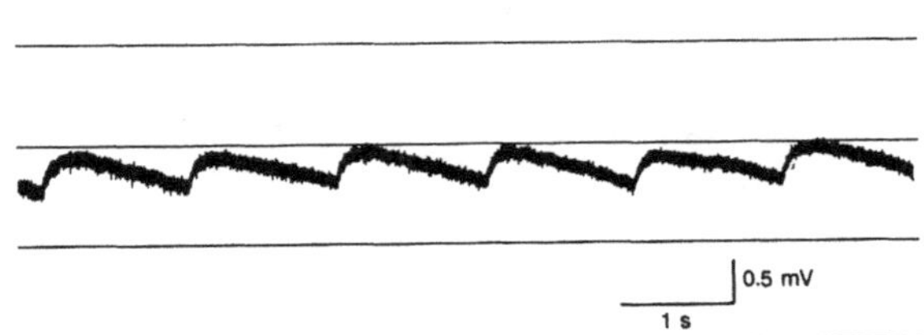

DISCUSSION AND CONCLUSIONS

1. Xenopus has different mucosae for smelling in air and water. The water mucosae showed more conspicuous effects to odorants. The strong response observed with food extract in the whole mucosa was completely suppressed by high concentrations of amiloride.

2. The response to BrcAMP in the whole mucosa of Rana catesbeiana is shown to be very similar in isolated cells of the air-mucosa of Xenopus. This allows the hypothesis that the response to BrcAMP might be a general phenomenon in amphibia olfactory mucosae and validates the whole mucosa preparation as far as the involvement of BrcAMP is concerned.

3. Isolated olfactory receptor cells react to odorants in the on-cell configuration of the patch-clamp technique. In this configuration action potentials can be detected by the mainly capacitative current though the patch which is associated with the voltage changes during an action potential. The spiking rate is increased following the application of odorants.

4. In the whole mucosa preparation, both BrcAMP as well as PDB were effective.

5. When the cAMP concentration was altered during an experiment, oscillations in the range of 1 Hz could often be observed. These oscillations might be similar or identical in nature with those described by Frings and Lindemann (1988). As they can be also observed "spontaneously", without voltage clamping or current injection, their basis must involve a cellular mechanism which influences a conductance which, in turn, is responsible for a voltage change. Among the currents measured in Xenopus (Schild 1989), one of the most probable candidates for a conductance which is altered by a cellular signal is the Ca-dependent potassium conductance.

6. Assuming that odours applied to the olfactory mucosa influence mainly olfactory receptor cells and given that BrcAMP has almost identical effects on single receptor cells and on the whole mucosa, one might, as a first order approximation, view the transmucosa voltage u_t as the difference of two voltages

$$u_t = u_{ms} - u_{mc}$$

whereby u_{ms} is the membrane potential across the soma and u_{mc} the membrane potential across the ciliar membrane of olfactory receptor cells. In the case that

$$u_{ms} \approx u_{mc} \qquad \text{or} \qquad u_{ms} \leq u_{mc}$$

one should expect that u_t is about zero at rest, possibly a little positive at the basolateral side. We usually measured a standing voltage over the whole mucosa of about 5 mV. The positive wave W_2 of u_t can generally (as far a receptor neurones are concerned) be interpreted in two ways: $u_t = u_{ms} - u_{mc}$ becomes more positive either if u_{mc} becomes less negative, i.e., if the ciliar membrane is depolarized, or if u_{ms} becomes more negative, i.e., if the soma of the cell is hyperpolarized. The opposite applies to the negative wave W_1 of u_t.

7. The general shape of the response to BrcAMP is the same in u_m (isolated cells) and u_t (mucosa); we never found the reversed response in single cells. It is clear that the second or late response is a depolarization in the single cells. In order that a depolarization in receptor neurones causes a more positive voltage U_t, the depolarization has to take place at the apical or ciliar side of the cells. This comparison thus suggests that BrcAMP depolarizes the ciliar membrane, which is in agreement with current hypotheses of cAMP-gated generator channels on the cilia.

8. An analogous argument applies to the early response, i.e., the hyperpolarization in single cells and the negative wave of u_t. Both effects can be interpreted by the apical hyperpolarization of the cells. This is a second, separate effect mediated by BrcAMP. The comparison of the results drawn from the patch-clamp approach and the whole-mucosa preparation seems thus to allow the localization of the channels that hyperpolarize the cell.

9. From the study of voltage-gated currents in Xenopus, we know that there are three different potassium currents. The major one is the Ca-dependent potassium current so that this outward current seems to be the most probable cause of the hyperpolarization of the ciliar membrane. In this case, it could be hypothesized that the increase of intracellular Ca^{2+} brought about by an increase in cAMP would open Ca^{2+}-dependent potassium channels on the apical side of the cells.

10. A tentative physiological explanation for the cAMP-induced hyperpolarization is the following: in low BrcAMP concentrations it is not seen so that these concentrations might exclusively lead to a depolarization. At higher concentrations, a potassium conductance might partly shortcircuit the subsequent generator inward current. This would be a powerful adaptation mechanism of the cell: it would partly compensate the ciliar generator potential without reducing the soma's sensitivity to the dendritic current. However, in order to better understand these processes, both currents and their kinetics have to be studied in much more detail.

11. If the hyperpolarization at the apical side of olfactory receptor cells is caused by potassium channels, there are two more consequences: first, one might expect that the ionic composition of the mucus is relatively potassium rich which was directly shown by Joshi et al. (1987). Second, if a part of the generator current is shortcircuited at the apical side of the cell, it would, of course, be impossible to measure the whole generator current by voltage-clamping the soma under normal physiological conditions.

REFERENCES

Firestein S, Werblin F (1987) Gated currents in isolated olfactory receptor neurons of the larval tiger salamander. Proc.Natl.Acad.Sci.USA 84:6292-6296

Firestein S, Werblin F (1989) Odor-induced membrane currents in vertebrate olfactory receptor neurons. Science 244:79-82

Frings S, Lindemann B (1988) Odorant response of isolated olfactory receptor cells is blocked by amilorid. J.Membrane Biol. 105:233-243

Hamill OP, Marty A, Neher E, Sakmann B, Sigworth FJ (1981) Improved patch-clamp techniques for high-resolution current recording from cells and cell-free membrane patches. Pflügers Arch. 391:85-100

Joshi et al.(1987) Spectrophotometric determination of cation concentrations in olfactory mucus. Neuroscience Letters 82:321-326

Kuba K, Nishi S (1976) Rhythmic hyperpolarizations and depolarizations of sympathetic ganglion cells induced by caffeine. J.Neurophysiol. 39:547-563

Lancet D (1986) Vertebrate olfactory reception. In: Cowan WM (ed) Annual Reviews of Neuroscience, 9. Annual Reviews, Palo Alto 329-355

Maue RA, Dionne VE (1987) Patch-clamp studies of isolated mouse olfactory receptor neurons. J.Gen.Physiol. 90:95-125

Persaud KC, Heck GL, DeSimone SK, Getchell TV, DeSimone JA (1988) Ion transport across the frog olfactory mucosa: the action of cyclic nucleotides on the basal and odorant-stimulated states. Biochim.Biophys.Acta 944:49-62

Restrepo D, Bruch RC (1989) Stimulus amino acids elicit rapid increase in intracellular calcium in dissociated olfactory neurons. Achems XI :150

Schild D (1989) Whole-cell currents in olfactory receptor cells of Xenopus laevis. Exp.Brain Res. 78:223-232

Schmiedel-Jacob I, Anderson PAV, Ache BW (1989) Whole-cell recordings from lobster olfactory receptor cells: responses to current and odor stimulation. J.Neurophysiol. 61:994-1000

Trotier D (1986) A patch-clamp analysis of membrane currents in salamander olfactory receptor cells. Pflügers Arch. 407:589-595

Zufall F, Stengl M, Hildebrand JG, Hatt H (1989) A patch clamp study of cultured olfactory receptor neurons from manduca sexta. Proceedings of the 17-th Göttingen Neurobiology Conference :75

RECEPTOR SELECTIVITY AND DIMENSIONALITY OF ODOURS AT THE STAGE OF THE OLFACTORY RECEPTOR CELLS

Gilles SICARD

Laboratoire de Physiologie Neurosensorielle, CNRS-URA 180
LYON I, 69622, Villeurbanne cedex FRANCE.

Several different approaches have been used to relate physico-chemical properties of odorants to their organoleptic characteristics such as judged by human subjects. Unfortunately, one get several difficulties if one tries a direct mapping of the whole perceptual olfactory space onto the physico-chemical space. The descriptions of odour rest always incomplete; correlations between individuals are low, probably due to the sociological conditions of learning (Berglund et al., 1971, O'Connell, 1989). In addition, the integration of peripheral information by physiological processes is complex, involving also non-olfactory - for instance trigeminal - pathways (Cain et al., 1980, Silver et al, 1988). As a consequence, a number of studies have considered restricted portions of the olfactory space in order to determine chemical similarities between members of "odorant families". Another interesting approach, is based on the assumption that there is a correspondence between specific anosmia known as a perceptual disfunction limited to an odorant family or note and genetic deficits of selective receptors in affected human subjects. Increasing numbers (several tenths) of different specific anosmias have been reported, thus suggesting the existence of several tenths of receptor sites (Amoore, 1982). Nevertheless, odorants are often described by multiple terms and thus can belong to several odorant families (Beets, 1982, Boelens, 1983). For example, beta-damascone shows a complicated odor profile in which fruity-flowery, exotic-spicy and chrysanthenum-like elements predominate (Olhoff, 1986). Moerover, correct analysis of a given odour requires several successive sniffs (Laing, 1983) what possibly means that time could be a pertinent factor in odour coding.

The peripheral messages and their modifications in higher centers can be analyzed using physiological methods. In Vertebrates, the discrimination of odours is based on the messages which are transmitted by the olfactory nerve after the interactions

NATO ASI Series, Vol. H 39
Chemosensory Information Processing. Edited by D. Schild

between odorous molecules and the receptor cells in the olfactory mucosa. In the olfactory mucosa, there are several millions of receptor cells so, we can study only a small portion of them in order to characterize the information which is transmitted to more central structures.

Considering that the initial olfactory message is defined as a spatio-temporal distribution of impulses in the multiple channels of the olfactory nerve (across-fiber activation pattern), three aspects of the functional organization of the periphery of the olfactory system could play a role in modeling the message: the selectivity of receptor cells, the spatial heterogeneity of the activation pattern throughout the olfactory mucosa, and also the organisation of the primary projections onto bulbar neurons, including spatial and temporal aspects which determine the response of the olfactory bulb.

SELECTIVITY OF RECEPTOR CELLS

Twenty-five years ago, the first electrophysiological records of single receptor cells indicated that the olfactory code would be a puzzling question for a long time (Gesteland et al., 1963). After a long decade of investigations, we have obtained a more satisfying view of the selectivity of olfactory receptor cells (Getchell, 1974, Duchamp et al., 1974, Revial et al., 1978, Revial et al., 1982, Sicard et al., 1984, Sicard, 1985).

In mammals, Gesteland et al. (1982) showed an increase of selectivity during development in rat and the selectivity of receptor cells was suspected to be higher in adult mouse than in frog (Sicard, 1986). However, it must be emphasized that our knowledge of selectivity is almost exclusively based on recordings of receptor cells in Amphibians or Fish.

In the frog, we recorded the responses of single receptor cells submitted to series of stimulations by different odorants in six separated studies (Duchamp et al, 1974, Revial et al., 1978, Revial et al., 1982, Sicard et al., 1984, Sicard, 1985). The outcome of this intensive investigation elucidates nervous and chemical aspects of the reception in the olfactory system but does not close the debates. It is possible to classify odorants on the basis of the receptor cell selectivity, but we never obtained an intelligible classification of the receptor cells themselves.

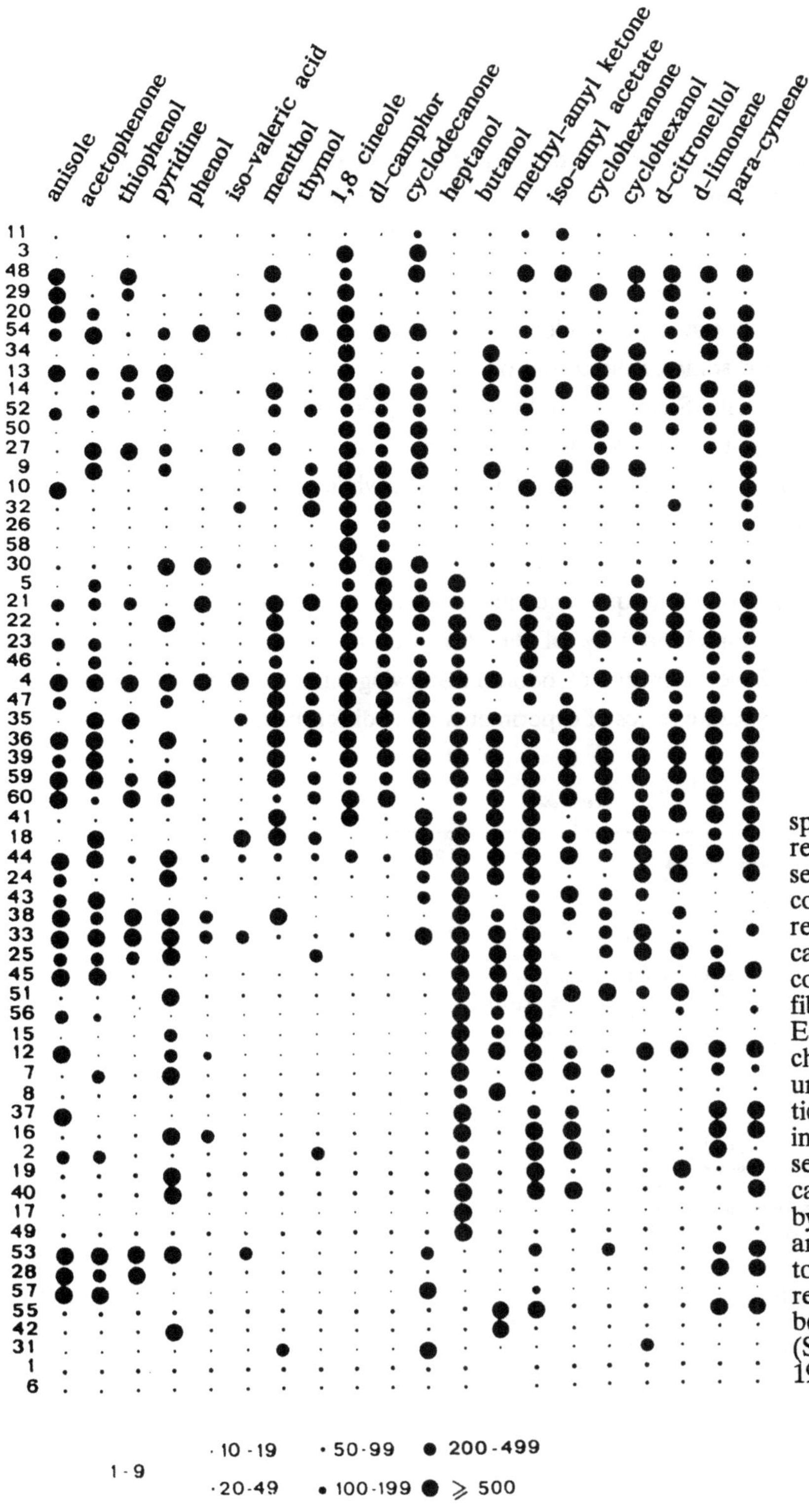

Figure 1. Responses of sixty receptor cells to a series of twenty compounds. This representation can illustrate the concept of across fiber pattern: Each odorant is characterized by a unique distribution of activations in a receptor cell set. Such a matrix can be processed by multimentional analysis in order to describe the relationships between odorants (Sicard et al., 1984).

PERIPHERAL OLFACTORY SPACE AND ODORANT GROUPS

In each experiment, we have studied the responses of 60 cells to 20 odorants. With 10-fold to 100-fold dilutions of saturated vapors, about 35% of the stimulations elicited responses of the tested cells. Cells could respond from 0 to 20 different stimuli (figure 1).

Systematically, all the tested odorants were distinguished, even enantiomeric compounds. However, some of the tested odorants elicited responses from very similar subsets of receptor cells. Significant links between pairs of odorants were shown by comparing the distribution of responses in a given set of receptor cells. For instance, acetophenone and anisole elicited similar responses while some other odorants, like cineole and heptanol did not. Multidimensional analysis allowed us to describe the relationships between all the tested odorants (figure 2). The concept of multidimentionality in olfactory quality coding is illustrated by the fact that five factors accounted generally for about 55% of the total variance of odorants/receptor cell response matrices. Several groups of odorants showing similar stimulative properties were confirmed by separate series of experiments using different chemical contexts.

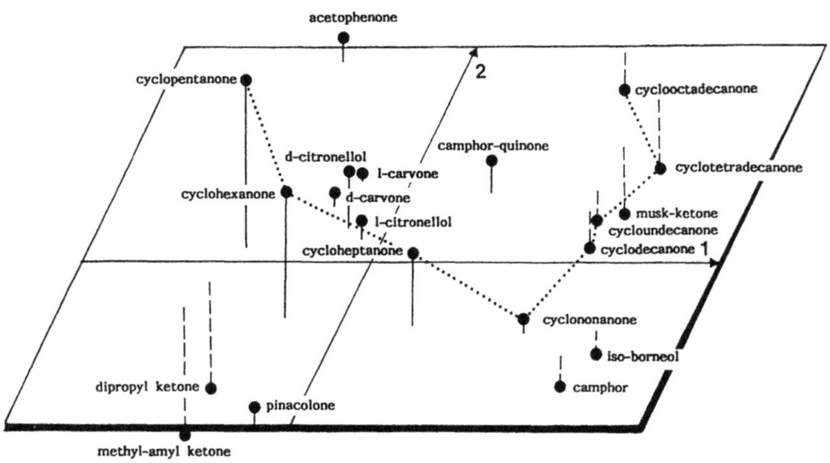

Figure 2. Description of the relationships between twenty odorants as revealed by the responses of a recetor cell set. First three factors support 51% of the variance of the original matrice which have been submitted to an analysis of correspondences. Cycloketones are spreading throughout the whole space. A group of odorants includes cyclooctadecanone, cyclotetradecanone, musk-ketone. All these compounds have a musky odour. (Revial et al., 1982)

Interestingly, in some cases, elements in a group share common odorant notes for human subjects (camphor group, terpene group) and thus groups could be also related to organoleptic families. This cannot be generalized: It is difficult to recognize a similar odour in chemicals which have been gathered in the "aromatic" group. However, chemists are not surprised by the revealed similarities between the odorants. Searching an explanation for the organization of the olfactory space, the chemical functionality is not sufficient to explain similarities. For example (figure 2), a weight-series of cycloketones was found to spread in the whole olfactory space including musk-ketone (musk odour), camphor (camphory odour) and d-carvone (minty odour). Only for some odorant groups (e.g. camphory odorants), it has been found that the shape and size of the molecules play an important distinctive role (Eminet et al., 1983). Additional investigations have established physico-chemical bases of the ressemblance of odorants belonging to the same group (Chastrette, 1981) and confirm the importance of shape, molecular weight, solubilities, electronic properties and polarisability (Dravnieks et al., 1972, Doving, 1974, Schiffman, 1974).

SELECTIVITY AND INTENSITY OF THE STIMULATIONS

Empirical human observations show that the quality of an odorant can change with the concentration of the stimulus. This point has been confirmed by a recent psychophysical study (Gross-Isserof and Lancet, 1989). Conversely, odorant can be often identified in spite of concentration changes.

Direct analyses of the responses of receptor cells indicated that the discrimination of the odorants can be affected by small variations of concentrations (Revial et al., 1982). These results are consistent with quality changes in perception. Alterations of across fiber activation patterns and discrimination by concentrations changes have been recently re-investigated (Duchamp-Viret et al., 1989): continuous recruitment of receptor cell responses were found over large concentration ranges (figure 3) and discrimination abilities were improved by increase of stimulation intensity.

To our opinion, until now the importance of the relation between quality and intensity of odour stimuli has been underestimated. It is difficult to evaluate common properties of discrete molecules.

Concentration-dependent changes of an odorant are determined by continuous variations of the vectors which are involved in the interactions of the odorant with the olfactory system. This can be of interest not only to study structure-odour relationships but also the principles of coding in the olfactory system.

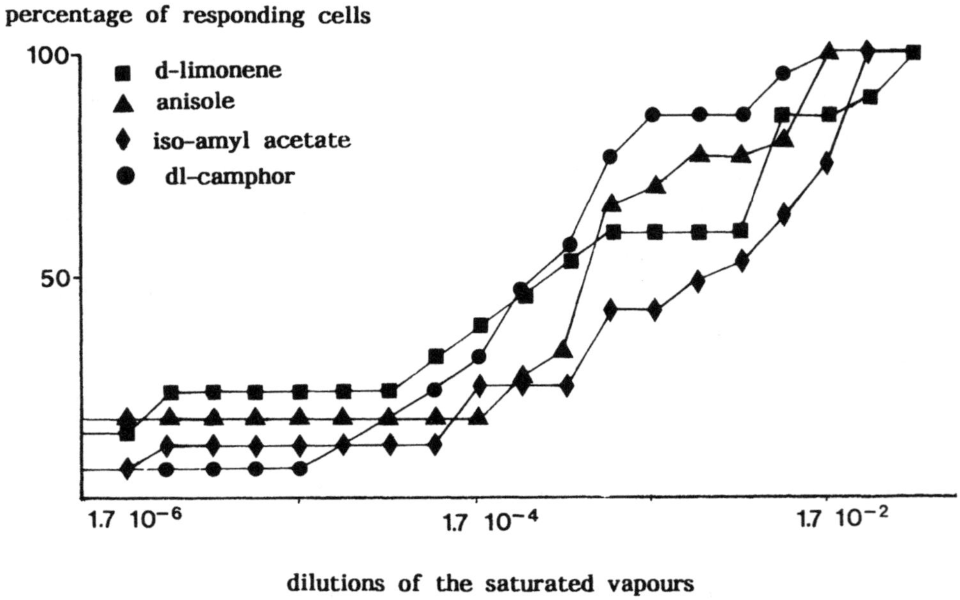

Figure 3. Recruitment of receptor cells by increasing concentrations of four different odorants (Duchamp-Viret et al., 1989).

BASIS FOR THE RECEPTOR CELL SELECTIVITY

It was not possible to classify the receptor cells themselves. This fact does not facilitate the description of the supports of the selectivity of the olfactory receptor cells. Several hypothesis have been emitted concerning the nature of the sites of interactions between odorants and receptor cells:

Interactions of the odorants with membrane lipids have been proposed (Davies, 1965, Kashiwayanagi et al., 1984, 1987). Differences in lipid composition of membranes might explain the diversity of the receptor cells. However, an interaction of odorants with membrane lipids does not exclude other discriminatory mechanisms.

Alternative propositions take ciliary proteins as main keys for the selectivity and have gained wide acceptance. A number of arguments support this idea.

Several experiments have shown that odorant stimulations can protect the olfactory mucosa against the blocking effects of protein-reagents (Gesteland et al., 1972, Delaleu et al.,1980, Shirley et al., 1983). Globally, these studies have emphasized the difficulty to make specific modifications of the olfactory mucosa responsiveness. Several proteins regarded as candidates for olfactory receptors have been isolated from the olfactory mucosa: camphor-binding protein (Fesenko et al., 1985), anisole-binding protein (Price, 1977), pyrazine derivative-binding protein (Pelosi et al., 1982, Pevsner et al., 1985), androstenone receptor (Gennings et al., 1977). Immunoreactivity tests have suggested that the receptors could share a common determinant (Goldberg et al., 1979). The olfactory receptors could belong to a family of glycoproteins analogous to rhodopsin or B-adrenergic receptors: in olfactory cells, transduction systems have been identified which involve a guanine nucleotide binding protein (G-protein) associated with an adenylate cyclase (Chen et al., 1984, Pace et al., 1985, Anholt et al., 1987, Fesenko et al., 1987, 1988, Shirley et al., 1987).

During the last fourty years, theoretical considerations have proposed a variety of descriptions of olfactory receptors. The results of the studies on receptor cell selectivity fit well in a balanced model which would include stereochemical and functional-group theories (Amoore, 1982, Beets, 1982, Polak, 1973). One of the prevalent ideas is that an odorant is generally recognized by several receptors. Therefore, an odorant is characterized by several properties and some of them can be shared by other odorants.

We can imagine that each receptor cell possesses a single receptor type. Our results and cross-adaptation properties of the receptor cells (Baylin et al., 1979) are more consistent with the assumption that the selectivity depends on multiple receptor sites with different specificities distributed on the membranes of the cilia of each receptor cell. This view introduces the complexity of receptor combinations on the receptor cells. In part, this complexity could explain why the number and the specificities of receptors are still unknown.

Now, isolation of receptors and direct analysis of their properties become accessible via molecular biological methods (Lee et al., 1988). We think that future studies will reveal much about the units which support the selectivity of receptor cells.

SPATIAL DIMENSION OF THE ODOUR

At the peripheral level of the olfactory system, the spatial dimension of the odour is not only described by a mosaic in which the cell is the unit. This dimension can also involve selectivity gradients which are "imposed" or "inherent" at a macroscopic level (Moulton, 1976). The word imposed refers to the distribution of odorants at the surface of the olfactory mucosa. This distribution depends on the stimulus flows and the diffusion of molecules within the mucus. The word inherent refers to a spatial segregation of the receptor cells with different selectivity profiles.

Electro-olfactogram recordings have been used to demonstrate regional variations in response to distinct odorants in the frog (Mustaparta, 1971, Daval et al., 1970), in the salamander (Mackay-Sim, 1982) and in the rat (Thomesen et al., 1977, Edwards et al., 1988). Biochemical arguments also indicate regional variations (Dreesen and al., 1982). These macroscopic observations of odorant-dependent differential sensitivities of the olfactory mucosa have not been confirmed at the single receptor cell level (Sicard et al., 1984), probably because of the small number of examined cells. Sometimes, we have got the opportunity to observe distinct selectivity-profiles in adjacent cells. Nevertheless, there is a long tradition of histological studies of olfactory receptor cell projections which demonstrate that a chemotopic organization of the periphery could be translocated to the bulbar level (Costanzo et al., 1978, Mackay-Sim, 1984, Astic et al., 1986, Saucier et al., 1986, Steward et al., 1985). Therefore, it has been speculated that the peripheral gradients of sensitivity could play a role in odour coding (see discussion by Moulton, 1976, Kauer et al., 1974).

However a conclusion about the actual importance of the epithelial chemotopy on quality coding is lacking. Peculiarly, consequences of limited lesions of the receptive sheets for quality coding are not well documented. Nevertheless, after destruction of the olfactory mucosa in mice by $ZnSO_4$ treatment partial biochemical and histological recovery occurs. These mice are able to detect food odour with less than 10% of the normal number of functional receptor cells (Harding et al., 1978). In frog, Duchamp-Viret et al. (1989) compared discrimination abilities of sets of bulbar neurons from animals with dorsal/ventral or only ventral intact olfactory mucosa. While some modifications of the neural sensitivity was observed, no important qualitative variation of the across fiber patterns was found.

CONCLUDING REMARKS

In order to understand completely the message which is transmitted by olfactory receptor cells, it is necessary to take the properties of the receiver into account. This is beyond the topic of this paper. However, the following comments are related to a temporal dimension of peripheral odour coding which might be important regarding the bulbar processes.

Modifications of across-fiber patterns can be obtained not only in experimental conditions as mentioned above, but also in physiological situations: Self adaptation and cross adaptation have been observed at the receptor cell level (Baylin et al., 1979). They can be explained by competition of odorants for receptors or saturations of transduction mechanisms. These properties demonstrate that the selectivity of receptor cells can be altered by their recent history. Such mechanisms could play a role in coding of odorant mixtures (Bell et al., 1987).

Signal transmission in the olfactory nerve is firstly modulated by the rythm of respiratory or sniffing air-flows. Intrinsic properties of receptor cells can also play a role. Phasic-tonic responses of receptor cells to long-lasting stimulations have been shown, bursts of impulses transmitted by these neurons can last longer than the stimulation and the response latencies depends on concentrations (Getchell et al., 1978). Little is known on the dynamic characteristics of the across fiber patterns, but these characteristics should play a role in the processing of messages within the olfactory system.

REFERERENCES

Amoore JE (1982) Odor theory and odor classification. In Theimer, ET (ed), Frangance chemistry. The science of the sense of smell, Academic Press, New York, London, pp 27-76.

Anholt RRH, Mumby SM, Stoffers DA, Girard PR, Kuo JF, Snyder SH (1987) Transduction proteins of olfactory receptor cells: Identification of guanine nucleotide binding proteins and protein kinase C. Biochem, 26:788-795.

Astic L, Saucier D (1986) Anatomical mapping of the neuroepithelial projection to the olfactory bulb in the rat. Br Res Bull, 16:445-454.

Baylin F (1979) Temporal patterns and selectivity in the unitary responses of olfactory receptors in the tiger salamander. J Gne Physiol. 74:17-36.

Baylin F, Moulton DG (1979) Adaptation and Cross adaptation to odor stimulation of olfactory receptors in the tiger salamander. J. Gen Physiol 74:37-55.

Beets MGJ (1982) Odor and stimulant structure. In Theimer, ET (ed), Frangance chemistry. The science of the sense of smell, Academic Press, New York, London, pp 77-122.

Bell, GA, Laing DG, Panhuber H (1987) Odour mixture suppression: evidence for a peripheral mechanism in human and rat. Br Res, 426:8-18.

Berglund, B, Berglund U, Ekman G, Engen T (1971) Individual psychophysical functions for 28 odorants. Percept Psychophys, 9, 379-384.

Boelens H (1983) Structure-activity relationships in chemoreception by human olfaction. TIPS, 421-426.

Cain WS, Murphy CL (1980) Interaction between chemoreceptive modalities od odour and irritation. Nature, 284:255-257.

Chastrette M (1981) An approach to a classification of odours using physico-chemical parameters. Chem. senses 6:157-163.
Chen Z, Lancet D (1984) Membrane proteins unique to vertebrate olfactory cilia: candidates for sensory receptor molecules. Proc. Natl. Acad. Sci., 81:1859-1863.
Costanzo RM, O'Connell RJ (1978) Spatially organized projections of hamster olfactory nerves. Br Res, 139:327-332.
Daval G, Leveteau J, Mac Leod P (1970) Electro-olfactogramme local et discrimination olfactive chez la grenouille. J Physiol Paris, 62:477-488.
Davies JT (1965) A theory of the quality of odours. J Theor Biol, 8:1-7.
Delaleu JC, Holley A (1980) Modification of transduction mechanism in the frog's olfactory mucosa using a thiol reagent as olfactory stimulus. Chem Senses, 3:205-218.
Doving KB (1974) Odorant properties correlated with physiological data. N.Y. Acad of Sci, 237:184-192.
Dravnieks A, Laffort P (1972) Physico-chemical basis of quantitative and qualitative odor discrimination in humans. In Schneider D (ed), Olfaction and Taste 4, Wissens-Verlag-MBH, Stuttgart, 142-148.
Dreesen TD, Koch RB (1982) Odorous chemical perturbations of (Na^+ + K^+) - dependent ATPase activities. Biochem J, 203:69-75.
Duchamp A, Revial MF, Holley A, Mac Leod P (1974) Odor discrimination by frog olfactory receptors. Chem Senses 1:213-233.
Duchamp-Viret P, Duchamp A, Vigouroux M (1989) Amplifying role of convergence in olfactory system: comparison of receptor cell and second order neuron sensitivities. J Neurophysiol 61:1085-1094.
Edwards DA, Mather RA, Dodd GH (1988) Spatial variation in response to odorants on the rat olfactory epithelium. Experientia, 44:208-211.
Eminet BP, Chastrette M (1983) Discrimination of camphoraceous substances using physico-chemical parameters. Chem Senses, 7:293-300.
Fesenko EE, Novoselov VI, Novikov, JV (1985) Molecular mechanisms of olfactory reception. VI Kinetic characteristics of camphor interaction with binding sites of rat olfactory epithelium. Biochem Biophys Acta 839:268-275.
Fesenko EE, Novoselov VI, Bystrova MF (1987) The subunits of specific odor-binding glycoproteins from rat olfactory epithelium. FEB, 219:224-226
Fesenko EE, Novoselov VI, Bystrova MF (1988) Properties of odour-binding glycoproteins from rat olfactory epithelium. Biochem. Biophys. Acta, 937:369-378.
Gennings JN, Gower DB, Bannister GH (1977) Studies on the receptor to 5 -androst-16-ene-3-one in sow nasal mucosa. Biochem. Biophys. Acta, 496: 547.
Gesteland RC, Lettvin JY, Pitts WH, Rojas A (1963) Odor specificities of the frog's olfactory receptors. In Zotterman (ed.), Olfaction and Taste, vol 1, Pergamon Press, London, pp 7-21.
Gesteland RC, Getchell, TV (1972) The chemistry of olfactory reception: stimulus specific protection from sulfhydryl reagent inhibition. Proc Nat Acad Sci USA, 69:1494-1498.
Gesteland RC, Yancey RA, Farbman AI (1982) Development of olfactory receptor neuroselectivity in the rat fetus, Neurosci., 7:3127-3136.
Getchell TV (1974) Unitary responses in frog olfactory epithelium to sterically related molecules at low concentration. J Gen Physiol, 64:241-261.
Getchell TV, Shepherd GM (1978) Adaptative properties of olfactory receptors analysed with odour pulses of varying durations. J Physiol, London, 282:541-560.
Goldberg SJ, Turpin J, Price S (1979) Anisole binding protein from olfactory epithelium: evidence for a role in transduction. Chem. Senses, 4:207-214.
Gross-Isserof R, Lancet D (1988) Concentration-dependent changes of perceived odor quality. Chem Senses, 13:191-204.
Harding JW, Getchell TV, Margolis FL (1978) Denervation of olfactory pathway in mice. V Long-term effect of $ZnSO_4$ intranasal irrigation on behavior, biochemistry and morphology. Br Res, 140:271-285.

Kashiwayanagi M, Sai K, Kurihara K (1987) Cell sustensions from porcine olfactory mucosa: Changes in Membrane Potential and membrane fluidity in response to various odorants. J.Gen. Physiol., 89:443-457.

Kauer JS, Moulton DG (1974) Responses of olfactory bulb neurons to odour stimulations of small nasal areas in the salamander. J Physiol, London, 243:717-737.

Laing DG (1983) Natural sniffing gives optimum odour perception for humans. Perception, 12:99-117.

Lee KH, Wells RG, Reed RR (1987) Isolation of an olfactory cDNA: Similarity to retinol-binding protein suggest a role in olfaction. Science, 235:1053-1056.

Mackay-Sim A, Shaman P, Moulton DG (1982) Topographic coding of olfactory quality: odorant-specific patterns of epithelial responsivity in the salamander. J Neurophysiol, 48:584-596.

Mackay-Sim A, Nathan MH (1984) The projection from olfactory epithelium to the olfactory bulb in the salamander, Ambystoma tigrinum. Anat Embryol 170:93-97.

Mustaparta A (1971) Spatial distribution of receptor-responses to stimulation with different odours. Acta Physiol Scand, 82:154-166.

O'Connel RJ, Stevens DA, Akers RP, Coppola DM, Grant AJ (1989) Individual differences in the qualitative responses of human subjects to various odors. Chem Senses 14:293-302.

Ohloff G, (1986) Chemistry of odor stimuli. Experientia, 42:271-279.

Pace U, Hanski E, Salomon Y, Lancet D (1985) Odorant-sensitive adenylate cyclase may mediate olfactory reception Nature 316:255-258.

Pelosi P, Baldaccini NE, Pisanelli AM (1982) Identification of a specific olfactory receptor for 2-isobutyl-3-methoxy pyrazine. Biochem. J,G. 201: 245-248.

Pevsner J, Trifiletti RR, Strittmatter S (1985) Isolation and characterisation of an olfactory receptor protein for odorant pyrazines. Proc Natl Acad Sci USA 82:3050-3054.

Polak EH (1973) Multiple profile-multiple receptor site model for vertebrate olfaction. J theor Biol, 40:469-484.

Price S (1978) Anisole binding protein from dog olfactory epithelium. Chem Senses, 3:51-55.

Revial MF, Duchamp A, Holley A (1978) Odour discrimination by frog olfactory receptors: a second study. Chem Senses, 3:7-21.

Revial MF, Sicard G, Duchamp A, Holley A (1982) New studies on odour discrimination in the frog's olfactory receptor cells. II Mathemathical analysis of electrophysiological responses. Chem. Senses, 8:179-194.

Saucier D, Astic L (1986) Analysis of the topographical organization of olfactory epithelium projection in the rat. Br Res Bull, 16:455-462.

Schiffman SS (1974) Physico chemical correlates of olfactory quality. Science, 185:112-117.

Shirley SG, Robinson CJ, Dodd GH (1987) The influence of temperature and membrane-fluidity changes on the olfactory adenylate cyclase of the rat. Biochem J 245:613-616.

Shirley SG, Polak E, Dodd GH (1983) Selective inhibition of rat olfactory receptors by concanavalin A. Biochem Soc Transact, 11:780-781.

Sicard G, Holley A (1984) Receptor cell responses to odorants: similarities and differences among odorants. Br Res, 292:283-296.

Sicard G (1985) Olfactory discrimination of structurally related molecules: receptor cell responses to camphoraceous odorants. Br Res, 326:203-215.

Sicard G (1986) Electrophysiological recordings from olfactory receptor cells in adult mice. Br Res, 397:405-408.

Silver WL, Arzt AH, Mason JR (1988) A comparison of the discrimininatory ability and sensitivity of trigeminal and olfactory systems to chemical stimuli in the tiger salamander. J Comp Physiol A, 164:55-66.

Steward WB, Pedersen PE, Greer CA, Shepherd GM (1985) The topography of olfactory epithelium to olfactory bulb projections in the rat. Chem Senses, 10:400

Thommesen G, Doving KB (1977) Spatial distribution of EOG in the rat: a variation with odour quality. Acta Physiol Scand, 99:270-280.

THE BIOCHEMISTRY OF ODORANT RECEPTION AND TRANSDUCTION

Richard G. Vogt
Robert Rybczynski*
Michael R. Lerner*

Section of Molecular Neurobiology
(*: Howard Hughes Medical Institute Research Laboratories)
Yale University School of Medicine
333 Cedar Street
P.O. Box 3333
New Haven, CT 06510

I. INTRODUCTION

Odor reception and signal transduction have for some years represented the two major foci within the "molecular mechanisms of olfaction" group. Significant findings of the past decade allow us to view olfaction in a new light. We are finally beginning to see how the primary step of odor recognition really does work. This passage alone frees us to consider a more limited set of possibilities at more central levels. A significant debate has been peripheral vs. central. Does the peripheral nervous system, i.e. the primary receptor cells, have a dynamic role in determining how we perceive odor, or are the dynamic aspects of odor perception the restricted domain of the central nervous system, i.e. the brain? The specificity of peripheral receptors has always been recognized, however the nature of this specificity has been questioned. Is peripheral odor discrimination accomplished using many receptors, each with high specificity, or only a few receptors, each with relatively low specificity? In the extreme many-receptor case, each receptor is activated by a different molecular component of a complex mixture. With increased specificity, fewer distinct molecules can be detected. There are thus certain volatile molecules that can not be smelled, and the number of distinct receptors should reflect the number of molecules that can be smelled. In the extreme few-receptor case, each receptor is activated by a broad spectrum of different

NATO ASI Series, Vol. H 39
Chemosensory Information Processing. Edited by D. Schild

component molecules. With decreased specificity, more distinct molecules can be detected. All volatile molecules can be smelled, but not necessarily distinguished, and the number of distinct receptors should be far fewer than the number of molecules that can be smelled. We have sought a solution somewhere between these two extremes, and it is now clear that more than one solution may exist within a single animal, correlating for example with the role certain odorant molecules play for that animal, e.g. sex pheromones vs. food odors. Our studies are showing us that the periphery does have a dynamic role in odor perception, as well as the nature of that dynamism.

In the above discussion, "receptor" has been used in its historically loose form to refer to the sensory cell in general. However, cells are only cells, made unique by which genes they express as proteins. It has long been assumed that the specificity of a given "receptor cell" is due to the presence of specific "receptor proteins" in the dendritic membrane of the cell. This has remained a cloudy issue because none of us has, as of this writing, offered a protein and its gene as an odorant receptor in a manner that convinces us that it is the beast.

The delay in producing a receptor protein has allowed the focus of molecular research interest to become more expanded, changing from simply odor reception and signal transduction to: (1) the molecular elements that are responsible for odor induced currents in olfactory cells and (2) the molecular events that occur to an odor molecule in the extracellular environment around the sensory dendrite. The past few years have yielded the identification of several transductory proteins from olfactory neurons, including ion channels, adenylate cyclase, phospholipase and G-proteins. Extracellular olfactory proteins have also been identified, including odorant binding proteins and odorant degrading enzymes. Also, several candidate olfactory receptor proteins have been described. These gene products allow us to view olfaction for the first time at its mechanistic level. From this level we can see that: (1) receptor proteins probably do exist and several are likely to be convincingly documented over the next few years; (2) there are probably several mechanisms utilized for odor transduction; (3) there are several potential peripheral odor filters which influence the kind of molecules that can be detected; (4) low and high specificity biochemical pathways coexist at the olfactory periphery; and (5) there is the genetic capacity to change the olfactory system generation to generation.

Based on anatomy alone, it appears that the complex organization of the olfactory periphery evolved quite early. Examining either vertebrate, arthropod or mollusc, one sees bipolar sensory neurons with cell bodies located in the epithelium and axons projecting to the brain (snail: Chase, 1986; insect: Sanes and Hildebrand, 1976; Steinbrecht, 1980; Gnatzy *et al.*, 1984; lobster: Grunert and Ache, 1988; vertebrate:

Lancet, 1986; Anholt, 1987). Primary synapses have a glomerular architecture in a bulb like appendage (snail: Chase and Tolloczko, 1986; insect: Matsumoto and Hildebrand, 1981; Homberg, *et al.*, 1989; vertebrate, Shepherd, 1972, 1981). Local, secondary processing occurs in this region, and output neurons carry this partially processed information to higher brain centers (snail: Chase, 1985; insect: Matsumoto and Hildebrand, 1981; Christensen and Hildebrand, 1987a,b; vertebrate: Greer, 1987; Shepherd and Greer, 1988). In insects, lobsters and vertebrates, the chemo-receptive membrane of the sensory neurons is on non-motile cilia (microtubules present without dynein arms) which project from a single dendrite emerging from the neuronal cell bodies. As much as the anatomy of olfaction appears to be conserved through these groups, we expect that the molecular mechanisms which underlie olfaction may also be conserved. This does not necessarily mean, for example, that only a single transductory pathway exists for all odorants in all animals. It does mean that we should expect examples of genetically homologous olfactory elements to be represented in each phyla. Thus, whether we study these molecular processes in vertebrate, arthropod or mollusc, the lessons learned should be cross applicable.

One reason to expect a maintenance of the olfactory machinery throughout this extent of evolution is the constancy of the olfactory environment. Terrestrial animals smell small, volatile lipophilic molecules. These molecules must pass through an aqueous medium in order to reach the receptive membranes of the olfactory cilia. This requirement for a hydrophobic molecule to pass through a hydrophilic barrier places a design constraint on terrestrial olfaction that all animals have in common. Aquatic animals have it somewhat easier, as they smell hydrophilic molecules, such as amino acids and nucleotides (e.g. ATP). In that terrestrialization developed in aquatic animals, either the aquatic olfactory mechanism needed to be modified to become a terrestrial mechanism, or the volatile terrestrial odorants needed to be modified to become aquatic odorants. We now see that the latter was apparently the case by evidence of the existence of odorant binding proteins (OBP) which are thought to solubilize the lipophilic molecules, thus allowing their transport through the aqueous barrier. The aquisition of OBPs may therefore be partly responsible for the maintenance of the basic organizational scheme of aquatic olfaction in a terrestrial life style. It may be, however, that a clear understanding of these evolutionary issues will not come until the relationships of taste and smell are understood in crustaceans and fish, for in taste and smell in both animal groups function against a more or less single chemical environment.

We currently view the process of odor detection as two sub-processes: (1) those events which occur to the odorant molecule both before and after it arrives at the receptor protein (perireceptor events); and (2) those transductional events which couple odorant-receptor binding to the generation of an action potential in the sensory axon. The actual interaction between odorant and receptor plays a key role in both sub-processes, and remains central in all studies of odor detection.

II. BIOCHEMISTRY OF THE OLFACTORY PERIPHERY: TRANSDUCTORY PROTEINS

In the detection of an odor molecule, the molecule is somehow recognized by the sensory cell, and the process of this chemical recognition leads to the induction of electrical currents in the cell. We assume that there is an initial generator current that is ligand induced, and that this generator current in turn induces a voltage activated current. The mechanism of the ligand activated current is presumably located in the sensory dendrite/cilium near the site of odor recognition, while the site of the voltage activated current is certainly located near the cell body in the axon, though it might also be located in the dendrite/cilium (see Evidence for Primary Generator Channels).

Numerous mechanisms for odor detection have been proposed over the decades, including infrared radiation (Wright, 1954, 1972, 1977; Wright and Brand, 1972) and microtubule perturbations (Atema, 1973, 1975). However, the 1970s showed us that many neurohormone and neurotransmitter systems function with specific membrane bound receptor proteins which are themselves ion channels, or are coupled to an ion channel by some "second messenger" process. Regardless of the popularity or plausibility of other olfactory mechanisms, by the mid-1970s, most molecular studies began to focus on a search for analogous systems operating in olfaction (see review by Margolis, 1977). Cagan and colleagues suggested that receptor proteins were present in fish olfaction by demonstrating amino acids binding to extracts of olfactory epithelium. Then Lancet and colleagues and others began to demonstrate the protein components for a cyclic AMP (cAMP) mediated pathway of transduction operating in isolated olfactory cilia. This was soon followed by the report of Nakamura and Gold (1987) of a cAMP gated current in olfactory cilia membranes. The result of these studies was a plausible model of olfactory transduction in which (1) odorant binding to a receptor protein induced (2) activation of an adenylate cyclase enzyme through G-protein coupling, and the resulting cAMP (3) acted directly on an ion channel inducing the generator current. There are as yet a number of problems with this view, however, most notably (1) the fact that no one has convincingly produced a receptor protein, let alone shown that it couples to a cyclase process, and (2) no one has demonstrated currents in cilia membranes which are induced by odorants acting via cAMP.

There remain several possible molecular mechanisms which fit our current view of olfactory transduction, including:

(1) ligand gated channels in which (i) the receptor and channel protein are the same (e.g. nicotinic acetylcholine channel), or receptor and channel are two distinct proteins either (ii) directly coupled or (iii) coupled through a G-protein (e.g. Codina *et al.*, 1987);
(2) some variant of a cAMP pathway, where odorant-receptor interaction induces cyclase activation through G-protein coupling and the resulting cAMP either acts directly on a channel or activates a protein kinase which phosphorylates a channel to an activated state;
(3) a pathway that activates a membrane bound or cytosolic guanylate cyclase where the resulting cGMP either activates a channel directly or activates a protein kinase which phosphorylates a channel to an activated state;
(4) an inositol triphosphate (IP_3) pathway where odorant-receptor interaction activates a phospholipase C (PLC) through G-protein coupling, and the PLC cleaves the IP_3 heads from the diacylglycerol tails (DAG) with at least three possible subsequent results: (i) IP_3 activates a channel directly (e.g. Kuno and Gardner, 1987), (ii) the IP_3 molecules move into the cytosol and interact with Ca^+ storage vesicles inducing Ca^+ release, and the Ca^{++} activates channels; or (iii) DAG migrates through the membrane to activate a protein kinase C (PKC), and the PKC activates a channels;
(5) some entirely novel pathway.

To date, evidence points strongly to the presence of the cAMP mediated pathway. However, this is largely because this general pathway is the best understood, is easily manipulated by pharmacological methods, and has thus been the most studied in olfaction. In their studies of odor induced cyclase activity in isolated cilia, Sklar *et al.* (1986) found that only certain odorants activated the adenylate cyclase, and concluded that those that did not must act via some other pathway. Also, in the lobster, individual sensory neurons have been recorded that show both hyperpolarizing and depolarizing responses to different odorants, suggesting that two or more transductory paths can exist in a given cell (McClintock and Ache, in press a). Furthermore, odorant induced activity in lobster neurons were generally unaffected by adenylate cyclase pathway perturbants, such as forskolin, IBMX, or cyclic nucleotide analogs, suggesting that in this system cAMP might not mediate odorant response (McClintock and Ache, in press b). These observations point strongly to the existence of at least one additional molecular transduction pathway in olfaction besides the cAMP path. It is noteworthy that in the odorant stimulated whole cell current studies of Firestein and Werblin (1989), the delay between odorant arrival at the cilia and current detection at the soma was so long (140-320 msec) as to suggest one of the multiple biochemical step processes rather than a direct ligand gated process, at least for a salamander cell responding to a mixture of acetophenone, amyl acetate, cineole, phenylethylamine and triethylamine.

A. Odorant Receptor Proteins

The convincing isolation of an odorant receptor protein has been elusive. Basic criteria for identifying a protein as an olfactory receptor might include: (1) the protein should be membrane bound; (2) it should be a component of the olfactory cilia; (3) it should bind odorant, and do so at physiologically reasonable odorant concentrations; (4) it should function as an olfactory receptor. Functionality could be demonstrated (i) by removing the protein from the system with a sole result that the system looses its ability to respond to odorant, or (ii) by placing the protein in an otherwise non-responsive system with a resulting gain of responsiveness. The first three points should be fairly easy to establish, though they have rarely been achieved. Demonstrating functionality is difficult, and generally requires that the mRNA or cDNA of the putative receptor be obtained. The nucleic acid can be transfected or injected into some foreign host where the nucleic acid is translated into protein, and in this foreign host the response to novel odorant and hence the functionality of the protein demonstrated. Knowing the nucleic acid sequence for a putative receptor also allows us to ask if it resembles or is homologous to a previously sequenced receptor of any type, and thus we might argue that by relationship it is a receptor. Hence, while several years ago it seemed that simply purifying and characterizing an olfactory receptor protein was sufficient, the goal now is clearly to obtain the nucleic acid sequence which encodes the receptor.

Isolations of candidate olfactory receptor proteins.

Several strategies have been utilized to isolate odorant receptor proteins over the past 15 years. Initially, ligand binding assays were developed in fish to follow the partial purification of membrane associated odorant binding proteins. Cagan and colleagues (Cagan & Zeiger, 1978; Rhein & Cagan, 1980, 1983; Cagan, 1984) demonstrated odorant binding sites in crude olfactory tissues as well as isolated olfactory cilia of rainbow trout. Amino acids (aquatic odorants) showed binding constants in the μM range and showed varying degrees of cross competition, suggesting different binding sites for different odorants. This binding activity could be sedimented by centrifugation, suggesting that it was membrane associated. Similar results were observed by Hara and colleagues (Brown & Hara, 1981), also in rainbow trout. None of these binding sites have since been purified, however.

Using a quite different approach, Price & Willey (1987, 1988) and Fesenko (Fesenko *et al.*, 1987, 1988) constructed affinity purification columns using two different odorants, made antibodies against the proteins which bound to these respective columns, and were then able to use these antibodies to block EOG responses *in vivo.* Price utilized olfactory epithelium from dogs and monoclonal antibodies, and isolated two different proteins with the same molecular mass, 61 kDa. Fesenko used a total epithelial membrane fraction from 800 rats and polyclonal antibodies, and isolated a protein of molecular mass ca. 140 kDa, a dimer comprised of an 88 kDa and a 55 kDa subunit. The 140 kDa protein was a glycoprotein, visualized by Con-A binding, and similar sized glycoproteins were observed in olfactory epithelium of rat, mouse, guinea-pig and hamster, but not frog or carp. Both the Price and Fesenko proteins have characteristics one would expect of a receptor protein, and one would hope now for a further purification of either, some sequence information, and some immunohistochemical evidence that the antigens are neuronal and located in the olfactory cilia.

In a more direct approach, Lancet (Chen *et al.*, 1986b) argued that a membrane bound receptor protein should be glycosylated, and visualized a 95 kDa glycoprotein from frog olfactory cilia using lectins. "gp95", as this protein became known, was membrane associated, was enriched in cilia homogenates relative to crude epithelial homogenates, and absent from the motile respiratory cilia of the same tissue. The protein, while very intriguing in its properties, has apparently proven difficult to purify, and unfortunately little more has been heard of it.

Moth antennae seem ideally suited to the purification olfactory proteins, due to their morphology and specificity of function. Vogt, Prestwich and Riddiford (1988) reported using a physiologically active photoaffinity analog of a sex pheromone to demonstrate pheromone binding to a membrane protein of the pheromone-specific dendrites in the moth *Antheraea polyphemus.* This 69 kDa protein was not present in membranes isolated from general nervous tissue or female antennae (only males smell this pheromone). Vogt and Lerner (unpublished data) have since observed that this protien is one of the major synthetic products appearing in the dendritic membranes at the time of final maturation of the sensory neurons. We have recently purified this protein from the isolated olfactory dendrites of 800 male antennae using preparative polyacrylamide gel electrophoresis, have obtained a partial amino acid sequence, and are in the process of obtaining the nucleic acid sequence which encodes this protein in the hopes of identifying it further.

One other approach that shows promise is homology probing of olfactory cDNA libraries. Jones and Reed (1989) very recently reported their isolation of an olfactory specific G-protein (G_{olf}, see below) by taking previously established nucleic acid fragments for other G-proteins, and using these fragments as screening probes in a cDNA library derived from rat olfactory mRNA. If any of the olfactory receptor proteins should be related to any other receptor proteins already sequenced, then one should similarly be able to use these other sequences to fish for an olfactory receptor protein. This approach relies heavily on histological verification of tissue location based on antibody binding or nucleic acid hybridization.

Chemoreceptor proteins of other systems.

While there is a mass of evidence, documented below, that supports an olfactory receptor coupled to a cAMP pathway, the possibilities are really still wide open. Not until several olfactory receptors are isolated and fully characterized will we be able to understand how olfactory detection and transduction works. Several chemoreceptors are now in fact known at the structural level, and they are quite different from each other. A class of *E. coli* chemoreceptors are transmembrane proteins with ligand (amino acid) binding sites on the outside and regulatory methylation sites on the inside (Russo & Koshland, 1983, Stewart & Dahlquist, 1987). The conformation state of the protein, and the resulting behavioral response of the bacterium, is apparently determined by whether or not ligand is bound and whether or not the protein is methylated.

Yeast mating factors (peptides) are recognized by receptor proteins with multiple transmembrane regions, reminiscent of the rhodopsin type receptors, and like the rhodopsin type receptors, the yeast receptors interact with the cell via G-protein coupling (Burkholder & Hartwell, 1985; Hagen *et al.*, 1986; Dietzel & Kurjan, 1987; Hall, 1987; Herskowitz & Marsh, 1987). Slime mold chemoreception (cAMP) is also mediated through a G-protein system, utilizing a rhodopsin type receptor protein that couples to an adenylate cyclase (Klein *et al.*, 1988).

The first eukaryotic chemoreceptor protein was recently cloned from the sperm of sea urchins by Garbers and colleagues (Ward *et al.*, 1985; Singh *et al.*, 1988). This peptide-receptor spans the membrane with a ligand binding site on the outside and a guanylate cyclase domain on the inside, such that peptide binding induces cGMP production. Furthermore, the inside portion shows the potential for regulatory phosphorylation, all of which is consistent with the chemotactic behavioral response shown by the sperm to species specific egg peptides (Miller, 1985).

On a final note, there is apparently a membrane bound sperm receptor protein in frog oocytes (Kline *et al.*, 1988). Frog oocytes are frequently used as expression systems for foreign mRNA. For example, injecting mRNA derived from rat brain produces both muscarinic acetylcholine receptors and serotonin receptors in the egg membrane. These receptors are detected electrophysiologically via their activation of a Ca^{+} dependent Cl^{-} channel by application of either transmitter, apparently through an IP_3 pathway (see above). Jaffee and colleagues (Kline *et al.*, 1988) recently found that this IP_3 pathway is apparently used in real life by sperm to induce a lifting of the fertilization membrane to block polyspermy. Thus, the expressed foreign receptors are apparently coupling into this IP_3 pathway by interacting with the G-protein that normally couples the sperm receptor to a phospholipase.

B. Evidence for Olfactory Adenylate Cyclase

The demonstration of odorant driven adenylate cyclase activity by Pace and Lancet (Pace *et al.*, 1985) has had profound influence on olfactory transduction research. Kurihara and Koyama (1972) had demonstrated that olfactory epithelium contained exceptionally high levels of adenylate cyclase activity, suggesting that this enzyme had a part to play in olfaction. Menevse *et al.* (1977) further implicated a role for cAMP in olfaction by showing epithelial depletion of phosphodiesterase (PDE) activity following bulbectomy and subsequent neuron death in the epithelium. PDE resets a cyclase pathway by degrading the cAMP messenger to inactive AMP.

Chen and Lancet (1984; Chen *et al.*, 1986a) developed a preparation of isolated olfactory cilia from frogs. Pace *et al.* (1985) demonstrated cyclase activity in this preparation by measuring produced cAMP in response to several pathway activators, including GTP-gamma-S, a non-hydrolizable analog of GTP which binds to and activates G-protein (see below). They then showed that this cyclase could be stimulated by a mixture containing 4 odorants (citral, L-carvone, 1,8-cineol and *n*-amyl acetate). They further suggested the presence of two alpha subunits of G-protein, implicated as G_S and G_I, based on ADP-ribosylation with cholera toxin and pertussis toxin.

The adenylate cyclase work was substantially elaborated on by Sklar *et al.* (1986) who examined odorant activation of adenylate cyclase in rat olfactory cilia. This study examined the response to over 40 odorants possessing a variety of different characteristic properties, i.e. fruity, floral, minty, herbaceous and putrid. Notably, while most of the odors tested induced cAMP production, not all did. Putrid odorants such as isobutyric acid did not, but a scattering of the other odorants also did not, such as vanillin. This suggested strongly that, while an adenylate cyclase pathway may mediate transduction for some odorants, there must be at least one additional pathway that is not mediated by cAMP.

These studies brought a rash of attention to olfactory transduction and to the cAMP pathway. Two weaknesses of these studies, overlooked by most, were the temporal aspects of the studies and the odor concentrations. As to the former, odor recognition occurs on a time scale of seconds, but the cAMP experiments required up to 10 minutes to accumulate enough cAMP to measure. Thus, the experiments could not resolve whether the cAMP response was one of primary transduction or of some secondary process. This issue might be resolved using some of the G-protein activators/inhibitors and forskolin in real time physiological experiments.

The concentration issue arose with a paper by Lerner *et al.* (1988) which documented an odorant induced adenylate cyclase response in cultured frog melanophores. These cells disperse there pigment in response to cAMP, a response thought normally to be due to a photoreceptor that is cyclase coupled. The cells seemed an attractive system for cloning a cyclase coupled receptor: (1) cultured cells could be transfected with mixed DNA, and DNA would be expressed as protein; (2) any cell receiving DNA for a cyclase driving olfactory receptor would disperse pigment when exposed to odorant. Thus, from a culture plate containing many thousand cells, the one or two which became pigmented could be easily observed, isolated, and cloned as containing olfactory receptors. In the course of trying to establish this system, Lerner *et al.* (1988) examined the cultured cells for their background, non-transfected response to odorants. Under similar conditions reported in the Sklar *et al.* (1986) study, many odorants stimulated cyclase mediated pigmentation in cultured frog melanophores. Although this led to an interesting, though brief, debate over the existence of receptor proteins (Dionne, 1988; Shirley and Robinson, 1988), the real message of this work was a caution to distinguish between specific (meaningful) and non-specific (non-meaningful) cyclase activation. Cyclase studies were being conducted at near mM odorant concentrations, and these concentrations could induce "meaningless" cyclase responses in non-olfactory cells. It would seem better to study a more specific pathway, one functioning at μM or lower concentrations.

C. Evidence for Olfactory G-Proteins

G-proteins are membrane bound GTP-binding proteins situated in the cytosolic face of the membrane. They are a heterotrimer, comprised of an alpha (39-52 kDa), beta (35-36 kDa) and gamma (7-10 kDa) subunit (see reviews by Sibley *et al.*, 1987; Gilman, 1987; and Neer & Clapham, 1988). Many receptors couple to other molecules via G-proteins, however while the beta and gamma subunits are common for all, there are several different alpha subunit types and these are associated respectively with specific receptor types. Among the better known alpha subunits are G_S and G_I which stimulate and inhibit adenylate cyclase, respectively, G_O which directly gates potassium channels in the brain (VanDongen *et al.*, 1988), G_K which gates K^+ channels in heart muscle (Codina *et al.*, 1987), and transducin or G_T which couples rhodopsin to guanylate cyclase in the visual system. There is also evidence that G_S has two forms based on alternative mRNA splicing, one form activating adenylate cyclase, and the other activating Ca^{++} channels (Mattera *et al.*, 1989).

In the G-protein OFF state, GDP is bound to the alpha subunit. Receptor activation induces GDP displacement and GTP binding and an ON state. The OFF state returns as the alpha subunit dephosphorylates the GTP to GDP, retaining the GDP until the cycle is repeated. During its ON, GTP bound state, the alpha subunit can activate its target protein. Some models suggest that the heterotrimer exists during GDP bound state, but that the complex dissociates and the alpha subunit migrates solo during the GTP bound state, and is involved in activating the target. However, there is at least one example of a beta-gamma dimer, rather than the alpha subunit, activating channels (Logothetis *et al.*, 1987).

Both Pace *et al.* (1985) and Sklar *et al.* (1986) were able to drive cyclase activity in isolated olfactory cilia using GTP-gamma-S, a non-hydrolyzable analog of GTP. Heldman and Lancet (1986) also demonstrated protein kinase (PK) activation by GTP-gamma-S. PK is a common cyclase pathway enzyme. These experiments suggested G-protein involvement in odor transduction, with GTP-gamma-S presumably binding to an alpha subunit and inducing long term activation.

Anholt utilized monoclonal antisera developed by S.M. Mumby (Mumby *et al.*, 1986) to directly demonstrate the presence of G-proteins in olfactory cilia of frogs (Anholt *et al.*, 1987a,b). Tissue homogenates were analyzed by immunoblot of polyacrylamide gels. Membranes of both cilia and deciliated epithelium showed only the presence of G_s, based on molecular weight (ca. 45 kDa). The deciliated olfactory epithelium membranes and brain membranes also showed additional strong titers of G_I and G_O, again based on molecular weight. Respiratory cilia of the olfactory mucosa showed no G-protein antigens. Immunohistochemical studies showed very strong staining with a beta-subunit antibody, but only comparatively weak staining with an general alpha-subunit antibody. The strong beta antibody staining argued convincingly for the presence of G-proteins in olfactory cilia. The weak staining of cilia by alpha antibody is confusing, as the immunoblot showed that the enriched cilia preparation contained an antigen of the correct mass to be a G_S subunit.

G_S and G_I can be ribosylated by cholera toxin and pertussis toxin respectively, using an ADP-ribose as the donor group. Both Pace and Lancet (1986) in rat and Bruch and Kalinoski (1987) in catfish demonstrated a G_S type alpha subunit in isolated olfactory cilia by using cholera toxin to ribosylate membrane proteins. The rat protein was 42 kDa while the fish protein was 45 kDa. The fish cilia also contained a pertussis toxin ribosylated protein, suggesting a G_I type alpha subunit as well. Today, these results are not so clear, as other alpha subunits are known to be ribosylated by these toxins. The rat antigen (Pace and Lancet, 1986) seems too small to be G_S (45 kDa), although the fish antigen is a correct size for a G_S (Bruch and Kalinoski, 1987).

A substantial offering has recently been made in a report by Jones and Reed (1989) of a new and olfactory specific alpha subunit which they have named G_{olf} (44 kDa). They screened a rat olfactory cDNA library with a nucleic acid probe based on a mRNA region highly conserved among all the known alpha subunits. They initially identified G_S, G_O, and G_I subunits. However, the G_S subunit, the only one of these subunits which drives cyclase, remained following bulbectomy, suggesting that it was non-neuronal and therefore not associated with olfactory cilia. They then rescreened with greater care and identified and sequenced a novel alpha subunit, G_{olf}. mRNA for G_{olf} disappeared from olfactory epithelium following bulbectomy. Antisera was generated against synthetic peptides corresponding to regions of G_{olf} having amino acid sequences unique from the other alpha subunits. G_{olf} antigen was then detected by immunoblot from isolated cilia and from normal extracted epithelium, but not from epithelium following neuron depletion by bulbectomy. Immunocytochemistry showed that this antisera cross reacted strongly with olfactory cilia. G_{olf} was finally shown to be capable of driving cyclase activity by transfecting a cell line deplete of G-proteins with G_{olf} DNA and then stimulating cyclase activity with GTP-gamma-S.

Jones and Reed (1989) suggest that G_{olf} is the real adenylate cyclase driver in olfactory cilia and that the previously observed G_S is non-neuronal, presumably located in support or gland cells. Anholt *et al* (1987a,b) may have visualized G_{olf} with the alpha-subunit antibody. G_S and G_{olf} have similar sequence, and similar molecular mass (Jones and Reed, 1989). The alpha-subunit antibody would presumably interact with G_S and G_{olf}, but not necessarily as strongly. Alternatively, G_{olf} might not be in frogs. Jones and Reed did not use the same antibodies used in the Anholt *et al.* (1987a,b) study, so it is difficult to compare these studies with certainty. Neither did Jones and Reed present any ribosylation data, so it is also difficult to compare G_{olf} with the 45 kDa protein observed by Bruch and Kalinoski (1987), although they are about the same size. The Pace and Lancet (1986) data is more difficult to fit to the Jones and Reed (1989) data. The ribosylated proteins of Pace and Lancet seem too small to be either G_S or G_{olf}. Consequently it is difficult to know what they visualized. Since cholera toxin is now known to ribosylate several alpha subunits (Logothetis *et al.*, 1987) it may be that what Pace and Lancet saw was not a G_S subunit at all, and possibly not even from the cilia (Jones and Reed, 1989).

D. Evidence for Odorant-Activated Channels.

We should expect at least two kinds of channels functioning in olfaction: (1) odorant-activated channels which are opened by odorant and which induce a local depolarization and a generator current, and (2) voltage activated channels which are induced to open by the generator current and which in turn generate the action potential. We assume that the odorant-activated channels are functionally located in the olfactory cilia while the voltage activated channels are functionally located near the soma/axon juncture. However, because we know nothing of where in these neurons membrane proteins are inserted, we do not really know where we should not find these channels. For example, patch recordings from the soma might capture odorant-activated channels, even though these are in a location that would not allow their activation by odorants *in vivo.*

Nakamura and Gold (1987) have presented the only really convincing recordings of possible odorant-activated channels. These were demonstrated as inside out patch recordings of ciliary membrane of dissociated olfactory neurons of frogs. These channels produced non-specific cation currents (Na^+, K^+) in response to cAMP and cGMP, based on their displaying a near 0 mV reversal potential. Channels of this character were also recorded in dendrite and soma membrane, suggesting perhaps that membrane protein insertion may occur near the soma. These recordings were extremely exciting when reported, as they suggested the first transduction pathway: odorant to receptor to cAMP to channel. However, the assumptions based on this study are seriously flawed by the lack of a demonstrable odorant connection.

Labarca and Anholt (Labarca *et al.*, 1988) reported a different kind of channel, but one that again possessed non-specific cation properties. They incorporated an isolated cilia preparation from frog into an artificial membrane and measured odorant induced conductances. Odorant concentrations were low, 40 nM bell pepper odor and 100 nM citralva, suggesting that they were looking at a ligand-protein interaction rather than membrane disruption. Because of the lack of any second messenger substrates and the physical dilution of second messenger enzymes in their preparation, they argued that ligand was acting directly on a channel, and not acting through a second messenger pathway.

Specificity, mechanisms, and choice of ligand.

The Labarca and Anholt study (Labarca *et al.*, 1988) raises highlights an interesting point. We've asked ourselves whether or not odorants in general act via receptor proteins, and we have visualized those receptor proteins as stand alone protein globules sitting in the membrane with a small odorant binding pocket oriented to the outside. Somewhere within the membrane, this receptor protein touches other membrain proteins, i.e. G-proteins. It seems likely that in aquatic olfaction, hydrophilic odorants work this way. It seems likely that in a highly specific terrestrial system such as moth sex-pheromone reception, that this receptor model fits. But in the terrestrial environment we smell lipophilic volatiles, and there are a lot of lipophilic volatiles. Must we expect that all terrestrial odorants activate channels through direct interaction with receptor proteins? Couldn't some, like a bell pepper odor, simply be extra good membrane protein perturbants? The question then arises, if a molecule is simply good at perturbing membranes or membrane proteins, how is sufficient specificity achieved so that it has a distinct odor? The answer could be that cells are unique based on which genes are expressed as proteins, and thus, not all olfactory neurons are a like. Perhaps only a subset of olfactory neurons can be "non-specifically" activated through channel or cyclase perturbation by a given volatile.

Odorants such as bell pepper odor are weird. Bell pepper odor (2-isobutyl-3-methoxypyrazine, or IBMP) binds to olfactory binding protein (OBP) with unusually high affinity relative to other odorants (Pevsner *et al.*, 1985). IBMP activates adenylate cyclase at 100 μM in frog and rat olfactory cilia (Sklar *et al.*, 1986). IBMP activates cation conductances in frog olfactory cilia in the absence of any second messenger (Labarca *et al.*, 1988). Perhaps we are being misled by unfortunate choices of ligands whose mode of action should not be expected to be identical. It is difficult to argue that olfactory systems evolved to be sensitively tuned to bell pepper odor in the way moth olfaction is tuned to sex-pheromone. Clearly, it is the property of bell pepper odor itself that makes it so potent. If evolution has acted on this system, it has acted on the plants which contain this compound, perhaps developing certain odors as cryptic feeding deterrents. Such an evolutionary process would not "care" which olfactory protein(s) were perturbed, so long as the end result was accomplished, i.e. action potentials were generated in some subset of olfactory neurons.

Perhaps the best criteria for selecting an odorant for use as a probe of molecular mechanisms is its behavioral significance to the animal. The extreme case is sex-pheromone reception/transduction by moths. The behavior is reproductive, and thus the evolutionary selective pressure on a well tuned process is extreme. Furthermore, both odor production by the females and sensitivity by the male are coevolved. The sex-pheromone is not an associatively learned odor. In this kind of system we expect a well defined biochemical pathway utilizing a receptor protein. That is, the initial protein in the transduction pathway, what ever else it is (i.e. cyclase or channel), is a membrane associated odorant binding protein with a specific, finely tuned odor-binding site.

Other transductory channels.

We began this section by assuming only non-voltage sensitive ion channels are present in the dendritic/ciliar membrane, however this may not be the case. de Kramer (de Kramer, 1985; de Kramer and Hemberger, 1987) made electrophysiological studies of transduction in moth pheromone detection, and suggested that action potentials might be induced in the dendritic membrane. In one of de Kramer's experiments, two extracellular electrodes were placed respectively at the tip and middle of a sensory dendrite. On odor stimulation, the electrodes detected coincident impulses, suggesting a spike had been initiated between, in the dendrite itself. Had a spike been initiated only in the soma/axon region, an antidromic impulse would have been detected first by the middle electrode, then by the tip electrode. de Kramer suggested that a spike generating mechanism in the dendrite would explain the apparent exquisite stimulus sensitivity of this very low noise insect neuron. Such a channel could be ligand and voltage activated so that only one species of channel need be represented in the cilium, or there might be voltage activated channels present in the dendrite in addition the odorant activated channels responsible for the primary generator current.

Labarca *et al.* (1988), in addition to the odorant activated channel, also detected a Ca^{++} sensitive K^+ channel and a cation channel (Na^+ or K^+). However, while their preparation was enriched in olfactory cilia, there is no assurance that these channels, or any of their observed channels for that matter, are neuronal.

Nature of the primary ionic conductance.

We still do not know what ion is responsible for the olfactory generator current. Consequently we do not know what kind of ion channel to expect. The channels observed so far appear to be non-discriminating cation channels. Given the choice they would pass K^+, Na^+ or Ca^{++}. Of course channels are not really given the choice *in vivo*, as transmembrane ionic distribution is generally based on factors such as ion pumps and membrane impermeant macromolecules. In olfaction, the ionic composition is considerably influenced by the glandular make up of mucus in vertebrates, and by neuronal support cell modification of mucus in vertebrates (sustentacular cells) and sensilla lymph in arthropods (tormogen and trichogen cells).

In insects, the extracellular fluid surrounding the sensory neurons (sensilla lymph) contains unusually high concentrations of K^+ (Steinbrecht and Zierold, 1987). Also, the support cell apical membranes, which face the sensilla lymph, contain structures having the appearance of K^+ pumps (Steinbrecht, 1986). Based an apparently similar situation in fly taste sensilla, Thurm and Kuppers (1980) suggested that a standing K^+ current existed between the odor activated channels in the dendrite membrane and the soma/axonal junction, and that ligand binding induced a modulation of this current. Kaissling and Thorson (1980) adapted this K^+ current model to moth olfactory sensilla. Thus it might be, at least in insect olfaction, that K^+ channels should be anticipated to be responsible for primary generator currents. The only report contrary to this is the study of de Kramer (1985) which suggested (1) that the combined membrane resistances of all the involved cells was too great for such a standing K^+ current loop to contribute significantly in transduction, and (2) that a dendritic spiking mechanism seemed to exist. Furthermore, the K^+ pumps observed by Steinbrecht (1986) are based on insect gut K^+ pumps which are used for osmotic H_2O transport out of the gut (Harvey *et al.*, 1981). It is likely that the sensilla K^+ pumps are functioning as osmotic H_2O pumps as well, maintaining the aquatic environment within the exposed sensilla, and may have little to do with sensory transduction directly.

E. Conceptual model of olfactory transduction: square pegs in round holes?

The studies described above have focused on an olfactory transductory pathway that is cAMP mediated. Taken together, the findings suggest a model where odorant-receptor interaction leads to adenylate cyclase activation, and the resulting cAMP acts directly on a cation channel to initiate a generator current. It is important to keep in mind that this scheme exists because we naturally try to create orderly explanations of our observations. It is just as likely that all of these steps do not fit together, that we are forcing together two or more independent pathways. Experiments are lacking which would allow us to connect the pieces. A cilia recording showing currents driven by odorant, $GTP_{gamma}S$, forskolin, and not inhibited by IBMX, a PDE inhibitor, would be nice. Even G_{olf} (Jones & Reed, 1989) does not have a clear function, as it stimulated cyclase activity only weakly compared to G_S in a transfected cell line. Perhaps G_{olf} activates a channel directly. Part of the interest in G-proteins was that they could be manipulated in order to in turn manipulate adenylate cyclase. Though we have forskolin that can drive adenylate cyclase, we have no probes that can specifically block adenylate cyclase. Our inability to uncouple adenylate cyclase makes it very difficult to understand the contribution of other pathways.

F. Evidence for other transductory pathways in olfaction: IP_3

There is as yet little information regarding the presence of other transductory pathways in olfaction. The most sought after has been the inositol triphosphate (IP_3) pathway, a search justified by Sklar *et al.* (1986) observing that not all odorants activated adenylate cyclase activity. The skeleton of this pathway was described earlier. Two components have been identified as being present in isolated cilia preparations. An initial step in the pathway is the activation of a phosphatidylinositol specific phosphodiesterase (PI-PDE). This enzyme is activated by odorant-receptor interaction through G-protein coupling. Huque and Bruch (1986) observed activation of a PI-PDE in isolated olfactory cilia from catfish. Activation was induced by both odorant (alanine) and $GTP_{gamma}S$.

Another process in the IP_3 pathway is the activation of Protein Kinase C (PKC), which might then activate a channel. Anholt *et al.* (1987a,b) observed PKC presence in isolated cilia using PKC antisera and phorbol ester binding. PKC was also present in respiratory cilia, and therefore was not as convincingly involved in olfaction as were the olfactory cilia-specific adenylate cyclase pathway proteins. Nevertheless, there appear to be hints of an olfactory IP_3 pathway, and further work may substantiate this. In particular, we are likely to see more studies utilizing phorbol ester. This probe activates PKC by mimicking diacylglycerol (DAG), one of the products of PI-PDE degraded PI. Phorbol ester also binds to PKC as a histological marker. Thus, phorbol ester can be a useful probe in demonstrating PKC function in tissue (Baraban, 1987).

III. BIOCHEMISTRY OF THE OLFACTORY PERIPHERY: PERIRECEPTOR PROTEINS.

Getchell *et al.* (1984) used the term "perireceptor events" to describe an odorant's entry into and passage through the mucus covering the olfactory epithelium. This was presented strictly as a vertebrate problem, and really concerned the fate of the stimulus molecule, rather than its transduction into a generator current. The problem perceived was that a lipophilic odorant needed to partition into the aqueous mucus, migrate through the mucus until it encountered a receptor protein, and then be gotten rid of. An analogous situation had previously been recognized for insects (Kaissling, 1974), where the sex-pheromone had to pass through a cuticular wall, then through an aqueous space to the receptor, and then be gotten rid of. Since 1980, a considerable amount of information has accumulated regarding the biochemistry of these passages, and with this a biochemical understanding has been achieved of the fate of an odorant in the extracellular space surrounding the sensory dendrite/cilium. The most complete story by far comes from the insect. However, the parallels seen in lobsters, snails and a number of vertebrates is so striking as to argue that both problem and solution are universal.

There are four specific aspects of dealing with odorants which concern the odor molecule itself:

1. the structure of the odor field between source and target; This concerns any animal involved in chemo-orientation, aquatic or terrestrial, and has probably been studied most in the context of moth pheromone tracking where behavioral strategies have evolved against the odor plume structure (see review by Baker, 1989).
2. the passage into the aqueous medium surrounding the sensory neurons; This is a problem unique to terrestrial olfaction, where lipophilic volatiles must go through a phase transition in order to get close to the sensory neurons.

3. migration through the aqueous medium surrounding the sensory neurons; This again is an adaptive problem for terrestrial olfaction, due to the hydrophobic nature of the odorants and their inherent insolubility in water.
4. inactivation. This ensures that the neurons only receive new signal and not residual signal. It is a signal to noise problem, and applies to both aquatic and terrestrial olfaction, though it has been studied most in arthropods.

Two biochemical solutions have evolved to deal with these four aspects: odorant binding proteins (OBPs) and odorant degrading enzymes (ODEs). OBPs act, in a sense, by converting a water insoluble odorant into a water soluble odorant. This is a rather simple solution to the problem of terrestrializing the olfactory system. ODEs are a bit different and seem to fit the life style of the animal. Flying moths make rapid assessments of external signal fluctuations, and have very potent and tissue unique enzymes for degrading signal rapidly. Slower moving lobsters have a similar though slightly less active enzyme system for degrading at least some odorants. No extracellular ODEs have been identified in vertebrates, but an intracellular enzyme, cytochrome P450, has recently been discussed as a possible odorant inactivator.

Comparative anatomy of olfactory epithelia: vertebrate, insect and snail.

There is remarkable similarity between the olfactory epithelia of vertebrate, insect and snail. The vertebrate epithelium is generally composed of two major cell types: sensory neurons and supporting or sustentacular cells (Menco, 1984; Menco and Farbman, 1985, 1987; Lancet, 1986; Getchell, 1986). The olfactory neurons are embedded among the sustentacular cells. The sustentacular cells have a microvillous apical membrane. The ends of the olfactory dendrites project to the surface, with their cilia extending beyond, into the olfactory mucus. A third cell type is present in the epithelium, the basal cells. However, these are restricted to the basal region of the epithelium, and function as progenitor olfactory neurons.

The olfactory neurons of insects are arrayed along the antennae. Two to several sensory neurons are encased within a cluster of 3 support cells, which sequentially surround the neurons like onion skin (Steinbrecht, 1980, 1986; Menco and van der Wolk, 1982; Gnatzy et el., 1984; Keil, 1984a,b). The sensory dendrite projects to the surface, and the cilia project outward. A difference between insect and vertebrate is the insect cuticle. The insect olfactory dendritic cilia are encased within a long hair-like structure made of cuticle. The neuronal cell bodies with their three support cells completely line the base of the hair. The entire structure is referred to as a sensillum, and the fluid within the sensillum and into which the dendritic cilia project is referred to as sensillum lymph (SL). The insect SL is analogous to the vertebrate olfactory mucus. The apical membranes of the support cells are highly folded, in a manner reminiscent of the apical microvilli of the vertebrate sustentacular cells.

The tentacle of the land snail serves as eye and nose (Chase, 1986). The eye is at the very tip of the tentacle, and the surrounding epithelium is olfactory. Columnar cells with microvillar like apical membranes surround ciliated olfactory neurons. The surface is wet with a non-mucus fluid rich in protein (hygroscopic?), which is distinctly different than the mucus covering the rest of the animal.

Thus, in all three phyla, primary neurons are surrounded by support cells with highly folded or microvillar apical membranes, and the neuronal dendrites/cilia project into a special fluid. Vertebrate and snail have epithelial glands which empty into and create and maintain the olfactory fluid while neuronal support cells presumably modify the fluid. Insects are a more closed situation, with the basal support cells acting to create and maintain the SL.

A. Odorant Binding Proteins

OBPs were independently identified in insects and vertebrates at about the same time (Vogt and Riddiford, 1981a,b; Pelosi *et al.*, 1982). They have generally been proposed to aid in the partitioning of volatile odorants from air into the aqueous environment surrounding sensory neurons. Also, once the odorant molecules are in the aqueous environment, the OBP continues to function by keeping the lipophilic odorant in solution. Thus, the OBPs aid in the solubilization of lipophilic odorants. The aquisition of odorant binding proteins may represent one of the adaptations animals evolved to deal with a terrestrial life style. This is suggested by the observation that insect and vertebrate OBPs show no genetic relationship (Gyorgyi *et al.*, 1988). In other words, while they are functional homologues, they are not genetic or structural homologues, a result that is consistent with the fact that insects became terrestrialized before vertebrates appeared. Thus, the insect and vertebrate OBPs seem to represent an evolutionary convergence. These animal groups solved a common problem in a similar manner but quite independently, utilizing different genetic source material to the same end.

Insect OBPs: multiple classes, specific and general.

The first moth OBP identified as such was the pheromone binding protein (PBP) of the large silkmoth *Antheraea polyphemus* (Vogt and Riddiford, 1981a,b). This soluble protein (1) was evident only in male antennae, (2) was localized to the extracellular fluid, the SL, which surrounds the sensory dendrite, and (3) bound pheromone. The *A. polyphemus* PBP is ca. 15 kDa (Vogt and Riddiford, 1981b), though some evidence suggests that it can form a 30 kDa homodimer, and therefore might *in vivo* (de Kramer and Hemberger, 1987). A striking feature of the PBPs is their high concentration in the SL, estimated at ca. 10 mM in both *A. polyphemus* (Vogt and Riddiford, 1981b; Klein, 1987) and the gypsy moth *Lymantria dispar* (Vogt *et al.*, 1989). While the ability of PBPs to bind sex-pheromone has been demonstrated (Vogt and Riddiford, 1981b; Vogt *et al.*, 1988; Vogt *et al.*, 1989), a binding constant has never been clearly established, due to problems of ligand (pheromone) solubility. Vogt and Riddiford (1986a) estimated the K_d might be in the range of 10^{-6} M, based on the concentration of PBP (10^{-5} M) required to retard enzymatic degradation of pheromone independent of pheromone concentration. Also, a glass competition assay suggested a K_d around 10^{-7} M (Kaissling *et al.*, 1985; de Kramer and Hemberger, 1987). In this study, pheromone alone in a glass vial containing H_2O quickly adsorbed to the glass wall. However, when PBP was added, a certain amount of pheromone remained in solution, and a K_d was estimated based on this amount. Unfortunately, the binding constant between pheromone and glass was never determined, so the 10^{-7} M K_d has unclear meaning. This study did show, though, that the degradative product of the pheromone (acetate ester pheromone, alcohol product) was not held in solution by PBP, indicating a certain degree of specificity with respect to which odorants can be solubilized.

PBP has been identified in a number of additional moth species, including *Antheraea pernyi, Hyalophora cecropia and Manduca sexta* (Kaissling and Thorson, 1980; Vogt and Riddiford, 1981a, Gyorgyi *et al.*, 1988), *L. dispar* (Vogt, 1987, Vogt *et al.*, 1989) *and Bombyx mori, Orgyia pseudotsuga* and *Heliothis verescins* (Vogt, unpublished data). All of these PBPs follow the general pattern described above. *M. sexta, B. mori* and *H. verescins* appear to have small amounts of PBP in female antennae, though the significance of this is unclear, as females are generally not known to respond to pheromone either physiologically or behaviorally. Also, *L. dispar* is peculiar in that it is the only member of this group that has two PBPs, both male antenna specific. The *M. sexta* PBP has been cloned and its entire sequence determined (Gyorgyi *et al.*, 1988).

The PBPs of *M. sexta, L. dispar* and *A. polyphemus* are first expressed during adult development, 2-3 days prior to adult emergence (Gyorgy *et al.*, 1988; Vogt *et al.*, 1989).

The switch which triggers this expression is apparently a declining titer in the steroid hormone 20-hydroxy-ecdysone (Vogt and Cruz, unpublished data). It is not yet known which cells synthesize PBP, but it is presumably the neuronal support cells. Once PBP expression begins, it continues at a high rate into the adult life (Vogt *et al.*, 1989). However, the absolute concentration of PBP remains constant, suggesting a high rate of PBP turnover, ca. 8×10^7 PBP molecules per hour per sensillum in *L. dispar* (Vogt *et al.*, 1989). This translates to complete PBP turnover every two days, and suggests an active process for the apical membranes of the support cells. Coated vesicles and pits, which might serve in PBP transport into and out of the SL have been observed among the apical membranes (Steinbrecht and Gnatzy, 1984).

The *N*-terminal amino acids have been compared between 7 PBPs from 6 species: *A. polyphemus* and *L. dispar* (Vogt *et al.*, 1989), *M. sexta* (Gyorgyi *et al.*, 1988) and *H. cecropia*, *B. mori*, and *O. pseudotsuga* (Vogt and Lerner, unpublished data). The PBPs can be characterized by several criteria.

(1) PBPs are antennal specific proteins which are either unique to or highly enriched in male antennae, depending on species. In several cases they have been confirmed to be present in SL and to bind pheromone.
(2) PBPs of different species are clearly homologous based on amino acid identities.
(3) PBP sequences are highly variable. Absolute identities range from ca. 30% to 70%, depending on which pairs are compared. The most similar PBPs are from species with the most similar pheromones.

A second group of moth OBP has recently been identified (Vogt and Lerner, unpublished data). This group has properties suggesting that it has general-odorant binding function, and is referred to as GOBP (general-odorant binding protein). Five GOBPs have been characterized from 5 species, those listed above with the exclusion of *O. pseudotsuga* due to insufficient material. The GOBPs have the following properties.

(1) GOBPs are antennal specific, but are expressed in female and male antennae. The *M. sexta* GOBP has been identified in isolated sensilla of female antennae. The *A. polyphemus* GOBP has been shown to bind pheromone, which is an odorant (Vogt and Riddiford, 1981b).
(2) GOBPs of different species are clearly homologous based on *N*-terminal sequence data. While they share rather low identity with the PBPs (ca. 30-40%), they are

clearly homologous to the PBPs based on several highly conserved regions which persist throughout the two groups.

(3) Unlike the PBPs, the GOBPs are nearly 100% identical among all 5 species.

Thus, at least two classes of OBP exist within moths: the PBPs which are sex specific and highly variable, and the GOBPs which are sex indifferent and highly conserved. We think that the PBPs are variable because they have become tuned to different ligands, the unique pheromone of each species. The GOBPs are conserved because they are broadly tuned to the general odorant world, which is more or less constant for all species. This touches on our initial peripheral coding question of whether there are many receptors each highly tuned to specific odorants, or relatively few receptors each broadly tuned to a group of odorants. In the moth, at least at the level of the OBPs, both alternatives exist, with the context of the odorant governing the "choice".

Vertebrate OBPs

Pelosi *et al.* (1982) identified a possible odorant receptor from the olfactory mucosa of cows, based on its ability to bind the potent bell pepper odor, 2-isobutyl-3-methoxypyrazine. They isolated and characterized the protein as a soluble, tissue specific homodimer, each subunit having a mass of 19 kDa (Bignetti *et al.*, 1985). They subsequently observed pyrazine binding in many vertebrate species, though generally absent from aquatic species (Baldaccini *et al.*, 1986). Pevsner *et al.*, (1985) similarly characterized the cow as well as a rat pyrazine binding protein. The protein showed two binding sites with different capacities: $K_{d1}=10^{-8}$ M, $B_{max1}=135$ pmol/mg protein; $K_{d2}=3x10^{-6}$ M, $B_{max2}=25$ nmol/mg protein.

Utilizing immunohistochemical techniques, Pevsner *et al.* (1986), demonstrated that the pyrazine binding protein was located in the olfactory mucus as well is in the nasal glands which secrete mucus. In addition to bell pepper odor, the protein was shown to bind methyldihydrojasmonate and 3,7-dimethyl-octanol. However, it bound amyl acetate and methoxypyrazine poorly, indicating a certain degree of specificity, a feature also supported by Topazzini *et al.* (1985). Because of its ability to bind several odorants and its extracellular location in the mucus, the authors suggested it had an odorant carrier function, and named it odorant binding protein, or OBP. The rat OBP was subsequently cloned, and its full length sequence determined (Pevsner *et al.*, 1988b). In addition, a frog OBP was cloned, identified as an OBP based on its sequence homology to the rat OBP (Lee *et al.*, 1987). The sequences of both proteins indicated a homology to a large family of small ligand-binding proteins (Cavaggioni *et al.*, 1987; Schofield, 1988).

Tissue *in situ* studies using nucleic acid probes derived from the rat OBP sequence demonstrated OBP synthesis in the lateral nasal gland (Pevsner *et al.*, 1988a). Because the duct from this gland empties at the nasal opening, the authors suggested that during an inhalation, mucus, containing OBP, becomes aspirated into a mist, and that this proteinaceous mist captures odors and then condenses on the olfactory mucosa. This seems a bit difficult to swallow, in as much as anyone smelling the potent bell pepper odor will notice that inhalation is hardly necessary for odor detection, and therefore inhalation velocity that could induce aspiration is unnecessary, at least for this odorant. Nevertheless, OBP is uniquely made in nasal glands, is clearly present in the olfactory mucus, and binds certain odorants. As it is located in a space where odors come the general aspect of its proposed function seems quite plausible.

Comparison of insect and vertebrate OBPs.

The vertebrate OBPs are homologous to a larger group of ligand binding proteins, sometimes referred to as a retinol binding protein family (Lee *et al.*, 1987; Cavaggioni *et al.*, 1987; Pevsner, 1988a,b; Schofield, 1988). However, the insect PBP is not homologous to the vertebrate OBP, based on full length sequence analysis (Gyorgyi *et al.*, 1988). This is very curious, especially since *M. sexta*, which provided the full length PBP sequence, contains a blood pigment protein, insecticyanin, which is a member of the retinol binding protein family (Riley *et al.*, 1984; Gyorgyi *et al.*, 1988). The fact that vertebrates and insects both evolved OBPs independently and from apparently different genetic material is consistent with our suggestion that OBPs represent independent adaptations for the terrestrial life style. It suggests a universal solution was achieved through convergent evolution.

Molluscan OBPs?

The olfactory epithelium of the land snail *Achatina fulica* is not covered with mucus nor does it contain mucus glands, unlike the rest of the body, but that it is wet possibly by virtue of glycoprotein secreting glands (Chase and Tolloczko, 1985; Chase, 1986). The protein is presumably hygroscopic, and therefore responsible in part for wetting the epithelium. However, it is likely that any protein with an accessible hydrophobic region can function to increase the aqueous solubility of an odorant. Chase (1986) suggests such a role. This snail protein may prove to be another OBP, another phylum's solution to the task of terrestrializing olfaction.

Do OBPs function as filters?

There are several examples where odorant selectivity has been demonstrated by OBPs. The A. polyphemus PBP solubilized the acetate ester pheromone but not the degradative product of the pheromone, which is the alcohol [E,Z]-6,11-decadienol (Vogt et al., 1985; Kaissling et al., 1985). Alcohols of this type are frequently pheromones, as is the case with the pheromone of the silk moth B. mori (Butenandt, 1959). Gypsy moth, L. dispar, has two PBPs, but the gypsy moth pheromone shows a much greater degree of association with one of them, PBP2 (Vogt et al., 1989). Also, the cow OBP showed widely differing affinities for a variety of pyrazine odorants, assayed by their ability to compete with or displace bell pepper odor binding (Pevsner et al., 1985). Thus some odorants are more readily solubilized by OBPs than are others, a property which suggests that OBPs act as a primary odor filter.

B. Odorant Degrading Enzymes

In order for an animal to maintain sensitivity to external odors, it requires a means of distinguishing what it just smelled from what it is about to smell. It needs to get rid of residual odorant molecules in order to recognize incoming odorant molecules. One way to do this is to rapidly degrade odorants soon after their arrival at the sensory neurons. Tissue specific, odorant degrading enzymes (ODE) have been described from moths and lobster (Vogt and Riddiford, 1981a,b; Vogt *et al.*, 1985; Rybczynski *et al.*, 1989, Trapido-Rosenthal *et al.*, 1987). In addition, a vertebrate intracellular enzyme has recently been discussed in the context of rapid odorant inactivation (Nef *et al.*, 1989). ODEs seem most desirable under conditions where it is important to closely monitor rapid fluctuations in external odor concentrations. Such is the case in chemo-orientation, such as flight orientation to sex-pheromone plumes for moths (reviewed by Baker, 1989). Moths are capable of dramatic changes in flight behavior 150-500 msec after loosing pheromone (Kennedy and Marsh, 1974; Marsh *et al.*, 1981; Baker & Haynes, 1987; Baker and Vogt, 1988). Such pheromone loss occurs frequently, due to wind shifts and fine scale turbulence (Murlis, 1986; David *et al.*, 1983). It may be that both the presence of an ODE and its potency in a given animal is highly dependent on the context of the odorant.

Insect ODEs: Antennal Pheromone Degrading Enzymes

The metabolic fate of pheromones in moth antennae has been a subject of interest since the 1970's (e.g. Kasang, 1971; Ferkovich *et al.*, 1973). It was early realized that because male moths could locate females from considerable distances, males must be able to detect small changes in pheromone concentration. Enzymatic modification of pheromone to inactive forms seemed a likely mechanism for maintaining sensitivity to such changes (Schneider, 1970). Over the years, data have accumulated from studies of several species of moths indicating that pheromone-degrading activity is common in antennae (Table 1). However, the majority of these studies were unable to demonstrate that the observed degradative activity was specifically associated with or near the sensory neurons.

Two pheromone degrading enzymes have been definitively identified: a 55 kDa esterase from the silk moth *A. polyphemus* (Vogt and Riddiford, 1981b, Vogt *et al.*, 1985; Klein, 1987), and a 150 kDa aldehyde oxidase (AOX) from the tobacco hawk moth *M. sexta* (Rybczynski *et al.*, 1989). Both enzymes are antennal specific, and have been localized to the SL, the fluid surrounding the sensory dendrites. While the *M. sexta* AOX is present in male and female antennae, the A. *polyphemus* esterase was only found in the male antennae. The pheromone of *M. sexta* is a mixture of two aldehydes, bombykal or (*E,Z*)-10,12-hexadecadienal (Starratt *et al.*, 1979) and (*E,E,Z*)-10,12,14-hexadecatrienal (Doolittle *et al.*, 1988). The *M. sexta* AOX degrades the bombykal component rapidly, the pheromone having an estimated *in situ* half-life of 0.6 msec (Rybczynski *et al.*, 1989). The pheromone of *A. polyphemus* is a 9:1 mixture of an acetate ester and an aldehyde, (*E,Z*)-6,11-hexadecadienyl acetate and (*E,Z*)-6,11-hexadecadienal (Kochansky *et al.*, 1975). The *A. polyphemus* esterase also degrades the acetate ester rapidly, with an estimated *in situ* half-life of 15 msec (Vogt *et al.*, 1985). The *M.sexta* AOX is known to first be expressed during adult development 2-3 days before adult emergence, at a time coinciding with initial PBP expression (Rybczynski *et al.*, 1989).

Table 1. Pheromone-degrading enzymes described from moth antennae. Only representative references are given for each species.

TYPE OF PHEROMONE	TYPE OF ENZYME	SPECIES	REFERENCES
Acetate ester	Esterase	*Trichoplusia ni*	Ferkovich *et al.*, 1973
		Anthereae polyphemus	Vogt and Riddiford, 1981
			Vogt *et al.*, 1985
		Plutella xylostella	Prestwich *et al.*, 1989
Aldehyde	Aldehyde dehydrogenase	*Choristoneura fumiferana*	Morse and Meighen, *1984*
		Heliothis virescens	Prestwich *et al.*, 1989
Aldehyde	Aldehyde oxidase	*Manduca sexta*	Rybczynski *et al.*, 1989
		Anthereae polyphemus	Rybczynski, Vogt and Lerner, unpublished data
		Bombyx mori	Rybczynski, Vogt and Lerner, unpublished data
		Heliothis virescens	Prestwich *et al.*, 1989
Epoxide	Epoxide hydrolase	*Lymantria dispar*	Kasang *et al.*, 1974
			Vogt, 1987
Alcohol	Alcohol oxidase or dehydrogenase	*Bombyx mori*	Kasang *et al.*, 1989

Until recently we have not known the fate of the second component of the *A. polyphemus* pheromone, the aldehyde. However, we have now identified an antennal specific AOX in this species, and have confirmed its location in the SL, along with the esterase (Rybczynski and Vogt, unpublished). Unlike the *A. polyphemus* esterase which is male specific, the *A. polyphemus* AOX is present in male and female antennae.

Although the females do not smell sex-pheromone, they do smell aldehydes such as trans-2-hexanal (Riddiford, 1967). Therefore the AOXs may do a double duty of degrading both pheromonal and non-pheromonal aldehyde odorants. *B. mori*, which utilizes an alcohol and an aldehyde for its pheromone (*E,Z*)-10,12-hexadecadienal (bombykol, Butenandt *et al.*, 1959) and (*E,Z*)-10,12-hexadecadienal (bombykal, Kaissling *et al.*, 1978), also possess an antennal specific AOX (Rybczynski and Vogt, unpublished). As in the other two species, the *B. mori* AOX is present in males and females. The *A. polyphemus* and *B. mori* AOXs are ca. 170 kDa and 130 kDa respectively. Taken together, these results support the view that specific and rapid pheromone degradation is important in these animals.

Lobster ODEs: Antennal Nucleotide Degrading Enzymes

Antennal ODEs have also been described from a non-insect arthropod, the spiny lobster *Panulirus argus* (Trapido-Rosenthal *et al*, 1987). Spiny lobsters utilize a number of odorants, including nucleotides (ATP, ADP, AMP), to detect food and assess its suitability (reviewed by Ache and Derby, 1985). The aesthetasc sensilla of these lobsters contain ectonucleotidase activities, probably membrane-bound, that extracellularly dephosphorylate nucleotides to adenosine. In these sensilla, there are receptors and ectonucleotidases for ATP, ADP and AMP respectively, as well as an uptake system for adenosine (Trapido-Rosenthal *et al.*, 1989). ADP and AMP are both antagonists on the ATP receptor, a feature consistent with the view that the animals are monitoring dead food for its freshness by assessing its ATP levels (Zimmer-Faust, 1987). An *in situ* half-life for AMP was estimated at ca. 0.5 sec, which corresponds to a behavioral and periodic antennal flick (Trapido-Rosenthal *et al.*, 1989). The antennal flicks are thought to be analogous to a vertebrate sniff, and also occur ca. every 0.5 sec. Thus, between flicks, the AMP ectonucleotidase is able to partially clear the sensilla of residual signal.

Insect ODEs: Non-antennal degradation.

Pheromones are generally hydrophobic molecules and as such can be adsorbed by the waxy cuticle covering most insects. If pheromone is adsorbed onto a male's body surface and then revolatilizes it could serve as a false source and complicate the behavior of locating a female, as well as attracting other males. If pheromone is absorbed by a female's body surface and then revolatilizes, its reemission will likely be in the wrong behavioral or temporal context, perhaps leading to complications in the mating process.

For these reasons, there has been interest in understanding the fate of pheromone on non-antennal surfaces (e.g., Kasang, 1971; Mayer, 1974). A number of studies have found, not surprisingly, that homogenates of many tissues can metabolize pheromone *in vitro.* Body surface enzymes have been reported from three species of moth: *Trichoplusia ni* (esterase: Ferkovich *et al.*, 1982); *Choristoneura fumiferana* (aldehyde dehydrogenase: Lonergan, 1986); and *A. polyphemus* (esterases: Vogt and Riddiford, 1981, 1986b). These enzymes include soluble enzymes of the integument (*T. ni*, Ferkovich *et al.*, 1982; *C. fumiferana*, Lonergan, 1986; and *A.polyphemus*, Vogt and Riddiford, 1981, 1986b) which may originate in epithelial cells underlying the cuticle, as well as a species specific and pheromone degrading esterase activity apparently cross-linked with the cuticle of body scales of *A. polyphemus* (Vogt and Riddiford, 1986b). The precise role of these surface enzymes has yet to be determined. However, as Vogt and Riddiford (1986b) point out a number of biochemical processes, both antennal and non-antennal, may have evolved to maintain low noise to signal ratios in pheromone-induced behaviors.

Vertebrate ODEs

A major difference between insect and vertebrate olfactory tissue is the cuticle covering the insect neurons. In contrast, the olfactory receptor cells of most vertebrates are in much more open contact with the external environment, being separated from it solely by a thin, flowing layer of mucus. In addition, these olfactory receptor cells are often associated with the respiratory system with its attendant flow of air or water. For these reasons, it has generally been assumed that the relevant fates of odorants in a vertebrate olfactory system are: (1) mucociliary transport to non-olfactory tissues, (2) internalization and transport by the circulatory system to non-olfactory tissues, or (3) desorption to the external environment (Hornung and Mozell, 1977). However, toxicologists have long been aware that the nasal apparatus of vertebrates is a rich source of catabolic enzymes, including some unique to that tissue e.g., nasal forms of cytochrome P-450 (Ding and Coon, 1988; for a review of enzymes in nasal tissue see Dahl, 1988).

Recently, interest has grown regarding the possibility that odorant metabolism plays a role in vertebrate odor perception (Getchell *et al.*, 1984) and a cDNA clone for a cytochrome P-450 specific to olfactory epithelium has been obtained from rat, with the specific aim of determining its role in olfaction (Nef *et al.*, 1989). Nevertheless, it presently seems likely that if enzymes play a role in vertebrate olfaction it is a different role than that assumed in invertebrates. Vertebrate olfactory enzymes are intracellular and hence removed from the immediate vicinity of the dendritic cilia, the presumed site of olfactory receptors and transduction. Thus it is unlikely that they can be responsible for very rapid attenuation of odor concentration in the receptor space in the same way as moth sensilla enzymes. Determining the exact role of vertebrate olfactory enzymes will require further, carefully focused biochemical and molecular studies.

C. Perireceptor Events: Models of Odorant Reception

Vogt (1987) proposed a model of sex-pheromone reception in moths based on biochemical studies:

1. adsorption onto waxy sensillum surface and migration to sensillum interior via lipophilic pore-tubules;
2. partitioning from lipophilic pore-tubule into hydrophilic sensillum lymph via interaction with PBP;
3. migration through sensillum lymph via continued interaction with PBP;

4a. interaction with membrane bound receptor protein in sensory dendrite to initiate generator current;

4b. degradation by appropriate sensilla enzyme.

Receptor interaction and degradation are both listed as step 4 to imply that degradation does not necessarily follow receptor binding. Rather, an odorant has a finite life time in the lymph, and kinetic parameters and protein concentrations are tuned such that degradation usually follows receptor binding. Furthermore we presume that it is the odorant itself, and not the odorant-PBP complex that binds to receptor. This latter view is biased by ligand-neuron specificity arguments.

Except for the initial step, this model is adaptable to non-insect olfactory systems as well, allowing for adjustments such as replacing degradative enzymes with mucociliary transport, internalization, desorption, etc. where appropriate. The first step, however, deals with an odorant's adsorption onto and passage through the sensillum cuticle. The sensilla walls are penetrated by small pores, ca. 1000 per sensillum in *A. polyphemus* (sensilla are 300 um long, pore density ca. $3/\mu m^2$)(Keil, 1984b). Each pore opens to the interior with small tubular intrusions, 3-4 per pore. These structures are remnants of wax canals, with the tubules presumably representing cell membrane extrusions torn away when the cell that cast the sensillum cuticle withdrew during morphogenesis. It has generally been assumed that these pore-tubules provide rapid entry of odor molecules into the sensillum. EM studies have shown a waxy surface layer on the sensillum which appears contiguous with pores (Steinbrecht, 1973). Thus, pheromone presumably adsorbs onto this lipophilic surface, and migrates into the sensillum via a pore-tubule.

Although early EM studies suggested that the pore-tubules made direct contact with the sensory neuron (Steinbrecht and Muller, 1971), these contacts were only rarely observed (Keil, 1982), and only when tissue was fixed in glutaraldehyde. Tisssue prepared in this manner generally looked distorted, dendrite profiles irregular and overly large, dendritic microtubules in disarray (Steinbrecht and Muller, 1971; Keil, 1982). When freeze substitution techniques were employed for tissue fixation, the tissue morphology was dramatically improved, dendrite profiles were small and round, microtubules were well ordered (Steinbrecht, 1976, 1980; Keil, 1984b). After freeze substitution, dendrites always were shown to be centered in sensilla profiles, pore-tubules never entered the lumen and dendrite-tubule contacts were never observed.

The above scheme differs from one proposed by Kaissling (1974, 1986) chiefly over the existence of functional dendrite-tubule contacts. The existence appears to be a function of improper fixation. Furthermore, the growing evidence of specific and aggressive pheromone degrading enzymes, the apparent tuning of PBPs to a species' pheromone and the indication of cross phyletic utilization of OBPs as a common solution to coping with volatile odorants would seem to support our model.

In the insect then, odorants are first captured in the waxy surface of a sensillum, migrate inward along a waxy conduit, and then face the aqueous receptor lymph. In the vertebrate and snail, the odorant confronts the aqueous mucus directly from the air. The movement of a lipophilic ligand from lipophilic environment into aqueous environment presumably requires far more energy than movement from air into water, and this may account for the extremely high concentration of PBP in the insect sensilla.

IV. PLASTICITY IN THE OLFACTORY PERIPHERY

Is there functional plasticity in the olfactory periphery that would influence its physiological the performance? The traditional view has held that the periphery has only a passive role in odor perception. Sensory neurons take in signal, and higher brain centers process the information. However, our biochemical view of the olfactory periphery shows at least three levels of odor discrimination existing already before the initial currents are generated: discrimination by receptors, discrimination by odorant binding proteins, and empirical discrimination by degradative enzymes (e.g. alcohols are not degraded by esterases, nor are all acetate esters degraded at equal rates (Vogt, *et al.*, 1985). However, while these represent examples of the contribution the periphery makes towards odor perception, they represent a relatively static contribution. Vogt (Vogt and Riddiford, 1986; Vogt, 1987) suggested that pheromone reception was dynamic because the overall kinetics of pheromone reception was dependent on pheromone concentration. In this situation, degradative enzyme and receptor were viewed as competing for pheromone from the PBP-pheromone pool. If enzyme and receptor represent separate paths from this pool, then there should be a path-bias that is dependent on the size of the pool. Thus, a seemingly fixed biochemistry would allow a dynamic response if it were constructed of parallel components. However, this is still not what we generally think of when we consider plasticity.

Plasticity is the capacity to change. As applied to the olfactory periphery, we usually consider plasticity to mean a directed change in sensitivity induced by some neuromodulatory mechanism. This is most likely to occur at three sites: (1) the olfactory cilia, where modulation might influence the sensitivity of odorant activated generator currents; (2) the soma/axon juncture, where modulation might influence the threshold for action potential generation; and (3) at the primary synapse, where modulation might influence the activity of the synapse. An important aspect of plasticity is the time scale in which it occurs. This can range from msec-sec for real time behavioral modulation to min-hours-days for behavioral state change modulation to generations for population level modulation.

The following are several examples of plasticity in the olfactory periphery. It is curious that these examples are all from arthropods. It may be that this represents some bias of this group. However, it most likely is a false bias, and more reflects the manner in which olfaction is studied in this group. We tend to study arthropod olfaction in the close context of behavior. This is less often the case in vertebrate studies. It is reasonable to expect peripheral modulation to occur among vertebrates, for example, where a seasonal use is made of the olfactory system for mating or migration.

Modulation by neurotransmitters and hormones

McClintock and Ache (in press c) have described a histamine gated chloride channel in the soma of some lobster olfactory neurons. These channels were isolated electrophysiologically in out-side out patches taken from dissociated neurons. The channel has characteristics of being directly gated by histamine, and the resulting Cl^- conductance is hyperpolarizing. The function of the channel is unclear. The authors suggest that it might be functional at or near the soma/axon junction, desensitizing the spike initiation zone to generator currents initiated in the dendritic cilia. Although histamine is an olfactory stimulant, it is generally excitatory and therefore, presumably, depolarizing. Thus the hyperpolarizing histamine gated Cl^- channel is probably not an olfactory receptor. However, as discussed earlier, nothing is known of the cellular sites of insertion of channel proteins in the membrane. This channel might actually be functional in the synaptic region as a synapse modulator, with its presence in the soma representing a stray appearance. While histamine is known to be present in the olfactory lobe, it is not known whether it circulates as a neurohormone in the lobster hemolymph. Synaptic modulation could occur on a very rapid time scale because histamine release would be local and rapidly transient. Modulation of spike initiation would act on a some what longer time scale. because circulating levels of histamine would change relatively slowly. There are no known synapses in the region of the olfactory somata, and therefore histamine levels would have to be regulated in the hemolymph which would require longer times due to the large volume.

Davis (1986) demonstrated a hormonal regulation of sensitivity to an odorant, lactic acid (LA) in olfactory sensory neurons in mosquitos. Female mosquitos seek out hosts by queuing on LA as a chemoattractant odorant. When they locate a host, they gorge themselves on a blood meal. Following the blood meal they cease their host seeking behavior and mature their oocytes. This continues until oviposition, and then host seeking behavior returns. Davis demonstrated, in electrophysiological recordings of single sensilla, that the periphery lost and gained sensitivity to LA exactly coincident with the loss and return of host seeking behavior. Furthermore, when blood of post blood meal females (no LA response) was injected into sugar fed control females, the control females showed a reduction in sensitivity to LA. Although the hormone was not identified, Davis suggested that this regulation of peripheral sensitivity was in part responsible for the regulation of host seeking behavior. This action required several hours to manifest, and therefore may have involved regulation of gene expression for some neural protein, conceivably the LA receptor protein.

Modulation by Inheritance

The precise ratio of individual components in a complex pheromone mixture is important for evoking a correct behavioral response in insects (Linn, 1986; Linn and Roelofs, 1983, 1985; Linn *et al.*, 1986). For example, male *A. polyphemus* moths respond best to a 9:1 mixture of the acetate:aldehyde components of their sex pheromone (Kochansky *et al.*, 1975). For a number of given species, isolated populations are known that utilize different ratios of the same basic mixture (Birch *et al.*, 1980; Mustaparta *et al.*, 1985; Bailey *et al.*, 1986; Guerin *et al.*, 1984). A species with a very interesting story is the European corn borer moth *Ostrinia nubilalis* (Klun, 1975).

The European corn borer utilizes a two component sex pheromone, (E)-11-tetradecenyl acetate and (*Z*)-11-tetradecenyl acetate (Klun and Maini, 1979; Roelofs *et al.*, 1987). This moth has been introduced into North America, and is known to have at least two populations which differ in that one utilizes an *E:Z* ratio of 99:1, referred to as the *E* morph, while the other population utilizes an *E:Z* ratio of 3:97, and is referred to as the morph. These ratios apply both to the mixture released by the females and to the mixture which best attracts the males. Hybrids of these two populations, which exist in nature and can be made in the lab, release and respond to a 1:1 *E:Z* ratio. It is stunning that sex pheromone synthesis in females and response in males should be so closely linked. What is even more extraordinary is that the differences in male response is detectable in the sensory neurons (Hansson *et al.*, 1987).

When a glass capillary electrode is slipped over a cut tip of an olfactory sensillum, extracellular spikes can be recorded which are not influenced by positional bias of the electrode. During pheromone stimulation, different amplitude spikes are elicited, dependent on which component of the pheromone is used as the stimulant. For example, the *A. polyphemus* sensilla, which each contain two sensory neurons, produce a large amplitude spike in response to the acetate component and a small amplitude spike in response to the aldehyde component of the pheromone (Kaissling and Thorson, 1980). This feature has been used to suggest that one sensory neuron is specifically sensitive to the acetate, while the other sensory neuron is specifically sensitive to the aldehyde (Kochansky *et al.*, 1975; Boeckh and Boeckh, 1979; Kaissling and Thorson, 1980).

In the European corn borer, Hansson *et al.* (1987) found that (1) the *Z* morph produces a larger amplitude spike in response to the *Z* isomer than to the *E* isomer , (2) the *E* morph produces large amplitude spike in response to the *E* compound than to the *Z* isomer, and (3) the *EZ* hybrid, which utilizes the 1:1 isomeric ratio, produces the same amplitude spike to both compounds. It is not clear what mechanism is responsible for generating large or small amplitude spikes, however the spike amplitude clearly reflects some property of the sensory neuron. This difference could relate to the relative sizes of the dendrites and cell bodies of the two neurons, or it could relate to some property of the spike generating mechanism of the respective neurons. But it is that this the property is consistently associated with a specific component of the pheromone, and must therefore be involved in some way in the coding of information conveyed by that component. It is extremely interesting that this regulation of information coding is occurring in the primary olfactory neurons, and apparently in the receiving end of those neurons, somewhere in the soma/axon juncture, the soma or the dendritic cilia . This is an example of genetic modulation occurring at the population level and on a generational time scale.

Another example of genetic modulation is suggested in the silk moth *A. polyphemus*, where Vogt (1987; Vogt and Prestwich, 1987) has observed what appears to be allelic variation among two olfactory proteins. One olfactory protein, the PBP, appeared as 2 electrophoretic morphs, with some individuals possessing only one or the other, and some individuals possessing both. Similarly, a second olfactory protein, a pheromone degrading esterase, appeared as at least 3 electrophoretic morphs of a basic quartet pattern, again with individuals appearing to have either different pair combinations or only a single morph. These patterns were suggestive of homozygous and heterozygous allelic inheritance. This in turn, led to the suggestion that the overall kinetic property of pheromone reception could vary between individuals, depending on which alleals the individual inherited, and assuming that the electrophoretic morphs represented kinetically distinct entities. The result of such differences would be individuals with slightly different behaviors with respect to performance. A correlation between morph type and behavioral performance has not as yet been attempted. However, allelic diversity, acted upon by selective pressures, would represent a population level plasticity operational over generational time scales.

V. THE CONTRIBUTION OF THE PERIPHERY IN OLFACTORY CODING

From a biochemical perspective, the olfactory periphery includes the non-synaptic regions of the primary olfactory neurons, and the extracellular environment that immediately surrounds the olfactory dendritic cilia. We have described several processes and biochemical entities which reside in this domain, and which influence the description received by the brain of the external world of air borne or water borne chemicals. At the very least, the olfactory biochemistry is empirically selective. Where there are parallel biochemical pathways, the periphery is also dynamic. Where there is modulation of the biochemical components, the periphery is plastic. We have presented examples of each of these aspects. Odorant binding proteins (OBP) and odorant degrading enzymes (ODE) provide initial filters to odorants en route to receptor proteins. Competition between ODEs and receptor proteins represents extracellular parallel pathways, while multiple transductory processes within a given olfactory sensory neuron represent intracellular parallel pathways. OBPs vary in their degree of specificity, as will likely be the case with receptors, with the degree of selectivity depending on context (e.g. pheromone vs. plant odorants in insects). Olfactory response can be modulated on short time scales by neuromodulators and hormones, and on long time scales by genetic means. This is not to say that, by any means, that integration is occurring at the periphery. This is empirically not true, unless some degree of cross talk can be demonstrated between sensory neurons in their distal regions. But it does say that not all olfactory inputs are equivalent, and that a given type of olfactory input might vary in its properties with time, and that time may be short or long. Before, we presumed these things to be likely, but we did not know these things to be true. Thus, the periphery has become more defined, and, as we stated at the beginning, this allows us to consider a more limited set of possibilities for the next level.

There is one aspect of biochemical studies of the olfactory periphery that is generally overlooked. Nowhere else in the nervous system is there so rich a dendritic tissue preparation available for direct study. Nor is their a piece of the nervous system so elaborated and accessible that is clearly involved in mediating specific behaviors. This tissue allows us to isolate a surprising number of protein species from a single source, and to understand them in the context against which they evolved. Because complex olfaction apparently evolved early, there seems to be a great commonality in the basic molecular and cellular strategies animals use to smell their environments. Yet within these constraints, olfactory mechanisms seem as diverse as the water and air borne chemicals which they processes, and the life styles of the animals which utilize the information conveyed by these chemicals. If there is one thing that future research in this field will accomplish, it will be the unveiling of the vast substance of this diversity.

ACKNOWLEDGEMENTS - We thank Drs T.S. McClintock and S. Firestein for helpful discussions and useful comments.
RGV was supported by NIH RO1-NS27086-01.

REFERENCES

Ache BW, Derby CD (1985) Functional organization of olfaction in crustaceans. Trends in Neurosci 8:356-360

Anholt RRH (1987) Primary events in olfactory reception. Trends Biochem Sci 12:58-62

Anholt RRH, Mumby SM, Stoffers DA, Girard PR, Kuo JF, Gilman AG, Snyder SH (1987a) Transduction proteins of olfactory receptor cells, identification of guanine nucleotide binding proteins and protein kinase C. In: Roper SD, Atema J (eds) Olfaction and Taste IX. Ann New York Acad Sci 510:638-641

Anholt RRH, Mumby SM, Stoffers DA, Girard PR, Kuo JF, Gilman AG Snyder SH (1987b) Transduction proteins of olfactory receptor cells, identification of guanine nucleotide binding proteins and Protein Kinase C. Biochem 26:788-795

Atema J (1973) Microtubule theory of sensory transduction. J Theor Biol 38:181-190

Atema J (1975) Stimulus transmission along microtubules in sensory cells: an hypothesis. In: Borgers M, deBrabander M (eds) Microtubules and Microtubule Inhibitors. North Holland, Amsterdam, pp. 247-257

Bailey JB, McDonough LM, Hoffmann MP (1986) Western avocado leafroller, *Amorbia cuneana* (Walsingham), (Lepidoptera: Tortricidae), discovery of populations utilizing different ratios of sex pheromone components. J Chem Ecol 12:1239-1245

Baker TC (1989) Sex pheromone communication in the Lepidoptera: New research progress. Experentia 45:248-262

Baker TC, Haynes KF (1987) Manoeuvers used by flying male oriental fruit moth to relocate a sex pheromone plume in an experimentally shifted wind-field. Physiol. Ent. 12:263-279

Baker TC, Vogt RG (1988) Measured behavioral latency in response to sex-pheromone loss in the large silk moth *Antheraea polyphemus*. J exp Biol 137:29-38

Baldaccini NE, Gagliardo A, Pelosi P, Topazzini A (1986) Occurrence of a pyrazine binding protein in the nasal mucosa of some vertebrates. Comp Biochem Physiol 84B:249-253

Barabab JM (1987) Phorbol esters: probes of protein kinase C function in the brain. Trends in Neurosci 10:57-58.

Bignetti E, Cavaggioni A, Pelosi P, Persaud KC, Sorbi RT, Tirindelli R (1985) Purification and characterization of an odorant-binding protein from cow nasal tissue. Eur J Biochem 149:227-231

Birch MC, Light DM, Wood DL, Browne LE, Silverstein RM, Bergot BJ, Ohloff G, West JR, Young JC (1980) Pheromonal attraction and allomonal interruption of *Ips pini* in California by the two enantiomers of ipsdienol. J Chem Ecol 6:703-717

Boeckh J, Boeckh V (1979) Threshold and odor specificity of pheromone-sensitive neurons in the deutocerebrum of *Antheraea pernyi* and *A. polyphemus* (Saturnidae) J Comp Physiol 132:235-242

Brown SB and Hara TJ (1981) Accumulation oh chemostimulatory amino acids by a sedimentable fraction isolated from olfactory rosettes of rainbow trout (*Salmo gairdneri*). Biochim. Biophys. Acta 675, 149-162

Bruch RC, Kalinoski DL (1987) Interaction of GTP-binding regulatory proteins with chemosensory receptors. J Biol Chem 262:2401-2404

Burkholder AC, Hartwell LH (1985) The yeast alpha-factor receptor: structural properties deduced from the sequence of the STE2 gene. Nuc Acids Res 23:8463-8475

Butenandt A, Beckmann R, Stamm D, Hecker E (1959) Uber den Sexuallockstoff des Seidenspinners Bombyx mori. Reindarstellung und Konstitution. Z Naturforsch 14b:283-284

Cagan RH (1984) Biochemical studies of physiological function: olfactory recognition on rainbow trout. In: Bolis L, Keynes RD, Maddrell SHP (eds) Comparative Physiology of Sensory Systems. Cambridge University Press, pp. 285-299

Cagan RH, Zeiger WN (1978) Biochemical studies of olfaction: binding specificity of radioactively labeled stimuli to an isolated olfactory preparation from rainbow trout (*Salmo gairdneri*). Proc Natl Acad Sci USA 75:4679-4683

Cavaggioni A, Sorbi RT, Keen JN, Pappin JBC, Findlay JBC (1987) Homology between the pyrazine-binding protein from nasal mucosa and major urinary proteins. FEBS Lett 212:225-228

Chase R (1985) Response to odors mapped in snail tentacle and brain by [^{14}C]-2-deoxyglucose autography. J Neurosci 11:2930-2939

Chase R (1986) Lessons from snail tentacles. Chem Senses 11:411-426

Chase R, Tolloczko B (1985) Secretory glands of the snail tentacle and their relation to the olfactory organ (Mollusca, Gastropoda) Zoomorph 105:60-67

Chase R, Tolloczko B (1986) Synaptic glomeruli in the olfactory system of a snail, *Achatina fulica*. Cell Tiss Res 246:567-573

Chen Z, Lancet D (1984) Membrane proteins unique to vertebrate olfactory cilia: candidates for sensory receptor molecules. Proc Natl Acad Sci USA 81:1859-1863

Chen Z, Pace U, Heldman J, Shapira A, Lancet D (1986a) Isolated frog olfactory cilia: a preparation of dendritic membranes from chemosensory neurons. J Neurosci 6: 2146-2154

Chen Z, Pace U, Ronen D, Lancet D (1986b) Polypeptide gp95: a unique glycoprotein of olfactory cilia with transmembrane receptor properties. J Biol Chem 261:1299-1305

Christensen TA, Hildebrand JG (1987a) Male-specific sex pheromone-selective projection neurons in the antennal lobes of the moth *Manduca sexta*. J Comp Physiol A 160:553-569

Christensen TA, Hildebrand JG (1987b) Functions, organization, and physiology of the olfactory pathways in the lepidopteran brain. In: Gupta AP (ed) Arthropod Brain: Its evolution, development, structure, and functions. Wiley, NY, pp 457-484

Codina J, Yatani A, Grenet D, Brown AM, Birnbaumer L (1987) The alpha subunit of the GTP binding protein G_K opens atrial potassium channels. Science 236:442-445

Dahl AR (1988) The effect of cytochrome P-450-dependent metabolism and other enzymes activities on olfaction. In: Margolis FL, Getchell TV (eds) Molecular Neurobiology of the Olfactory System Plenum Press, NY, pp 51-70

David CT, Kennedy JS, Ludlow AR (1983) Finding of a sex pheromone source by gypsy moths released in the field. Nature (London) 303: 804-806.

Davis EE (1986) Peripheral chempreceptors and regulation of insect behaviour. In: Payne TL, Birch MC, Kennedy CEJ (eds) Mechanisms in Insect Olfaction. Clarendon Press, Oxford, pp 243-251

de Kramer JJ (1985) The electrical circuitry of an olfactory sensillum in *Antheraea polyphemus*. J Neurosci 5:2484-2493.

de Kramer JJ, Hemberger J (1987) The Neurobiology of pheromone reception. In: Prestwich GD, Blomquist GJ (eds) Pheromone Biochemistry. Academic Press, Orlando, pp 433-472

Dietzel C, Kurjan J (1987) The yeast SCG1 gene: a G_{alpha}-like protein implicated in the a- and alpha-factor response pathway. Cell 50:1001-1010

Ding X, Coon MJ (1988) Purification and characterization of two unique forms of cytochrome P-450 from rabbit nasal microsomes. Biochem 27:8330-8337

Dionne VE (1988) How do we smell? Principle in question. Trends Neurosci 11:188-189

Doolittle RE, Tumlinson JH, Mitchell ER, Brennan MM (1988) The spectroscopic identification and stereoselective synthesis of the sex pheromone blend of the tobacco hornworm *Manduca sexta*. Abstract AGRO 38, Third Chemical Congress of North America.

Ferkovich SM, Mayer MS, Rutter RR (1973) Conversion of the sex pheromone of the cabbage looper. Nature 242:53-55

Ferkovich SM, Oliver JE, Dillard C (1982) Pheromone hydrolysis by cuticular and interior esterases of the antennae, legs, and wings of the cabbage looper moth *Trichoplusia ni* (Hübner). J Chem Ecol 8:859-866

Fesenko EE, Novoselov VI, Bystrova MF (1987) The subunits of specific odor-binding glycoproteins from rat olfactory epithelium. FEBS Lett 219:224-226

Fesenko EE, Novoselov VI, Bystrova MF (1988) Properties of odour-binding glycoproteins from rat olfactory epithelium. Biochim Biophys Acta 937:369-378

Firestein S, Werblin F (1989) Odor-induced membrane currents in vertebrate olfactory receptor neurons. Science 244:79-82

Getchell TV (1986) Functional properties of vertebrate olfactory receptor neurons. Physiol Rev 66:772-817

Getchell TV, Margolis FL, Getchell ML (1984) Perireceptor and receptor events in vertebrate olfaction. Prog Neurobiol 23:317-345

Gilman AG (1987) G proteins: transducers of receptor-generated signals. Ann Rev Biochem 56:615-649

Gnatzy W, Mohren W, Steinbrecht RA (1984) Pheromone receptors in *Bombyx mori* and *Antheraea pernyi*: II morphometric analysis. Cell Tiss Res 235:35-42

Greer CA (1987) Golgi analysis of dendritic organization among denervated olfactory bulb granule cells. J Comp Neurology 257:442-452

Grunert U, Ache BA (1988) Ultrastructure of the aesthetasc (olfactory) sensilla of the spiny lobster, *Panulirus argus*. Cell Tiss Res 251:95-103

Guerin PM, Baltensweiler W, Arn H, Buser H-R (1984) Host race pheromone polymorphism in the larch budmoth. Experentia 9:9-18

Györgyi TK, Roby-Shemkovitz AJ, Lerner MR (1988) Characterization and cDNA cloning of the pheromone-binding protein from the tobacco hornworm, *Manduca sexta*: a tissue-specific developmentally regulated protein. Proc Natl Acad Sci USA 85:9851-9855

Hagen DC, McCaffrey G, Sprague Jr GF (1986) Evidence the yeast STE3 gene encodes a receptor for the peptide pheromone a-factor: gene sequence and implications for the structure of the presumed receptor. Proc Natl Acad Sci USA 83:1418-1422

Hall Z (1987) Three of a kind: the beta-adrenergic receptor, the muscarinic acetylcholine receptor, and rhodopsin. Trends Neurosci 10:99-101

Hansson BS, Lofstedt C, Roelofs WL (1987) Inheritance of olfactory response to sex pheromone components in Ostrinia nubilalis Naturwissenschaften 74:

Harvey WR, Cioffi M, Wolhersberger MG (1981) Portasomes as coupling factors in active ion transport and oxidative phosphorylation. Am Zool 21:775-791

Heldman J, Lancet D (1986) Cyclic AMP-dependent protein phosphorylation in chemosensory neurons: identification of cyclic nucleotide-regulated phosphorylated phosphoproteins in olfactory cilia. J Neurochem 47:1527-1533

Herskowitz I, Marsh L (1987) Conservation of a receptor/signal transduction system. Cell 50:995-996

Homberg U, Christensen TA, Hildebrand JG (1989) Structure and function of the deutocerebrum in insects. Ann Rev Entomol 34:477-501

Hornung DE, Mozell MM (1977) Odorant removal from the frog olfactory mucosa. Brain Res 128:158-163

Huque T, Bruch RC (1986) Odorant- and guanine nucleotide- stimulated phosphoinositide turnover in olfactory cilia. Biochim. Biophys Res Comm 137:36-42

Jones DT, Reed RR (1989) G_{olf}: an olfactory neuron specific-G protein involved in odorant signal transduction. Science 244:790-795

Kaissling K-E (1974) Sensory transduction in insect olfactory receptors. In: Jaenicke L (ed) Biochemistry of Sensory Functions. Springer-Verlag, Berlin, pp 243-273

Kaissling K-E (1986) Chemo-electrical transduction in insect olfactory receptors. Ann Rev Neurosci 9,121-145

Kaissling KE, Kasang G, Bestmann HJ, Stransky W, Vostrowsky O (1978) A new pheromone of the silkworm moth *Bombyx mori*, sensory pathway and behavioral effect. Naturwissenschaften 65:382-384

Kaissling K-E, Klein U, de Kramer JJ, Keil TA, Kanaujia S, Hemberger J (1985) Insect olfactory cells: electrophysiological and biochemical studies. In: Changeux JP,

Hucho F, Maelicke A, Neumann E (eds) Molecular Basis of Nerve Activity. Walter de Gruyter, Berlin, pp 173-183

Kaissling K-E, Thorson J (1980) Insect olfactory sensilla: Structural, chemical and electrical aspects of the functional organization. In: Sattelle DB, Hall LM, Hildebrand JG (eds) Receptors for Neurotransmitters, Hormones and Pheromones in Insects. Elsevier/North-Holland, Amsterdam, pp 261-282

Kasang G (1971) Bombykol reception and metabolism on the antennae of the silkmoth *Bombyx mori.* In: Ohloff G, Thomas AF (eds) Gustation and olfaction. Academic, New York, pp 245-253

Kasang G, Knauer B, Beroza M (1974) Uptake of the sex attractant ^{3}H-disparlure by male gypsy moth antennae (*Lymantria dispar*) [= *Portheria dispar*]. Experentia 30:147-148

Kasang G, Nicholls M, van Proff L (1989) Sex pheromone conversion and degradation in antennae of the silkworm moth *Bombyx mori.* L. Experentia 45:81-87

Keil TA (1982) Contacts of pore tubules and sensory dendrites in antennal chemosensilla of a silkmoth: demonstration of a possible pathway for olfactory molecules. Tissue Cell 14:451-462

Keil TA (1984a) Very tight contact of tormogen cell membrane and sensillum cuticle: ultrastructural basis for high electrical resistance between receptor-lymph and subcuticular spaces in silkmoth olfactory hairs. Tissue Cell 16:131-135

Keil TA (1984b) Reconstruction and morphometry of silkmoth olfactory hairs: a comparative study of sensilla trichodea on the antennae of male Antheraea polyphemus and Antheraea pernyi (Insecta, Lepidoptera) Zoomorph 104:147-156

Kennedy JS, Marsh D (1974) Pheromone-regulated anemotaxis in flying moths. Science 184:999-1101

Klein PS, Sun TJ, Saxe III CJ, Kimmel AR, Johnson RL, Devreotes PN (1988) A chemoattractant receptor controls development in *Dictyostelium discoideum.* Science 241:1467-1472

Klein U (1987) Sensillum-lymph proteins from antennal olfactory hairs of the moth *Antheraea polyphemus* (Saturniidae) Insect Biochem 17:1193-1204

Kline D, Simoncini L, Mandel G, Maue RA, Kado RT, Jaffe LA (1988) Fertilization events induced by neurtransmitters after injection of mRNA in *Xenopus* eggs. Science 241: 464-467

Klun JA (1975) Insect sex pheromones: intraspecific pheromonal variability of *Ostrinia nubilalis* in North America and Europe. Environ Entomol 4:891-894

Klun JA, Maini S (1979) Genetic basis of an insect chemical communication system: the European corn borer. Environ Entomol 8:423-426

Kochansky J, Tette J, Taschenberg EF, Carde RT, Kaissling K-E, Roelofs WL (1975) Sex pheromone of the moth *Antheraea polyphemus.* J Insect Physiol 21:1977-1983

Kuno M, Gardner P (1987) Ion channels activated by inositol 1,4,5-triphosphate in plasma membrane of human T-lymphocytes. Nature 326:301-304

Kurihara K, Koyama N (1972) High activity of adenyl cyclase in olfactory and gustatory organs. Biochem Biophys Res Commun 48:30-34

Labarca P, Simon SA, Anholt RRH (1988) Activation by odorants of a multistate cation channel from olfactory cilia. Proc Natl Acad Sci USA 85:944-947

Lancet D (1986) Vertebrate olfactory reception. Ann Rev Neurosci 9:329-355

Lee K-H, Wells RG, Reed RR (1987) Isolation of an olfactory cDNA: similarity to retinol-binding protein suggests a role in olfaction. Science 235:1053-1056

Linn Jr CE (1986) Book review of Lewis T (ed) Insect Communication. Academic, New York, 1984 (with explanatory comments by Linn of his own work). J Chem Ecol 12:1311-1317

Linn Jr CE, Roelofs WL (1983) Effect of varying proportions of the alcohol component on sex pheromone blend discrimination in oriental fruit moths. Physiol Entomol 8:291-306

Linn Jr CE, Roeloffs WL (1985) Response specificity of male pink bollworm moths to different blends and dosages of sex pheromone. J Chem Ecol 11:1583-1590

Linn Jr CE, Campbell MG, Roelofs WL (1986) Male moth sensitivity to multicomponent pheromones: critical role of female-released blend in determining the functional role of components and active space of the pheromone. J Chem Ecol 12:659-668

Logothetis DE, Kurachi Y, Galper J Neer EJ, Clapham DE (1987) The beta/gamma subunits of GTP-binding proteins activate the muscarinic K^+ in heart. Nature 325:321-326

Lonergan GC (1986) Metabolism of pheromone components and analogs by cuticular enzymes of *Choristoneura fumiferana*. J Chem Senses 12:483-496

Margolis FL (1977) Biochemical studies of the primary olfactory pathway. Soc Neurosci Symp 3:167-188

Marsh D, Kennedy JS, Ludlow AR (1981) Analysis of zigzagging flight in moths: a correction. Physiol Ent 6, 225

Mattera R, Graziano MP, Yatani A, Zhou Z, Graf R, Codina J, Birnbaumer L, Gilman AG, Brown AM (1989) Splice varients of the alpha subunit of the G protein G_S activate both adenyl cyclase and calcium channels. Science 243:804-807

Matsumoto SG, Hildebrand JG (1981) Olfactory mechanisms in the moth *Manduca sexta*: response characteristics and morphology of central neurons in the antennal lobes. Proc R Soc London Ser B 213:249-277

Mayer MS (1974) Hydrolysis of sex pheromone by the antennae of *Trichoplusia ni*. Experentia 31:452-454

McClintock TS, Ache B (to be published a) Hyperpolarizing receptor potentials in lobster olfactory receptor cells: implications for transduction and mixture suppression. Chem Senses

McClintock TS, Ache B (to be published b) Failure to implicate cAMP in transduction in lobster olfactory receptor cells. Chem Senses

McClintock TS, Ache B (to be published c) Histamine directly gates a chloride channel in lobster olfactory receptor neurons. Proc Natl Acad Sci USA

Menco BPM (1984) Ciliated and microvillous structures of rat olfactory and nasal respiratory epithelia. Cell Tissue Res 235:225-241

Menco BPM, Farbman AI (1985) Genesis of cilia and microvilli of rat nasal epithelia during prenatal development. I. Olfactory epithelium, qualitative studies. J Cell Sci 78:283-310

Menco BPM, Farbman AI (1987) Genesis of cilia and microvilli of rat nasal epithelia during prenatal development. III. Respiratory epithelium surface, including a comparison with the surface of the olfactory epithelium. J Anat 152:145-160

Menco BPM, van der Wolk FM (1982) Freeze-fracture characteristics of insect gustatory and olfactory sensilla. I. A comparison with vertebrate olfactory receptor cells with special reference to ciliary components. Cell Tissue Res 223:1-27

Menevse A, Dodd G, Poynder TM (1977) Evidence for the specific involvement of cyclic AMP in the olfactory transduction mechanism. Biochem Biophys Res Commun 77:671-677

Miller RL (1985) Demonstration of sperm chemotaxis in Echinodermata: Asteroidea, Holothuroidea, Ophiuroidea. J Exp Zool 234:383-414

Morse D, Meighen E (1984) Detection of pheromone biosynthetic and degradative enzymes in vitro. J Biol Chem 259:475-480

Mumby SM, Kahn RA, Manning DR, Gilman AG (1986) Antisera of designed specificity for subunits of guanine nucleotide-binding regulatory proteins. Proc Natl Acad Sci USA 83:265-269

Murlis J (1986) The structure of odour plumes. In: Payne TL, Birch MC, Kennedy CEJ (eds) Mechanisms in Insect Olfaction. Clarendon Press, Oxford, pp 27-38

Mustaparta H, Tommeras BA, Lanier GN (1985) Pheromone receptor cell specificity in interpopulational hybrids of *Ips pini* (Coleoptera: Scolytidae). J Chem Ecol 11:999-1007

Nakamura T and Gold GH (1987) A cyclic nucleotide-gated conductance in olfactory receptor cilia. Nature 325:442-444

Nef P, Heldman J, Lazard D, Margalit T, Jaye M, Hanukoglu I, Lancet D (1989) Olfactory-specific cytochrome P-450. J Biol Chem 264:6780-6785

Neer EJ, Clapham DE (1988) Roles of G protein subunits in transmembrane signalling. Nature 333:129-134

Pace U, Hanski E, Salomon Y, Lancet D (1985) Odorant-sensitive adenylate cyclase may mediate olfactory reception. Nature 316:255-258

Pace U, Lancet D (1986) Olfactory GTP-binding protein: signal-transducing polypeptide of vertebrate chemosensory neurons. Proc Natl Acad Sci USA 83:4947-4951

Pelosi P, Baldaccini NE, Pisanelli AM (1982) Identification of a specific olfactory receptor for 2-isobutyl-3-methoxypyrazine. Biochem J 201:245-248

Pevsner J, Hwang PM, Sklar PB, Venable JC, Snyder SH (1988a) Odorant-binding protein and its mRNA are localized to lateral nasal gland implying a carrier function. Proc Natl Acad Sci USA 85:2383-2387

Pevsner J, Reed RR, Feinstein PG, Snyder SH (1988b) Molecular cloning of odorant-binding protein: member of a ligand carrier family. Science 241:336-339

Pevsner J, Sklar PB, Snyder SH (1986) Odorant-binding protein: localization to nasal glands and secretions. Proc Natl Acad Sci USA 83:4942-4946

Pevsner J, Trifletti RR, Strittmatter SM, Snyder SH (1985) Isolation and characterization of an olfactory receptor protein for odorant pyrazines. Proc Natl Acad Sci USA 82:3050-3054

Prestwich GD, Graham S, Handley M, Latli B, Streinz L, Tasayco MJ (1989) Enzymatic processing of pheromones and pheromone analogs. Experentia 45:263-270

Rhein LD, Cagan RH (1980) Biochemical studies of olfaction: isolation, characterized, and odorant binding activity of cilia from rainbow trout olfactory rosettes. Proc Natl Acad Sci USA 77:4412-4416

Rhein LD, Cagan RH (1983) Biochemical studies of olfaction: binding specificity of odorants to cilia preparation from rainbow trout olfactory rosettes. J Neurochem 41:569-577

Riddiford LM (1967) *trans*-2-Hexenal: mating stimulant for *Polyphemus* moths. Science 158:139-140

Riley CT, Barbeau KB, Keim PS, Kezdy FJ, Heinrikson RL, Law JH (1984) The covalent protein structure of insecticyanin, a blue biliprotein from the hemolymph of the tobacco hornworm, *Manduca sexta* L. J Biol Chem 21:13159-13165

Roelofs W, Glover T, Tang X-H, Sreng I, Robbins P, Eckenrode C, Lofstedt C, Hansson BS, Bengtsson BO (1987) Sex pheromone production and perception in European corn borer moths is determined by both autosomal and sex-linked genes. Proc Natl Acad Sci USA 84:7585-7589

Russo AF, Koshland Jr DE (1983) Separation of signal transduction and adaptation functions of the aspartate receptor in bacterial sensing. Science 220:1016-1020

Rybczynski R, Reagan J, Lerner MR (1989) A pheromone-degrading aldehyde oxidase in the antennae of the moth *Manduca sexta*. J Neurosci 9:1341-1353

Sanes JR and Hildebrand JG (1976) Origin and morphogenesis of sensory neurons in an insect antenna. Dev Biol 51:300-319

Schneider D (1970) Olfactory receptors for the sexual attractant (Bombykol) of the silk moth. In: Schmitt FO (ed) The Neurosciences: Second study program. Rockefeller Univ Press, NY, pp 511-518

Schofield PR (1988) Carrier-bound odorant delivery to olfactory receptors. Trends Neurosci 11:471-472

Shepherd GM (1972) Synaptic organization of the mammalian olfactory bulb. Physiol Rev 52:864-917

Shepherd GM (1981) The olfactory glomerulus: its significance for sensory processing. In: Katsuki Y, Norgran R, Sato M (eds) Brain Mechanisms of Sensation. Wiley, New York, pp 209-223

Shepherd GM, Greer CA (1988) The dendritic spine: adaptations of structure and function for different types of synaptic integration. In: Intrinsic determinants of neuronal form and function. Alan R. Liss, pp 245-262
Shirley SG, Robinson CJ (1988) Are we really studying olfactory transduction? Trends Neurosci 11:532-533
Sibley DR, Benovic JL, Caron MG, Lefkowitz RJ (1987) Regulation of transmembrane signaling by receptor phosphorylation. Cell 48:913-922
Singh S, Lowe DG, Thorpe DS, Rodriguez H, Kusng W-J, Dangott LJ, Chinkers M, Gorddel DV, Garbers DL (1988) Membrane guanylate cyclase is a cell-surface receptor with homology to protein kinases. Nature 334:708-712
Sklar PB, Anholt RRH, Snyder SH (1986) The odorant-sensitive adenylate cyclase of olfactory receptor cells. J Biol Chem 261:15538-15543
Starratt AN, Dahm KH, Allen N, Hildebrand JG, Payne TL, Roller H (1979) Bombykal, a sex pheromone of the sphinx moth *Manduca sexta*. Z Naturforsch 34c:9-12
Steinbrecht RA (1973) Der Feinbau olfaktorischer Sensillen des Seidenspinners (Insecta, Lepidoptera). Z Zellforsch Mikrosk Anat 139:533-565
Steinbrecht RA (1976) Freeze substitution and freeze fracturing of insect sensilla without cryoprotectants. In: Ben-Shaul Y (ed.) Proc. 6th Eur. Congr. Electron Microscopy Jerusalem, Intertal, Givatayim, Vol II, 111-113
Steinbrecht RA (1980) Cryofixation without cryoprotectants: freeze substitution and freeze etching of an insect olfactory receptor. Tiss. Cell 12:73-100
Steinbrecht RA (1986) ATPase particles (portasomes) on mitochondrial cristae and the plasma membrane of an insect as demonstrated by freeze substitution. Naturwissenschaften 73:275-276
Steinbrecht RA, Gnatzy W (1984) Pheromone receptors in Bombyx mori and *Antheraea pernyi*. Cell Tissue Res 235:25-34
Steinbrecht RA, Muller B (1971) On the stimulus conducting structures in insect olfactory receptors. Z Zellforsch 117:570-575
Steinbrecht RA, Zierold K (1987) The electrolyte distribution in Insect olfactory sensilla as revealed by X-ray microanalysis. In: Roper SD, Atema J (eds) Olfaction and Taste IX. Ann New York Acad Sci 510:638-641
Stewart RC, Dahlquist FW (1987) Molecular components of bacterial chemotaxis. Chem Rev 87:997-1025
Thurm U, Kuppers J (1980) Epithelial physiology of insect sensilla. In: Locke M, Smith DS (eds) Insect Biology in the Future. Academic, New York, pp 735-763
Topazzini A, Pelosi P, Pasqualetto PL, Baldaccini NE (1985) Specificity of a pyrazine binding protein from cow olfactory mucosa. Chem Senses 10:45-49
Trapido-Rosenthal H, Carr WES, Gleeson RA (1987) Biochemistry of an olfactory purinergic system: dephosphorylation of excitatory nucleotides and uptake of adenosine. J Neurochem 49:1174-1182
Trapido-Rosenthal H, Carr WES, Gleeson RS (1989) Biochemistry of purinergic olfaction: the importance of nucleotide dephosphorylation. In: Teeter J, Cagan RH, Kare MR, Brand JG (eds) Receptor and Transduction Mechanisms in Taste and Olfaction. Marcel Dekker, New York, pp 243-262
VanDongen AMJ, Codina J, Olate J, Mattera R, Joho R, Birnbaumer L, Brown AM (1988) Newly identified brain potassium channels gated by the guanine nucleotide binding protein G_o. Science 242:1433-1437
Vogt RG (1987) The molecular basis of pheromone reception: its influence on behavior. In: Prestwich GD, Blomquist GJ (eds) Pheromone Biochemistry. Academic Press, Orlando, FL, pp 385-431
Vogt RG, Koehne AC, Dubnau JT, Prestwich GD (to be published) Expression of pheromone binding proteins during antennal development in the gypsy moth *Lymantria dispar*. J Neurosci

Vogt RG, Riddiford LM (1981a) Pheromone deactivation by antennal proteins of Lepidoptera. In: Sehnal F, Zabza A, Menn JJ, Cymbrowski B (eds) Insect Development and Behaviour. Polytechnical University of Wroclaw Press, Wroclaw, Poland, pp 955-967

Vogt RG, Riddiford LM (1981b) Pheromone binding and inactivation by moth antennae. Nature 293:161-163

Vogt RG, Riddiford LM (1986a) Pheromone reception: a kinetic equilibrium. In: Payne TL, Birch MC, Kennedy CEJ (eds) Mechanisms in Insect Olfaction. Clarendon Press, Oxford, pp 201-208

Vogt RG, Riddiford LM (1986b) Scale esterase: a pheromone-degrading enzyme from scales of silk moth *Antheraea polyphemus.* J Chem Ecol 12:469-482

Vogt RG, Riddiford LM, Prestwich GD (1985) Kinetic properties of a pheromone degrading enzyme: the sensillar esterase of *Antheraea polyphemus.* Proc Natl Acad Sci USA 82:8827-8831

Vogt RG, Prestwich GD (1988a) Variation in olfactory proteins: evolvable elements encoding insect behavior. In: Roper SD, Atema J (eds) Olfaction and Taste IX. Ann New York Acad Sci 510:689-691

Vogt RG, Prestwich GD, Riddiford LM (1988b) Sex-pheromone receptor proteins: visualization using a radiolabeled photoaffinity analog. J Biol Chem 263:3952-3959

Ward GE, Brokaw CJ, Garbers DL, Vacquier VD (1985) Chemotaxis of *Arbacia punctulata* spermatozoa to resact, a peptide from the egg jelly layer. J. Cell Biol. 101, 2324-2329

Wright RH (1954) Odor and molecular vibration. I. Quantum and thermodynamic considerations. J Appl Chem 4:611-616

Wright RH (1972) Stereochemical and vibrational theories of odour. Nature 239:226

Wright RH (1977) Odor and molecular vibration: neural coding of olfactory information. J Theor Biol 64:473-502

Wright RH, Brand J (1972) Correlation of ant alarm pheromone activity with molecular vibration. Nature 239:225-226

Zimmer-Faust RA (1987) Crustacean chemical perception: tuning to energy and nutrient reward? In: Roper SD, Atema J (eds) Olfaction and Taste IX. Ann New York Acad Sci 510:735-738

OLECULAR ELEMENTS OF OLFACTORY SIGNAL TRANSDUCTION IN INSECT ANTENNAE

H.Breer, I. Boekhoff, J. Strotmann,
K. Raming, E. Tareilus

University Stuttgart-Hohenheim
Institute of Zoophysiology
7000 STUTTGART 70
West Germany

TRODUCTION

ıe sense of smell is a key for survival and adaptation in the animal world and has been e subject of intense interest and research, however, the mechanisms underlying its nction are still largely unknown (Lancet, 1986). In many ways, insects are ideal models r exploring the details of olfactory reception and transduction; they are easy to obtain large numbers, and the collection of olfactory tissue is mechanically straightforward. ost insects show strong male-female dichotomy in the production of and response to ıeromone blends and the relevant odorants are simple lipids, often of closely related ımponents. Furthermore, the olfactory organs are relatively large and accessible for perimental work, these advantages as well as a continuous refinement of ectrophysiological and neuroantomical techniques has led to numerous studies which ıve developed a large body of information on the structure and function of insect factory receptor cells (Kaissling, 1986) and the field of research has continued to clude elucidation of the mechanisms of central integration of olfactory information ıd its translation into the motor responses of behaviour (Boeckh and Ernst, 1987; ildebrand, 1987). Thus, the analysis of the olfactory signal and its neuronal processing quite advanced in insects, however these biological events are the output of molecular ocesses in receptor cells which are completely unknown.

The chemo-electrical transduction in antennal cells includes a chain of events ading from specific binding of odorants to the generation of receptor potentials and

NATO ASI Series, Vol. H 39
Chemosensory Information Processing. Edited by D. Schild

nerve impulses and, eventually, to the inactivation of the stimulus molecules. Thus, chemosensory transduction appears to be similar to other neuronal reception systems, with the odorant or pheromone play a role analogous to that of neurotransmitter, reacting with specialized receptors to activate conductance changes; these changes give raise in turn to action potentials. In most cases the transduction mechanisms involves the production of second messengers; these transduction systems are initiated by the binding of a ligand to the receptor, the activated receptor catalyzes the binding of GTP to the G-protein which in turn is activated. The activated G-protein turns on a catalytic element (e.g. adenylate cyclase, phospholipase C), and this activation ceases upon hydrolysis of GTP to GDP, leaving the G-protein inactive and ready for a new cycle. This chain of events involves a high degree of amplification, in particular at the catalytic unit level (Gilman, 1986). Based on the observation that the high level of adenylate cyclase in the olfactory epithelium was activated by certain classes of odorants in a GTP-dependent manner (Sklar et al., 1986), considerable evidence has accumulated indicating that the olfactory transduction systems in vertebrates involve adenylate cyclase and cAMP (Lancet et al., 1988).

There is no direct biochemical evidence regarding the mechanism of transduction in insect olfactory systems. However, data from single unit recordings from olfactory cells of Antheraea polyphemus indicate that there is a significant delay between stimulus presentation and electrophysiological response (Kaissling, 1986). This finding is consistent with the activation of a second messenger cascade by odorant or pheromone. Furthermore, the extreme sensitivity of pheromone-reactive cells may be due to the involvement of an amplification mechanism of the kind afforded by a second messenger cascade.

SECOND MESSENGER REACTION CASCADES

Experiments on antennae from various insects have indicated that the activity of adenylate cyclase is rather low in these preparations and was not affected by odorants and pheromones (Raming, 1988). A moderate accumulation of cGMP in insect antennae upon exposure to pheromones has recently been observed (Ziegelberger and Kaissling, 1988). However, the elevated cGMP-level lasted for several minutes; it was therefore suggested that cGMP may be involved in adaptation processes rather than playing a pivotal role in the transduction cascade. Phospholipase C, which acts on a membrane lipid (phosphatidylinositol bisphosphate) to produce two different second messengers, inositol trisphosphate (IP_3) and diacylglycerol (Berridge and Irvine, 1984) may be considered as an alternative for an odorant-activated transduction mechanism. Comparing the specific activity of phospholipase C (PLC) in various tissues of different insect species it was found that the antennal preparation contains PLC at a much higher activity than any other tissue. GTP and its non-hydrolysable analogues significantly enhanced the antennal PLC activity, suggesting that the enzyme is under the control of a G-protein.

ANTENNAL G - PROTEINS

In order to prove if suitable GTP-binding regulatory polypeptides are present in insect sensory tissue, antennal preparations from various insect species were probed for G-proteins by ADP-ribosylation and specific immunoreactivity. For most species a 45 kDa substrate was found for cholera toxin and a 40 kDa substrate for pertussis toxin. In immunoblotting experiments using antisera reactive of the α-subunits of all G-proteins revealed that the major immunoreactivity was located at a position corresponding to a molecular mass of 40 kDa, i.e. the relative mobility of the pertussis toxin substrate (Breer et al., 1988). It is interesting to note that a block of pheromone-induced IP_3 production by pertussis toxin was observed (see below); this observation is reminescent to the toxin-inhibition of IP_3 release induced by neurotransmitter agonist as described in several cases (Miller, 1988).

STIMULATION OF PHOSPHOLIPASE C BY PHEROMONES

An important criterion for a functional role of PLC in antennal chemo-electrical signalling would be its modulation by odorous compounds in a cell-free system. Using antennal preparations from cockroaches it was found that several compounds (periplanon B, bornyl acetate), which have been identified as specific pheromones (Sass, 1983; Bowers and Bodenstein, 1971) significantly stimulated the antennal PLC in a GTP-dependent manner; the periplanon effect was already observed in nanomolar concentrations. Stimulation of PLC in locust antennae was detected with grass odorants (hexenic acid, hexenal) (Boekh, 1967), however, the cockroach pheromone, periplanon B did not affect the locust enzyme indicating the specificity of the stimulatory effects (Boekhoff et al., 1989).

PHEROMONE - INDUCED IP_3 - RELEASE

Inositol trisphosphate is one of the second messengers produced by phospholipase C; in order to explore if the pheromone-induced stimulation of PLC leads to a significant elevation of the IP_3 level, the concentration of IP_3 was determined using a radioreceptor assay. In unstimulated antennal preparations a basal concentration of 53 pmol/mg protein was detected; this value is very close to data reported for rat brain tissue (Challis et al, 1988). Incubation with 1 μM GTP or 6 nM periplanon B for 30 sec induced a 3-4 fold increase of IP_3. A simultaneous application of GTP and pheromone gave a synergistic effect, causing an almost 20 fold increase of IP_3 (fig. 1). The notion that the stimulatory action of pheromone is mediated via specific G-proteins was further supported in experiments demonstra-ting that pertussis toxin completely prevented the induced PLC activation.

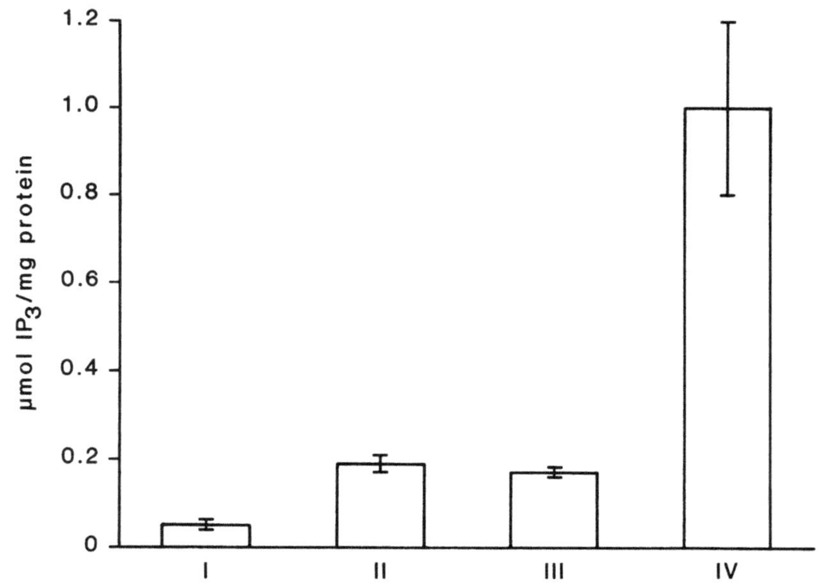

Fig. 1. IP_3 concentrations in antennal preparation from cockroaches.

I. basal level;
II. concentration after incubation with 1μM GTP;
III. 6 nM periplanon B;
IV. 6 nM periplanon B plus 1μM GTP.

Incubation time: 30 sec.

A key question concerning the possible role of IP_3 in the reaction cascade of olfaction is whether changes in IP_3 level elicited by pheromones occur rapidly enough for IP_3 to be the transmitter. In time course experiments it was found that the level of IP_3 was quite constant in non-stimulated preparations; after application of periplanon, however, the maximal concentration of IP_3 (about 70 times of control) was determined already after 2 sec, the shortest time point that could be measured. In order to approach the kinetics of IP_3 formation in the msec time range, 'stopped flow' experiments were performed. It was found that the level of IP_3 very rapidly increased after pheromone application reaching a peak (more than 100 times of control) at about 50 msec; thereafter the concentration rapidly declined. This time course is reminescent of the changes in cGMP concentrations in vertebrate rod outer segments after illumination (Stryer, 1986). Thus, the increase of IP_3 accumulation seems to be rapidly enough for IP_3 to be a transmitter in olfactory excitation.

CALCIUM - DEPENDENT K^+- CHANNELS

The primary effects of IP_3 are supposed to be mediated via Ca^{2+} ions released from intracellular storage compartments into the cytoplasm. In recent studies it has been shown that many neurons respond to the formation of inositol phosphates, or to direct intracellular microinjection of IP_3, with a transient opening of Ca^{2+}-dependent K^+ channels in the plasma membrane. An elevation of Ca^{2+} concentration on the cytoplasmic side leading to an increase in K^+ conductance.

The ionic currents which generate receptor potentials in pheromone-sensitive cells have been studied (Kaissling, 1986). The data indicate that the application of pheromone causes opening of single channels. The specificity of these channels are not known. However, since the sensillum lymph has a very high K^+ concentration ($\geq$ 200 mM), it has been suggested that the initial depolarizing currents of signal transduction are mediated by an inward K^+ current. In cell attached and excised patch clamp analysis on antennal cells Ca^{2+}-dependent K^+-channels were detected. In inside-out patches applying different Ca^{2+} concentrations, we have found that elevated calcium concentrations greatly increased the opening probability of these channels.

IDENTIFICATION OF ANTENNAL MEMBRANE PROTEINS

A number of approaches have been used to identify olfactory receptors (Margolis and Getchell, 1988). Assuming that the receptors for odorants and pheromones function very much like neurotransmitter receptors, one would expect the molecules to be membrane glycoproteins localized selectively in antennae. Using the Western blot technique and applying various lectins a number of highly reactive polypeptides were found in antennal preparations, however, no obvious candidates for antennae-specific glycoproteins could be detected. One approach to investigate the specificity of proteins is through the use of specific, site-directed agents. Because antibodies are ideally suited for this, we have used the hybridoma technique to generate monoclonal antibodies against antennal membrane preparations from locust and analysed these antibodies for antennae-specificity in a differential screening assay. One of the antibodies (B14-6D2) was further characterized. Under native conditions the epitop was localized on a large membrane protein complex (≥ 150 kDa); however after SDS-PAGE followed by immunoblotting two polypeptides (40 kDa, 33 kDa) were detected. Interestingly these two polypeptides are also identified using G-protein specific antisera. It is now a working hypothesis that the monoclonal antibody may recognize a membrane protein associated with G-proteins. In immunocytochemical approaches on cell suspensions it was demonstrated that only a subpopulation of antennal cells was reactive (fig. 2); the identity of the cell type could not be determined yet.

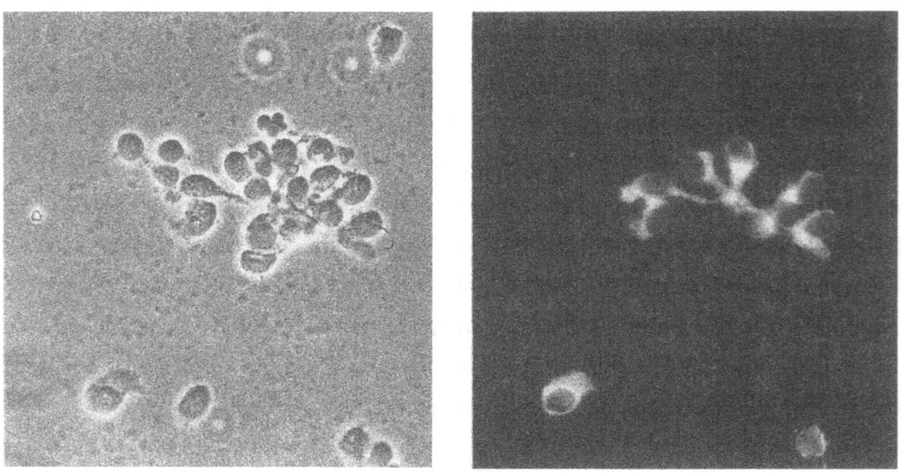

Phase contrast B14-6D2

Fig. 2. Cytochemical localization of B14-6D2 immunoreactivity in antennal cells from locust. Only a specific population of cells is selectively labelled.

Investigations on various developmental stages indicated that the epitop was only accessible just before hatching. In order to explore if this was due to a stage-dependent unmasking of the epitop or if the antigen expression is developmentally regulated, experiments were performed using protein synthesis inhibitors. It was found that in the presence of inhibitors, like cycloheximide, no additional immunoreactivity accumulated in antennal cells as was observed in controls. These results indicate that the antigen is actually expressed during this critical period of development. Identification of the cells which express this antigen as well as the exploration of its molecular properties may contribute to unravel the functional role of this antennae-specific protein.

ANALYSIS OF PHEROMONE BINDING PROTEINS

The sensory hair lumen contains an proteinaceous fluid, the sensillum lymph, providing the characteristic ionic milieu (high potassium level) for the receptor dendrites and contains 30% soluble protein, mainly the 15,000 dalton pheromone-binding protein (PBP) which is present in 10 mM concentration and sequesters (K_D of about 6 x 10^{-8}) the odour molecules within the hair (Vogt and Riddiford, 1981; Kaissling et al., 1985). Recently it has been suggested that the PBP may play a multifunctional role in the process of pheromone perception, operating as solublizer, carrier, and protector of the extremely hydrophobic pheromone molecules (Vogt, 1987). Some partial N-terminal amino acid sequences have been obtained recently for PBP from _Antherea polyphemus_ and _Lymantria dispar_ showing some degree of homology (Vogt, 1987). To better understand the structural and functional properties of PBP we have started to unravel the complete primary structure of PBPs from two moth species _Antheraea polyphemus_ and _Antheraea pernii_ using recombinant DNA cloning techniques. cDNA libraries, derived from poly (A^+) RNA from the antennae of newly eclosed males, were screened with a redundant, 15-base oligonucleotide probe, designed according to the N-terminal sequence of the PBP from amino acid 3 to 7. The oligonucleotide probe was found to hybridize to several plaques formed by recombinant phage. Five clones were plaque purified, and their cDNA insert were subcloned and sequenced. The open reading frame, which starts with a methionine codon preceded by 30 nucleotides of 5' untranslated sequence and ends with a TGA codon, encodes a protein of 163 amino acids. The region 5' of the amino-terminal residue in the native protein codes for 21 amino acids which form the signal peptide, rich in hydrophobic amino acids, and characteristic of exported protein precursor (Walter et al. 1984). The predicted mature PBP consists of 142 amino acids giving a calculated molecular mass of 15,826 dalton and a theoretical isoelectric point of pH 4.54; this value correlates very well with the pI of pH 4.7 determined by isoelectro-focusing (Klein, 1987). The N-terminal amino acid sequence of the mature protein deduced from the nucleotide sequence (Fig. 3) almost precisely matches the one obtained in microsequencing experiments (Klein et al., 1987; Vogt, 1987).

```
                   1        10        20        30
                   |        |         |         |
A. polyphemus      SPEIMKNLSNNFGKAMDQCKDELSLPDSVV
A. pernii          ----I----Q--C--------Q--NI----I

A. polyphemus*     T-------------
A. polyphemus**    ------------------S----------
                                (L)             (N)
```

Fig. 3. Alignment of the N-terminal amino acid sequence of pheromone binding proteins from two Antheraea species. The sequence in the upper rows was predicted from the encoding cDNA; in the lower rows sequences derived from protein sequencing.
* Klein et al., 1987; ** Vogt, 1987.

On inspection, the sequences of A. polyphemus and A. pernii show a very high degree of amino acid homology (95 %), and even the sequence of PBP from Manduca sexta (Györgyi et al., 1988) is quite homologous (64%). This interspecies sequence similarity suggests that particular stretches of amino acid residues are conserved to maintain the function of these proteins; the degree of difference may reflect both the degree of relatedness of the different species and the differences in the chemical structure of their pheromones. An alignment of the amino acid sequence from different species indicates that hydrophobic domains at the C-terminal end of the protein are particular conserved. It is conceivable that these domains may play an important role in forming odorant-binding pockets.

CONCLUSION

There is clearly a continuing need for a multidisciplinary approach using refined physiological and biochemical techniques to explore molecular mechanisms involved in the chemo-electrical transduction of olfactory signals in insects. The recruitment of molecular genetics may give some new and powerful impetus to answer fundamental questions related to receptor function and specificity; the recently reported odorant-specific olfactory-deficient mutants in Drosophila (Siddiqi, 1987) may be particular useful in this respect.

As many insects play an important role as pests of serious agricultural or medical importance, unraveling the olfactory mechanisms appears to be a necessity for the development of behaviour-modifying agrochemicals in the future.

ACKNOWLEDGEMENT - This work was supported by the Deutsche Forschungsgemeinschaft and the Ministerium für Wissenschaft und Kunst. We are most grateful to Prof J. Boekh for providing Periplanon B. K.R. acknowledges the support of Schering AG.

REFERENCES

Berridge, M.J., and Irvine, R.F. (1984) Inositol trisphosphate, a novel second messenger in cellular signal transduction. Nature 312: 315-321.

Boeckh, J. (1967) Reaktionsschwelle, Arbeitsbereich und Spezifität eines Geruchsrezeptors auf der Heuschreckenantenne. Z. Vergl.Physiol. 55: 378-406.

Boeckh, J. and Ernst, K.D. (1987) Contribution of single unit analysis in insects to an understanding of olfactory function. J Comp. Physiol A 161: 549-565.

Boekhoff, I., Strotmann, J., Raming, K., Tareilus, E. and Breer, H. (1989) Odorant-sensitive phospholipase C in insect antennae. (submitted).

Bowers, W.S., and Bodenstein, W.G, (1971) Sex-pheromone in lobster, Homarus americanus. Nature 232: 259-261.

Breer, H., Raming, K. and Boekhoff, I. (1988) G-proteins in the antennae of insects. Naturwissenschaften 75: 627.

Challis, R.A.J., Batty, I.H. and Nahorski, S.R. (1988) Mass measurements of inositol (1,4,5) trisphosphate in rat cerebral cortex slices using a radioreceptor assay: effect of neurotransmitters and depolarization. Biochem. Biophys. Res. Comm. 157: 684-691.

Gilman, A.G. (1986) Receptor-regulated G proteins. Trends Neurosci. 9, 460-463.

Györgyi, T.K., Roby-Shemkovitz, A. and Lerner, M.R. (1988) Characterization and cDNA cloning of the pheromone-binding protein from the tobacco hornworm Manduca sexta: A tissue specific developmentally regulated protein. Proc. Natl. Acad. Sci. U.S.A. 85: 9851-9855.

Hildebrand, J.G. (1987) From semiochemical to behavior: olfaction in the sphinx moth Manduca sexta. In: Molecular Entomology, ed. J.H. Law. Alan R. Liss, New York, pp. 21-31.

Kaissling, K.E. (1986) Chemo-electrical transduction in insect olfactory receptors. Annu Rev. Neurosci. 9: 121-145

Kaissling, K.E., Klein, U., DeKramer, J.J., Keil, T.A., Kanaujia, S., and Hemberger, J. (1985) Insect olfactory cells: Electrophysiological and biochemical studies. In: Molecular Basis of Nerve Activity. (eds. J.P. Changeux, F.Hucho, A. Maelicke, and E. Neumann) de Gruyter, Berlin. pp 173-183.

Klein, U. (1987) Sensillum-lymph proteins from antennal olfactory hairs of the moth Anteraea polyphemus (Saturniidae). Insect Biochem. 17: 1193-1204.

Klein, U., Kehl, M., Hemberger, J. and Günzel, A. (1987) Sensillum lymph proteins in insect olfactory hairs. (ECRO Conf. 1986). Chem. Senses 12: 211

Lancet, D. (1986) Vertebrate olfactory reception. Ann. Rev. Neurosci. 9: 329-355.

Lancet, D., Lazard, D., Heldman, J. Khen, M. and Nef, P. (1988) Molecular transduction in smell and taste. In: Cold Spring Harbor Symposia of Quantitative Biology LIII. pp.343-348.

Margolis, F.L. and Getchell, T.V., eds. (1988) Molecular Neurobiology of the Olfactory System. Plenum Press, New York.

Miller, R. J. (1988) G-proteins flex their muscles. Trend Neurosci. 11, 3-6.

Raming, K. (1988) Vergleichende Studien zur Identifizierung von G-Proteinen im neuronalen und sensorischen Gewebe. Diploma Thesis, University Osnabrück.

Sass, H. (1983) Production, release and effectiveness of two female sex pheromone components of Periplaneta americana. J. Comp. Physiol. 152: 309-317.

Siddiqi, O. (1987) Neurogenetics of olfaction in Drosophila melanogaster. Trends Genet. 3: 137-142.

Sklar, P.B., Anholt, R.H., and Snyder S.H. (1986) The odorant sensitive adenylate cyclase of olfactory receptor cells: Differential stimulation by distinct classes of odorants. J. Biol. Chem. 261: 15538-15543.

Stryer, L. (1986) Cyclic GMP cascade of vision. Ann. Rev. Neurosci. 9: 87-119.

Vogt, R.G. (1987) Molecular basis of pheromone reception. In: Pheromone biochemistry. (eds. G.D. Prestwich & G.J. Blomquist). Academic Press, Orlando pp. 385-431.

Vogt, R.G. and Riddiford, L.M. (1981) Pheromone binding and inactivation by moth antennae. Nature (Lond.) 293: 161-163.

Walter, P., Gilmore, R. and Blobel, G. (1984) Protein translocation across the endoplasmic membranes. Science 230: 400-407.

Ziegelberger, G. and Kaissling E.E. (1988) Cyclic nucleotides in the olfactory system of the moths Antherea polyphemus and Bombyx mori. In: Sense organs, Proceedings of the 16th Göttingen Neurobiology Conference, eds. Elsner, N. and F. G. Barth, Thieme, Stuttgart, p. 58.

THE *XENOPUS* OOCYTE AS AN *IN VITRO* TRANSLATION AND EXPRESSION SYSTEM FOR CHEMOSENSORY - SPECIFIC GENE PRODUCTS

Thomas V. Getchell[1], Frank L. Margolis[2]

1) Department of Physiology & Biophysics
College of Medicine
University of Kentucky
Lexington, KY 40536 - 0084

2) Department of Neuroscience
Roche Institute of Molecular Biology
Roche Research Center
Nutley, NJ 07110

INTRODUCTION

In 1971, Gurdon and coworkers published a seminal paper reporting the first successful *in ovo* translation of exogenous mRNA into protein. The poly (A^+) RNA that was isolated biochemically from red blood cells and microinjected into *Xenopus* oocytes directed the synthesis of protein globin. Their experiments not only established a system in which to study the *in vitro* translation of genetic material but also, identified a system in which the functional activity of the translation products could be assayed after expression. In this system, the injected mRNA effectively competes with the endogenous mRNA for the translation machinery to synthesize a variety of proteins. This background synthesis and post-translational modification of functional proteins from endogenous mRNA is particularly advantageous when investigating the functional expression of a multicomponent transduction system that utilizes the G-protein, phospholipase C, inositol triphosphate or the adenylate cyclase 2nd messenger intermediates. More recently, the *Xenopus* oocyte has been utilized to investigate the expression of neurotransmitter receptors, neuromodulators and voltage-gated ion channels (Dascal, 1987; Snutch, 1988) and the pharmacology of 2nd messengers in the B-adrenergic receptor system (Smith et al., 1987). This technique has also been

NATO ASI Series, Vol. H 39
Chemosensory Information Processing. Edited by D. Schild

combined with techniques of recombinant DNA technology to investigate the molecular biology of ion channels (Claudio, 1986), the substance-K receptor (Masu et al., 1987) and visual transduction (Khorana et al., 1988). Recently, several investigators habe been exploring the use of the *Xenopus* expression system to study chemosensory systems. Following a preliminary report from Lancet's laboratory (Dascal et al., 1986), Getchell reported that total RNA isolated from the frog olfactory mucosa directed the synthesis of odorant receptor molecules (Getchell, 1988), Getchell and Margolis reported that poly (A^+) RNA isolated from catfish taste barbels directed the synthesis of amino acid receptors (Getchell et al., 1989) and Kurihara reported that poly (A^+) RNA isolated from the carp olfactory epithelium directed the synthesis of a K^+ channel (Yoshi & Kurihara, 1989). In each report, the investigators utilized the two electrode voltage clamp technique as an assay to monitor the functional expression of a newly synthesized induced membrane molecule. These studies together with contemporary advances in molecular biological investigations of the olfactory system (Margolis, 1988) and the molecular cloning of chemosensory-gene products (Jones & Reed, 1987, 1989) suggests that substantial progress is forthcoming in our understanding of transduction in chemosensory systems. In this communication, we summarize our results on the functional expression of odorant and amino acid receptors.

METHODS

The sequence of steps in the isolation of oocytes for microinjection of exogenous RNA in preparation for the electrophysiological assay is shown in Figure 1. Stage V and VI oocytes are isolated from the ovarian lobes of oocyte positive South African clawed frogs, *Xenopus laevis*, using mild enzymatic treatment and mechanical agitation (Dumont, 1972; Colman, 1984). Healthy oocytes have a diameter of about 1 mm with clearly identified animal (dark) and vegetal (light) poles that are well-demarcated at the equatorial region. Oocytes that can exclude $^{45}Ca^{++}$ generally have an intact membrane and are suitable for microinjection (Getchell et al., 1989). It is crucial to the successful outcome of the experiments, i.e., functional expression of receptor molecules, that objective as well as subjective criteria be utilized to select suitable oocytes.

We have isolated total RNA from the olfactory mucosae of 4-6 week old CD rats and from the barbels of 12-14 inch long catfish using a variation of the RNAzol method (Chomczynski and Sacchi, 1987) and prepared poly (A^+) RNA from this material by the procedure of Badley, et al. (1988). The resultant RNA pellet was dissolved in water at a

Preparation of *Xenopus laevis* Oocytes

ovarian lobes

↓

sterile ND-96-Py-PS

↓ tease lobes
visual evaluation

Ca^{++}-free OR-2

↓

Ca^{++}-free OR-2 collagenase

↓ shaker table, 2-3 hrs. and
visual evaluation every 20 min.

↓ defolliculated oocytes Ca^{++}-free OR-2 (1x) | ↓ discard tissue

↓

ND-96-Py-PS (3x)

↓ select Stage V/VI oocyes | ↓ discard earlier Stage and sick oocytes

↓

monitor $^{45}Ca^{++}$ uptake

↓

ND-96-Py-PS

↓ 2-24 hrs., room temp.,
shaker table

↓ Stage V/VI oocytes | ↓ discard sick oocytes

↓

monitor $^{45}Ca^{++}$ uptake

↓

microinjection of A^{+} RNA $\equiv$ d_O

↓

1-3d <u>in ovo</u> translation

↓ Assay for protein expression | ↓ discard sick oocytes

Figure 1. Preparation of *Xenopus laevis* oocytes for microinjection of exogenous RNAs and elctrophysiological assay of membrane conductances.

concentration of 1 μg/ml in preparation for microinjection. The oocytes are injected with approximately 50 ng poly (A^+) RNA / 50 nl sterile water within 24 hours of isolation and this time period is defined as day 0. Initial experiments indicated that the volume of solution injected can be controlled more precisely using the Eppendorf 5242 pressure injection system than the Dumont mechanical system. The oocytes are maintained in sterile ND-96-pyruvate-pencillin-streptomycin solution at room temperature. Representative oocytes are selected on days 1 through 5 following microinjection to examine their membrane properties in response to a variety of ligands under voltage clamp conditions. The incubation period is related to the time required for the translation and insertion of functional membrane proteins.

RESULTS

In earlier experiments (Getchell, 1988), ligands in solution that were delivered to the voltage clamped oocyte were mixed with the slowly flowing (0.5 to 1.0 ml/min) perfusate. This procedure resulted in long latency, long duration, low magnitude influxes of current that were difficult to manipulate experimentally due to their physical characteristics and receptor desensitization. We now utilize the Picospritzer II that permits for greater stimulus control including small volumes with sharp onsets, short durations and abrupt terminations. The magnitude and time course of the conductance changes that one records in response to ligand stimulation is also strongly influenced by the orientation of the delivery pipette to the oocyte. As is shown in Figure 2, we standardized the orientation of the pipette to the oocyte in order to compare responses among ligands recorded from a single oocyte, since the delivery pipette is not multi-barreled, or among oocytes. Under normal recording conditions (Figure 2a), the delivery pipette (~35μm in diameter) is oriented under visual control normal to and approximately 400 μm from the membrane. As shown in Figures 3 and 4 the resultant membrane current transients have sharp onsets and are of sufficient magnitude and duration that their properties can be investigated without substantial recording artifacts. When the delivery pipette is moved closer to the oocyte (Figure 2b) into the "slam position" considerable indentation of the membrane occurs with concomitant movement of the recording microelectrodes (M1, M2) when a ligand pulse is delivered. The actual concentration of the ligand in the perfusate surrounding the oocyte was not determined but has been estimated to range from the prepared concentration of the ligand to approximately 10^3 times more dilute. Therefore, the methodology of stimulus delivery and control, the orientation and distance of the delivery pipette to the oocyte and the concentration of the ligand are three critical variables that determine the response properties recorded from the oocyte membrane. In turn, careful implementation of these controls will facilitate the electrophysiological assay of the oocyte membrane for the expression of gene products.

Figure 2. Two electrode voltage clamp recording and ligand delivery systems showing the orientation of the ligand delivery pipettes in the normal (A) and slam (B) positions.

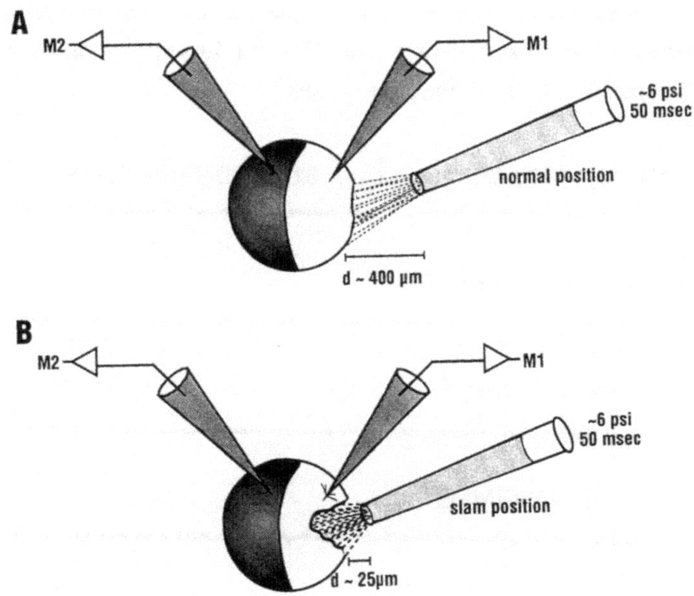

We utilized the catfish barbel taste system as a rich source of chemoreceptor RNA because of the high density of taste receptor cells, electrophysiological characterization of amino acid taste responses, binding studies of ligand-receptor interactions and evidence suggesting that two independent classes of amino acid binding sites exist (reviewed in: Getchell et al., 1989). We examined electrophysiologically 21 oocytes microinjected with poly (A^+) RNA isolated from catfish barbels. The results from 3 oocytes were set aside because of unstable membrane currents under voltage clamp; 5 (28 %) of the remaining 18 oocytes were responsive to amino acid stimulation on days 2 or 3 following microinjection. The amino acids ranged in stock concentration from 0.1 to 10 mM.

Representative control and experimental data are seen in Figure 3. They were obtained from a single oocyte on the second day following microinjection of 50 ng catfish barbel poly (A^+) RNA / 50 nl sterile Water. When aliquots of the perfusate ND-96 were pulsed (50 msec, 6 psi) toward the oocyte membrane in the normal configuration (a) and the "slam position" (b) small magnitude transients were recorded in the latter configuration that are probably of mechanical origin (see Figure 2b). Examination of the responsiveness of the oocyte to amino acids indicated that it was more responsive to L-arginine that L-alanine and non-responsive to L-lysine at the same concentrations. The current transients elicited by L-arginine showed progressive stages of receptor desensitization when the ligand pulses were separated by about 5 sec and partial to complete resensitization at longer interpulse intervals of about 15 to 45 sec. A maximum desensitization of about 75 % in 40 sec was observed in these studies with short duration (50 msec) iterative pulses.

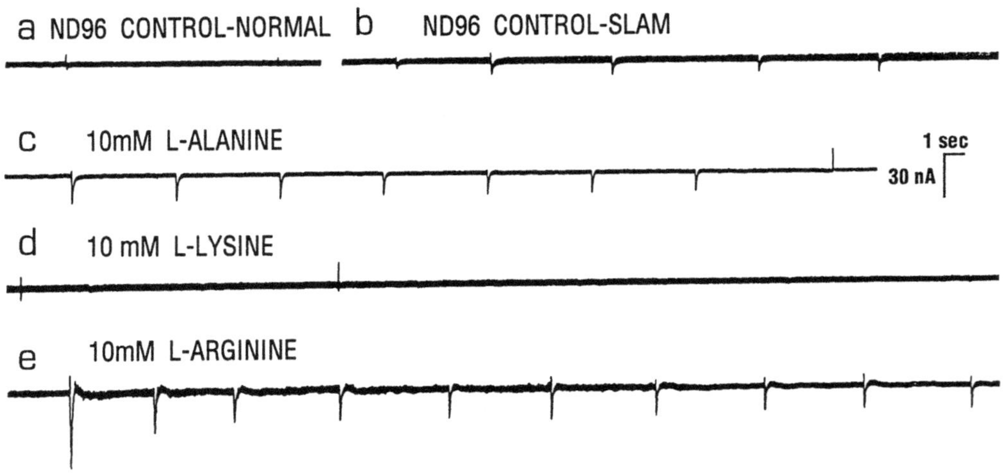

Figure 3. Membrane conductances recorded from a voltage clamped oocyte (-60 mV) microinjected with catfish barbel poly (A^+) RNA are shown in response to L-alanine (c), L-lysine (d) and L-arginine (e) under normal recording conditions; controls for the normal (a) and slam (b) positions.

We also examined the membrane properties of 10 oocytes in response to an odorant cocktail that were microinjected with poly (A^+) RNA isolated from rat olfactory mucosa. The cocktail consisted of the odorants citral, cineole, i-amylacetate, a- and B-ionone that ranged in stock concentration from 0.05 to 5 mM each. The results from 2 oocytes were set aside because of unstable membrane currents under voltage clamp; 4 (50 %) of the 8 remaining oocytes were responsive to odorant stimulation. The results shown in Figure 4 are an example of current transients evoked by pulses of the odorant cocktail in which the concentration and pulse duration were varied. A relatively brief pulse (50 msec) of 50 μM cocktail elicited a rapid influx of current about 50 nA in magnitude (a). When the pulse duration was extended to 150 msec (b), the response consisted of an initial current transient followed by a long latency (10 sec) and a relatively slow influx of current. When the ligand concentration was increased to 500 μM, a long duration pulse (150 msec) elicited a complex current transient that consisted of a large magnitude initial influx of current (90 nA) immediately followed by a short latency (~1 sec), long duration component.

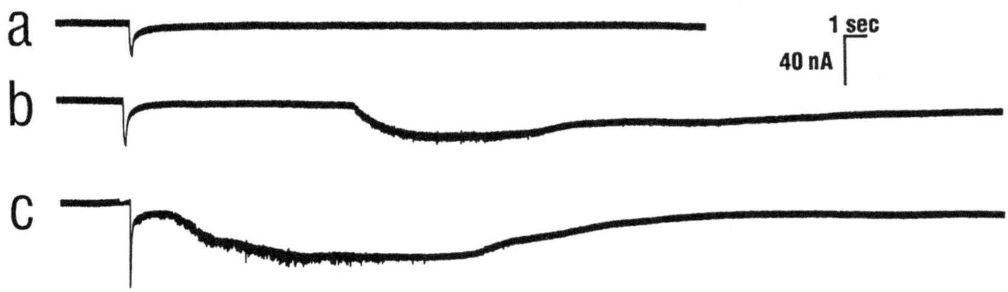

Figure 4. Membrane conductances recorded from a voltage clamped oocyte (-60 mV) microinjected with rat olfactory mucosa poly (A^+) RNA are shown in response to odorant cocktail stimulation; responses evoked by 50 μM aliquots of cocktails 50 msec (a) and 150 msec (b) in duration, and a 150 msec pulse of 500 μM cocktail.

In our studies, oocytes that were microinjected with catfish barbel poly (A^+) RNA were non-responsive to the odorant cocktail and those microinjected with rat olfactory mucosal poly (A^+) RNA were non-responsive to amino acid stimulation.

DISCUSSION

The results of our studies confirm the use of the *Xenopus* oocyte as an mRNA translation and expression system and extend its use to chemosensory systems. The functional assay utilized suggests that the full complement of active components participating in taste and olfactory transduction were synthesized from endogenous as well as exogenous mRNA from chemosensory tissue. They were apparently organized in functional assemblies permitting transduction to proceed from the initial binding event through membrane conductance changes. Four lines of evidence indicate that poly (A^+) RNA isolated from catfish barbels directed the synthesis of an amino acid receptor. First, control oocytes microinjected with sterile water or poly (A^+) RNA from rat olfactory mucosa were non-responsive to amino acid stimulation; second the functional expression of amino acid receptors required one to two days to develop, an interval that is presumably related to the time required for protein translation, processing and assembly into the transduction sequence; third, the responses recorded electrophysiologically were amino acid selective and specific and fourth, they show desensitization with iterative stimulation at concentrations that evoke neural impulses from catfish taste neurons. The selectivity and specificity of odorant responses is less clear in oocytes microinjected with rat poly (A^+) RNA and is presumably related to the molecular events underlying the complex current transients shown in Figure 4.

Thus, we have demonstrated that the *Xenopus* oocyte can be used as a functional assay to study expression of chemoreceptor mRNAs. We have isolated and characterized chemosensory poly (A^+) RNA, defined the ligand delivery and recording parameters and characterized certain basic response properties of the injected oocytes. The use of the one-by-one electrophysiological assay of oocytes to study the functional expression of chemosensory gene products is a definitive, but severely rate limiting, step in the overall experimental design. Although 28-50 % of the oocytes that we examined electrophysiologically for functional expression respond to ligands, we need to understand why the remaining oocytes are non-responsive or alternatively, to develop additional selection criteria. The next phase of this study will use this functional assay to identify, isolate and characterize functionally relevant cloned chemosensory mRNAs. To achieve this, it will be necessary to increase the efficiency of this assay and to develop companion screening procedures to optimize the selection of appropriate clones for electrophysiological characterization. Efforts in these directions are already underway. In summary, we have established a functional assay that will facilitate the biochemical, electrophysiological and molecular biological characterization of the components of eucaryotic chemosensory transduction.

ACKNOWLEDGEMENT - Partially supported by NIH NS-16340 and F33 NS-08486 (TVG); assistance of M. Grillo, S. Tate, R. Urade, V. Corondan and J. Teeter who participated in various aspects of these experiments is appreciated.

REFERENCES

Badley JE, Bishop GA, St. John T, Frelinger JA (1988) A simple, rapid method for the purification of poly (A^+) RNA. Biotechniques 6:114-116.

Chomczynski P, Sacchi N (1987) Single step method of RNA isolation by acid guanidinium-thiocyanate-phenol-chloroform extraction. Anal Biochem 162:156-159.

Claudio T (1986) Recombinant DNA technology in the study of ion channels. Trends Pharmacol Sci 7:308-312.

Colman A (1984) Translation of eukaryotic messenger RNA in *Xenopus* oocytes. In: Transcription and Translation, a practical approach. BD Hames, SJ Higgins (eds.). IRL Press, Washington, pp 271-300.

Dascal N (1987) The use of *Xenopus* oocytes for the study of ion channels. CRC Critical Rev Biochem 22:317-387.

Dascal N, Heldman J, Gershon E, Lancet D (1986) Possible expression of odorant receptor proteins in *Xenopus* oocytes injected with rat olfactory epithelial mRNA. Soc Neurosci Abstr. 13:369.

Dumont JN (1972) Oogenesis in *Xenopus laevis* (Daudin) I. Stages of oocyte development in laboratory maintained animals. J Morph 136:153-180.

Getchell TV (1988) Induction of odorant-evoked current transients in *Xenopus* oocytes injected with mRNA isolated from the olfactory mucosa of *Rana pipiens*. Neurosci Lett 91:217-221.

Getchell TV, Grillo M, Tate SS, Urade R, Teeter J, Margolis FL (1989) Expression of catfish amino acid taste receptors in *Xenopus* oocytes. Science, submitted.

Gurdon JB, Lane CD, Woodland HR, Marbaix G (1971) Use of frog eggs and oocytes for the study of messenger RNA and its translation in living cells. Nature 233:177-182.

Jones DT, Reed RR (1987) Molecular cloning of five GTP-binding protein cDNA species from rat olfactory neuroepithelium. J Biol Chem 262:14241-14249.

Jones DT, Reed RR (1989) G_{olf}: An olfactory neuron specific-G protein involved in odorant signal transduction. Science 244:790-795.

Khorana HG, Knox BE, Hasi E, Swanson R, Thompson DA (1988) Expression of a bovine rhodopsin gene in *Xenopus* oocytes: Demonstration of light-dependent ionic currents. Proc Natl Acad Sci USA 85:7917-7921.

Margolis FL (1988) Molecular cloning of olfactory-specific gene products. In: Molecular Neurobiology of the Olfactory System. FL Margolis, TV Getchell (eds.) Plenum Press, New York, pp. 237-265.

Masu Y, Nakayama K, Tamaki H, Harada Y, Kuno M, Nakanishi S (1987) cDNA cloning of bovine substance-K receptor through oocyte expression system. Nature 329:836-838.

Smith AA, Brooker T, Brooker G (1987) Expression of rat mRNA coding for hormone-stimulated adenylate cyclase in *Xenopus* oocytes. FASEB J 1:380-387.

Snutch TP (1988) The use of *Xenopus* oocytes to probe synaptic communication. Trends Neurosci 11:250-256.

Yoshi K, Kurihara K (1989) Inward rectifier produced by *Xenopus* oocytes injected with mRNA extracted from carp olfactory epithelium. Synapse 3:234-238

STIMULUS PROPERTIES AND BINDING TO RECEPTORS

Maurice CHASTRETTE

Laboratoire de Chimie Organique Physique, UA CNRS 463
Université Claude Bernard Lyon 1
43 Bd du 11 Novembre 1918
69622, Villeurbanne, France

Structure-odour relationships (SOR) are important in the design of new odorants as well as in the understanding of interactions between molecules and receptor sites. Ideally interpretations of SOR should be related to the structure of the receptors. Unfortunately no receptor structure is presently known, so we have to rely on indirect approaches.

The activity of odorants is generally interpreted by assuming that intact molecules interact at sites with specific requirements for optimum interaction. Accordingly some conception of the structure of the receptor can be inferred by discussing common features in the structure of the most active molecules. Discussing pertinent common features, however, necessitates some hypotheses on the interaction mechanisms which are involved.

Amoore (1952) considered molecular shape as the dominating influence on the olfactory activity of molecules. Within his theory, an experimental study of the relationship between molecular shape and odour has to be based on some measures of morphological similarity. This was done using comparison methods (such as the shadow-matching method, and later more sophisticated procedures) which measure the similarity between the shape of a molecular model and that of a reference object, for instance the model of a receptor site or the model of a reference molecule.

Amoore (1964) postulated highly specific receptor sites embedded in the olfactory epithelium, with structural characteristics which are complementary to those of the odorants to be accomodated. For example molecules of musks fit in an oval site with specific dimensions. The odours which correspond well to specific types of receptors are considered as primary odours, the number of which was determined to be seven or eight, later extended to thirty two.

NATO ASI Series, Vol. H 39
Chemosensory Information Processing. Edited by D. Schild

Beets, (1978 and in a series of papers starting in 1957) developed a different approach to the receptor-site interaction.

While Amoore seems to consider that a primary odour results from the summation of identical interactions, Beets considers the odour from a statistical point of view. A population of identical molecules will arrive in the vicinity of receptors with different orientations, and different conformations if not completely rigid. The orientation pattern will be more or less homogeneous but perfect homogeneity will never be attained. These differently oriented molecules interact with a rich variety of sites on the epithelium to produce a heterogeneous set of interactions, finally resulting in a given odour.

According to this approach, morphological similarity between two molecules cannot be measured by seeking maximum coincidence between centres of gravity and axes of minimum and maximum dimensions of the two molecular models. The molecules should be compared from the point of view of their dipolar interactions with the receptor. Thus, for dipolar molecules, their dipole axes should be made to coincide. Then, maximum superposition should be sought by suitable modifications of the conformations, and by rotations and translations. In the case of non polar molecules such as saturated hydrocarbons, one should expect a random orientation pattern, resulting in heterogeneous interactions and finally in a complex odour, which is in fact observed.

Other aspects of the functional group such as its steric accessibility and its contribution to the direction of the molecular dipole were also considered by Beets. Molecular shape, polar groups and dipoles are different aspects of a same molecular structure and should not be discussed separately.

A consequence of Beets' model is that the number of different sites a molecule can interact with should follow a roughly gaussian distribution, varying from a relatively high number in the case of saturated hydrocarbons to a very small number in the case of molecules having some particular odours. To find those particular odours we searched for descriptors which appear alone in the descriptions of the olfactory properties of pure substances (isolated descriptors).

We obtained from Arctander's (1969) book a data bank containing olfactory descriptions of nearly 2500 pure substances. Using statistical methods we were able to measure the degree of "isolation" of the 74 most frequent words used to describe the odours of these substances. The words musk, amber, sandalwood and anisic (Chastrette et al. 1986, 1988) appeared as the most isolated. This was taken as indicating a relative homogeneity of the corresponding interaction types and suggested that these odours correspond to well defined receptor sites. Two of them, musk and sandalwood odours, were chosen as targets for structure-odour relationships.

To gain an understanding of molecule-receptor interactions we followed a four-step strategy presented in detail in the case of musks. Strategy and results obtained with sandalwood odorants will be only briefly decribed.

In the first step, a sample of pure substances possessing well defined odours and chemical structures was obtained from the literature. From this sample were drawn a training set and a test set.

In the second step, pertinent structural elements were identified using data from the training set. The corresponding knowlege is usually stated in the literature under the form of a set of rules. Using statistical methods we were able to assess the validity of these rules and to produce new rules and sets of favourable and unfavourable structural elements.

In the third step, the relative positions and orientations of these structural elements were determined, using stereochemical considerations and hypotheses on the interaction types associated with each of them. As a result an interaction model, valid for the training set of molecules, was designed.

In the fourth step, the model was tested on a test set, different from the training set, and modified if necessary.

THE INTERACTION MODEL FOR MUSKS

To design an interaction model, we had to rely on a sample of pure substances (musks and non-musks) with well-defined odours and structures. The musk odour is found in several classes of molecules possessing very different chemical structures. Thus, the nitro musks are relatively rigid benzene derivatives while the macrocyclic musks show a considerable conformational freedom.

We first constituted a sample of 244 pure substances pertaining to the classes of nitrobenzenes, indans or tetralins. From this population 104 compounds were described as musks and 140 as non musks (either odourless or odorous but not musk). This sample was our training set, used to design the interaction model.

We also constituted a test sample, including 43 musks and 17 non-musks, of pure substances with different chemical structures.

A critical examination of the empirical rules found in the literature resulted in a first set of pertinent structural elements. This set was refined using a combination of three methods of description of molecular structures. The DARC system and the autocorrelation method, both already well-known in pharmacology, were used for the first time in the field of olfaction. A third original method, named GESDEP (Generation and Selection of Descriptors and Elaboration of Patterns), consisting in a combination of stepwise generation of progressively more complex descriptors and selection of the best descriptors at each degree of complexity, was designed and used.

As a result three muskophore patterns (i.e. patterns which are necessary for the presence of the musk odour) were found, two for the nitrobenzenes and one for indans and tetralins (fig. 1).

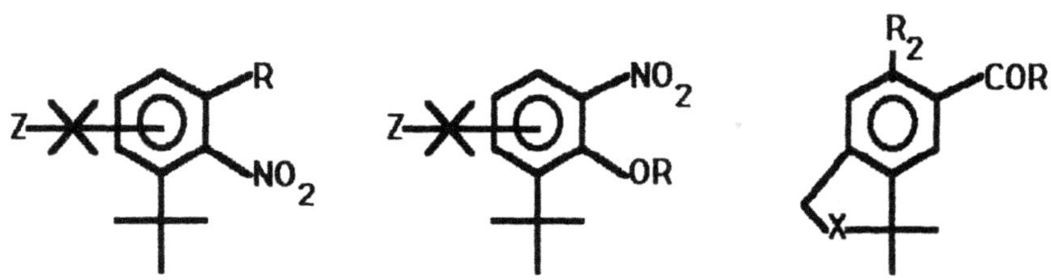

P_1 P_2 P_3

R = Me, Et   x = C or C-C   R1 = H, Me, Et   Z = NH_2, OH, COOH

R_2 = Me, Et $R_2 \neq$ OR, COOH, CH_2OR

Figure 1. Muskophore patterns

These patterns are very similar to ortho-musks and meta-musks (Beets 1978) except that they include explicit conditions for the absence of unfavourable substituants such as Z in P_1 and P_2. An unique site cannot account for all of the three patterns, as long as interactions are thought in terms of dipole orientations. The different olfactory qualities of known musks should then be related to the different patterns found in their molecular structures.

Nevertheless, we decided to test the possibility of interactions via hydrogen bonding to the functional group, as suggested by Ohloff (1986) and others. The problem of finding an unique pattern was reformulated as: is it possible to find acceptable conformations of patterns P_1 to P_3 in which associated hydrogen atoms, hydrogen-bonded to the functional group, would occupy the same position relative to the aromatic ring and to the t-butyl group?

To answer this question one has to introduce the geometrical characteristics of hydrogen bonds to nitro and carbonyl groups. Fortunately such characteristics were recently determined in a statistical study of X-ray cristallographic data from the Cambridge Crystallographic Data File (Murray-Rust and al. (1984), Dunitz and al. (1985)). The important points are (fig. 2) that the bonded hydrogen atom should lie close to the plane of the carbonyl, at a distance of 1,7 to 1,8 Å from the oxygen atom.

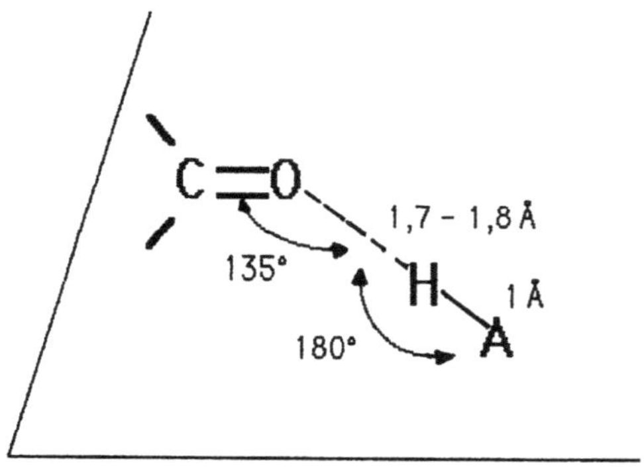

Figure 2. Geometrical constraints of hydrogen bonds at a carbonyl group.

The COH angle has an optimum value of 135° with a possible deviation of ± 15 to 20°. The OHA angle has an optimum value of 180 ± 10°.

As no information was found concerning the hydrogen bonds to a nitro group we decided to use as an approximation the same geometrical constraints as in the case of the carbonyl group. Once this point is accepted, the similarity between patterns P_2 and P_3 becomes obvious and we only habe to consider P_1 and P_2.

If P_1 and P_2 interact with the same site through an hydrogen bond, the associated hydrogen-bonded atoms H_1 and H_2 should lie at the same position relative to the aromatic ring and to the t-butyl group (fig. 3).

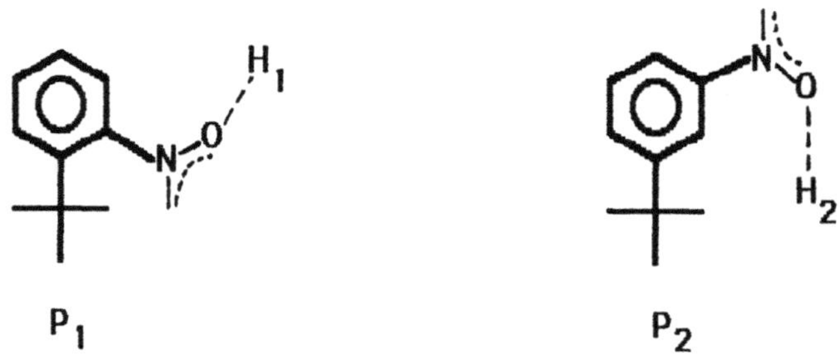

R = Me, Et

Figure 3. Hydrogen bonding to P_1 and P_2

Let β_1 and β_2 be the angles between the ring and the nitro group in P_1 and P_2. The question now becomes: is it possible to find a pair of β_1 and β_2 values such as the hydrogen atoms H_1 and H_2 coincide (fig. 4) ?

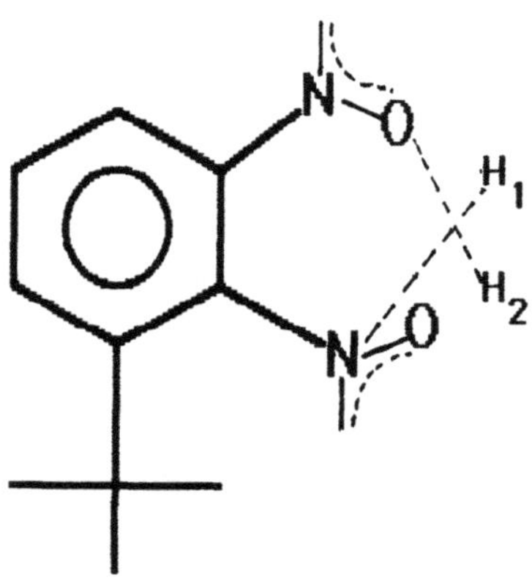

Figure 4. Compatibility of P_1 and P_2.

Taking into account the geometrical constraints of hydrogen bonding, it was possible, using the Alchemy II programme, to obtain a distance $d(H_1\text{-}H_2)$ of ≈ 0,04 Å for β_1 = - 52° and β_2 = 51°. The distance between the associated hydrogen atom and the quaternary carbon atom (centre of the hydrophobic part) is about 5 Å. This suggests a probable interaction with an unique site of molecules possessing different patterns, resulting in a musk odour. Interaction of musk molecules with other sites, beside this specific one, produce secondary odours.

The high values of the β angles mean that the nitro group has lost a good part of its conjugation with the aromatic ring. A confirmation of the somewhat surprising values of β was sought in three different ways.

The first way consists in an estimation, using conformational energy calculations, of the lower energy conformations of molecules possessing patterns P_1 and P_2 (Chastrette and Zakarya, 1989). The results (fig. 5) show that the β values are in good agreement with β_1 and β_1.

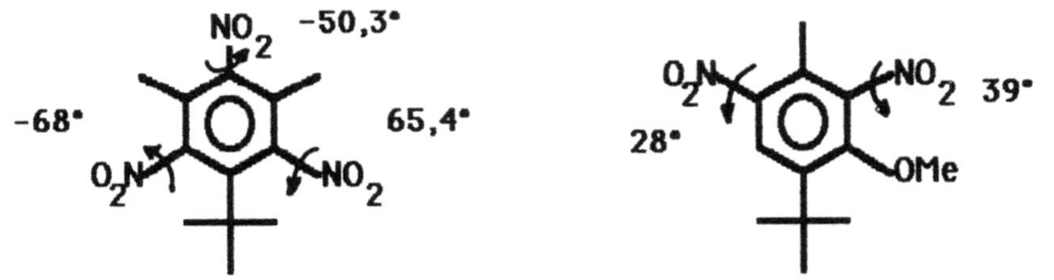

xylol musk (P_1) ambrette musk (P_2)

Figure 5. Calculated β angles for xylol an ambrette musks.

A second confirmation comes from crystallographic data (Bryden, 1972; Havere et al., 1982) of substances the structures of which are very similar to patterns P_1 and P_2. High β values are quite common in the crystals of these substances (fig. 6).

Figure 6. β angles in some nitrobenzenes, from crystallographic data.

Finally, our model implies that an optimum β angle should be found and that too large or too small substituants should produce too large or too small β angles. This was found to be the case in two series of indans and tetralins (fig. 7) where the odour intensity is related to the size of the substituant (Chastrette and Zakarya 1988).

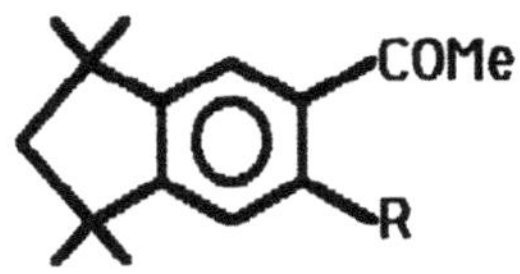

R = H	non-musk
R = Me	strong musk
R = Et	fairly strong musk
R = iPr	weak musk

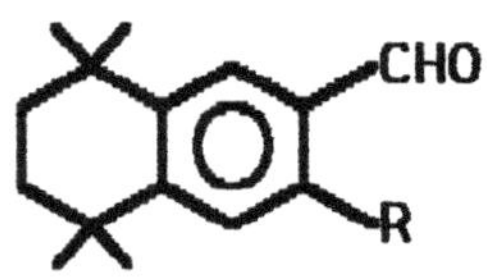

R = H	non-musk
R = Me	musk
R = Et	musk
R = iPr	weak musk

Figure 7. Intensity of musk odours as a function of substituant size.

Thus the interaction model consists in a set of structural elements, geometrical characteristics and interactions. The t-butyl group interacts with the hydrophobic part of the site through dispersion forces. The functional group interacts with the hydrophilic part of the site through an hydrogen bond, within the geometrical constraints of hydrogen bonding. The intensity of odour can be ascribed, at least partly to the quality of fit between molecules and sites. The unfavourable effect of hydrogen bonding substituants is easily explained.

The interaction model gives a quite satisfactory percentage of correct predictions (89,4 % for musks and 93,6 % for non-musks) on the training set and excellent predictions for the classical strong musks.

The last step was a test of the model on the test set of 60 compounds. The predictions were again quite acceptable (100 % for the musks and 82,4 % for the non-musks). For structures including carbonyl or nitro groups it was easy to use the model. For structures devoided of these groups but including an oxygen atom, the known characteristics of hydrogen bonding to ethers (Murray-Rust, 1984; Dunitz, 1985) were used.

Macrocyclic musks were a more difficult case as they have highly variable conformations and no t-butyl group. The lower energy conformation of muscone was calculated using a MM2 program and the molecular model of muscone and P_2 were superposed, using the Alchemy II program. It was found that when the carbonyl groups are superposed, the methylene groups corresponding to carbons numbered 5, 6 and 7 are superposed on the t-butyl group and are probably those involved in the hydrophobic interaction with the site.

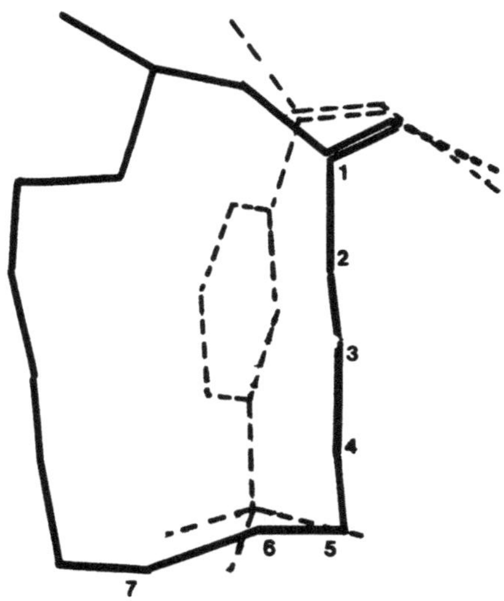

Figure 8. Superposition of P_2 on muscone.

To test further the limits of the interaction model the constraints on the hydrophobic part were also examined. The optimum size of the hydrophobic group, which seems to correspond to a t-butyl or t-amyl group, was more precisely determined using results from Wrobel and Wannagat (1982) on a series of analogs of musk xylol in which the carbon atom of the t-butyl group was replaced by a heteroatom (fig. 9).

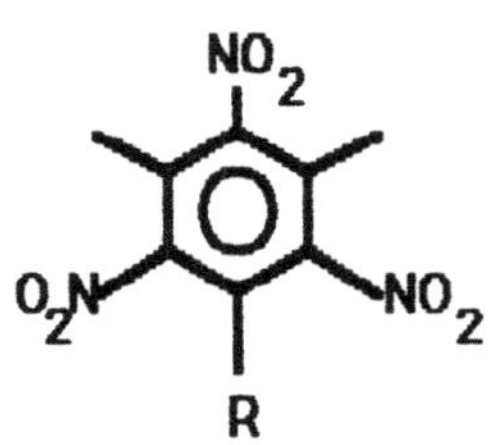

R = t-butyl ===> strong musk

R = Si Me_3 ===> very strong musk

R = Ge Me_3 ===> weak musk

Figure 9. Intensity of the musk odour and size of the hydrophobic substituant.

Substitution by a silicon atom gives a musk described as more brilliant than musk xylol while the germanium derivative is a weaker musk.

However, the musk odour cannot be entirely ascribed to the quality of the fit between molecules and sites. Other molecular properties such as vapor pressure and solubility in the mucus are clearly playing an important role. Furthermore the complexity of the odours of some musks suggests that their interaction with other sites produces secondary odours. When for some reason (e.g. presence of unfavourable substituents such as an hydroxyl group or steric effects) the musky character fades away the secondary notes may appear stronger and become dominant.

THE INTERACTION MODEL FOR SANDALWOOD ODOURS

When the same strategy was applied to the sandalwood problem it was possible to define a set of sandalophore patterns leading to correct predictions of the odours of a sample of 139 compounds (57 sandalwood and 82 non-sandalwood). It was impossible to fuse these patterns in an unique pattern as in the case of musks, even taking hydrogen bonding into account.

However, we were able to design an interaction model including a bulky alkyl group and an hydrogen bonding functional group with specific relative positions and orientations. When the functional groups of the molecules with strong sandalwood odours are superposed, the hydrophobic parts are not completely overlapping and different molecules seem to interact with different parts of the same hydrophobic site. To account for this possibility we introduced the concept of superpattern, defined as the envelope of several patterns. No actual molecule is expected to possess all the features of the superpattern but the most active compounds were found to possess almost all of them.

Using the concept of superpattern, excellent predictions (100 %) were made for all active and inactive compounds in the sample.

REFERENCES

Amoore EJ (1952) The Sterochemical Specificities of Human Olfactory Receptors. Perf. and Ess. Oil Rec. 43:321-30

Amoore EJ (1964) Current Status of the Steric Theory of Odour. Ann. N.Y. Acad. Sc. 116:457-76

Arctander S (1986) Perfume and Flavor Chemicals. Arctander, Monclair

Beets MGJ (1978) Structure-activity Relationships in Human Chemoreception. Applied Science Publishers, London

Bryden JH (1972) The Crystal Structure of 2,4,6 Trinitro-m-xylene (TNX). Acta Cryst. B28: 1395-1398

Chastrette M Elmouaffek A Zakarya D (1986) Etude statistique multidimensionnelle des similarités entre 24 notes utilisées en parfumerie. C.R. Acad. Sci. t303:1209-1214

Chastrette M Elmouaffek A Sauvegrain P (1988) A Multidimensional Statistical Study of Similarities Between 74 Notes Used in Perfumery. Chem. Senses 13:295-305

Chastrette M Zakarya D (1989) unpublished results

Chastrette M Zakarya D (1988) Sur le rôle de la liaison hydrogène dans l'interaction entre les récepteurs olfactifs et les molécules à odeur de musc. C.R. Acad. Sci. t307:1185-1188

Havere WV Lenstra ATH Geise HJ (1982) 4-ter-Butyl-3,5-dinitroanisole. Acta Cryst. B38:3119-3120

Murray-Rust P Glusker JJ (1984) Directional Hydrogen Bonding to sp^2 and sp^3 Hybridized Oxygen Atoms and its Relevance to Ligand-Macromolecule Interactions. J. Am. Chem. Soc. 106:1018-1025

Ohloff G (1986) Chemistry of Odor Stimuli. Experientia 42:271-279

Vedani A Dunitz JD (1985) Lone-Pair Directionality in Hydrogen Bond Potential Functions for Molecular Mechanics Calculations: the Inhibition of Human Carbonic Anhydrase II by Sulfonamides. J. Am. Chem. Soc. 107:7653-7658

Wrobel D Wannagat U (1982) Sila-Substituted Perfumes, Sila Derivatives of Some Musk Scents. J. Organomet. Chem. 225:203-210

PRINCIPLES AND PROPERTIES OF SOME SOLID STATE CHEMICAL SENSORS

G. Heiland

RWTH Aachen
5100 Aachen
FRG

I. INTRODUCTION

In medicine and biology sensory phenomena are studied already for a long time. A review has been given some years ago (Stiewe 1983). In the field of technique the demand for sensors has been induced by the fast development of semiconductor electronics. Often the computers require transducers for communication with the external world. Sensors are small electronic devices observing physical or chemical quantities and giving out a corresponding electrical signal. The computers accept this signal and turn out their result via read outs or actuators.

Of special interest are at present the solid state chemical sensors. They serve mainly for the detection of gases in air and of ions in fluids. In many cases a quantitative determination of low concentrations is required. Here the sensors are in competition with complex and expensive equipment like optical spectrometers, mass spectrometers and gas chromatographs. However, quite different properties are expected for the sensors: They should be small, rugged and inexpensive. This holds also for the connected electronics and power supply. Some applications require that the whole apparatus can be carried easily, or, that many sensors can be distributed in a building. Very small sensors are developed for invasive monitoring and for implantation. For such advantages a price has to be paid consisting of lower analytical ability and longtime stability. In some applications the stability or regeneration problems may be solved by replacing a cheap sensor sufficiently often.

For medical and biological applications biosensors can serve (Turner et al. 1987). Besides the International Conferences on Transducers having at present a repetition

NATO ASI Series, Vol. H 39
Chemosensory Information Processing. Edited by D. Schild

period of 2 years (e.g. De Rooij 1989) there are now established also special Meetings on Chemical Sensors (e.g. Aucuturier et al. 1986) and on biosensors (e.g. Schmid 1989). Fig. 1 shows the scheme of a biosensor employing a biologically sensisitive layer. The output of this layer consists of intermediate products, which are observed by means of a transducer which can be a solid state chemical sensor.

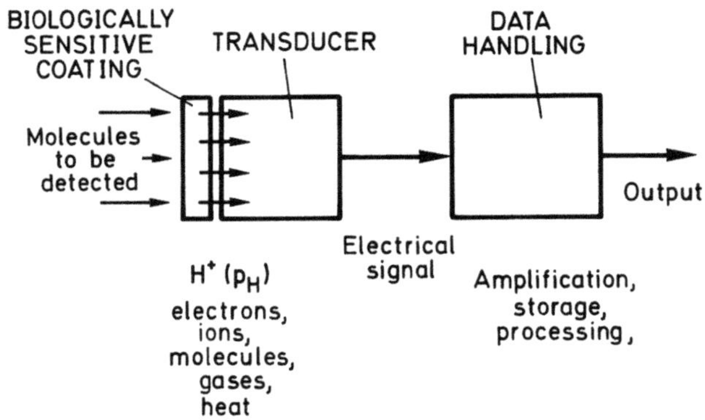

Fig. 1. A generalized biosensor configuration. The biologically sensitive coating generates an intermediate product in response to the analyte. This product causes an electrical signal in the transducer. After Lowe (1985).

The intention of the present article is to give an overview of some important solid state chemical sensors and to explain the principles. With regard to the required brevity not many details are mentioned. However, the more deeply interested reader finds an extended list of references.

II. SOLID STATE CHEMICAL SENSORS

II. 1 Homogeneous semiconductor sensors

Table 1 (at the end of the text) gives a list of some solid state chemical sensors including short informations about materials and properties. The sensors employing electronic semiconductors are separated in two groups: The conductivity based sensors consisting of a homogeneous sensitive layer with two or four electrodes for measurement of the conductance, and the structured sensors like diodes, capacitors and field effect transistors. The conversion of a chemical input into an electrical signal exhibits remarkable differences between the two groups.

The homogeneous semiconductor sensors are apt for the detection of noxious or dangerous gases in air. In many cases the sensitivity reaches down to a few ppm allowing the detection of the maximal work place concentration. Besides warning against air pollution also control and steering of technical processes require such gas sensitive devices. An idealized response to a gas increasing the conductance is sketched in Fig. 2. It can be advantageous to use for a signal not the stationary increase $g\text{-}g_0$ but the initial slope dg/dt upon admission of the active gas. Then a shorter exposure is sufficient for the measurement resulting in a faster read out of the signal and in a shorter recovery time. Furthermore, the linear range of the calibration curve is extended to higher concentrations.

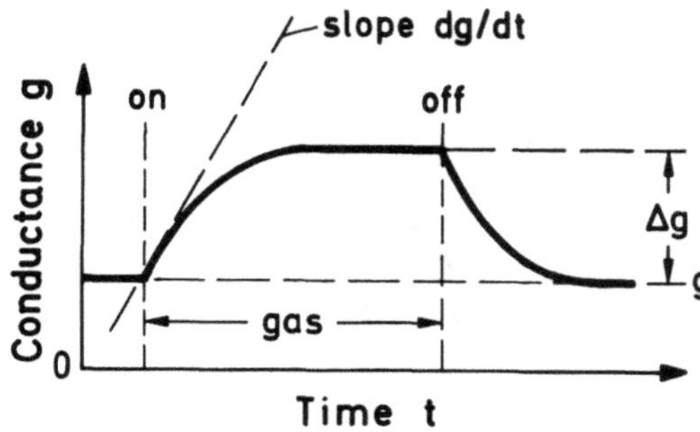

Fig. 2. Homogeneous, conductivity based semiconductor sensor. Schematic drawing of the conductance g = I/U during transient exposure to a gas, that increases the conductance. g_0 conductance in air without an active gas.

II. 1.1 Sensors employing anorganic semiconductors

The surface of a semiconducting n-type metal oxide is partially covered in air by chemisorbed oxygen. The negative oxygen ions bind electrons and decrease the conductance within a space charge layer to a depth of about 10^{-6} to 10^{-5} cm (Heiland 1982, Heiland and Kohl 1985, Kohl 1989). Some gases can increase the conductance by adsorbing as donors injecting electrons into the oxide. Reducing gases can also react with the ionosorbed oxygen freeing the bound electrons from the surface acceptors, or with the lattice bound oxygen generating oxygen vacancies. The vacancies act as a more stable kind of surface donors after annealing at sufficiently high temperatures.

The reduction of the surface by the active gas is connected with a catalytic oxidation of the gas. This process can be influenced by the addition of a few percent of a noble metal to the oxide (Kohl 1989 Montreux). For the recovery of the sensors after the exposure the reoxidation of the surface by the oxygen of the air plays a decisive role. A temperature of several hundred centigrade is required for a sufficiently fast response and recovery.

In polycrystalline samples the conductance can depend on the grain boundary resistance which changes by surface reactions of the ambient gas. In thin films with small grains the space charge layer penetrates the whole grains.

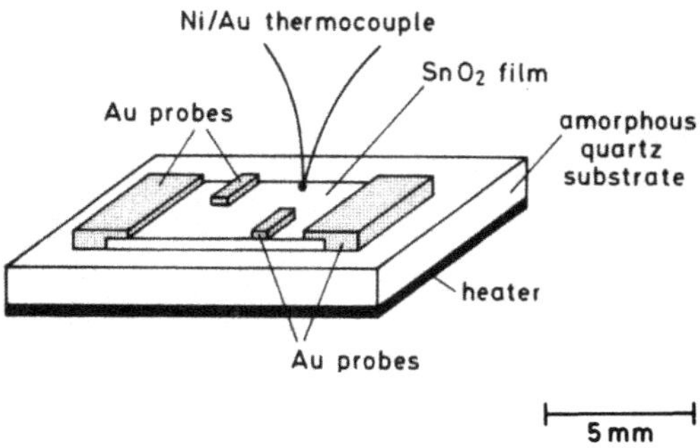

Fig. 3. Thin film gas sensor. To exclude the influence of contact resistance there are two probes for the current and two potential probes. After Mokwa et al. (1985).

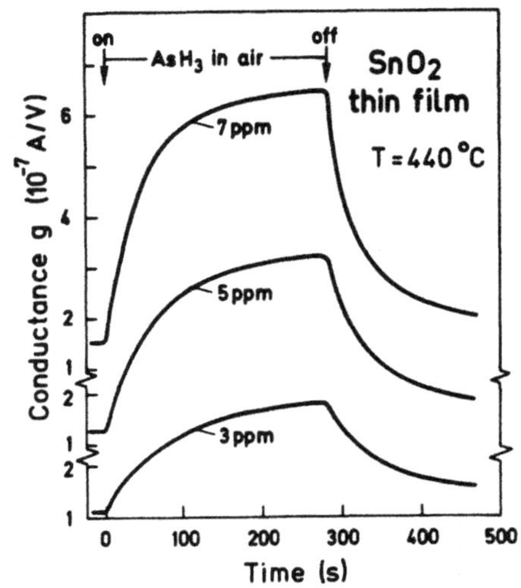

Fig. 4. Thin film sensor as shown in Fig. 3. Increase of conductance g during transient exposure to arsine in air, The permanent ambient was air without any admixture. After Mokwa et al. (1985).

An example of a gas sensor with a sensitive film of SnO_2 can be seen in Fig. 3. The response to a small content of arsine in air is shown in Fig. 4. The increase of conductance has been found proportional to the arsine concentration from 0.5 to 7 ppm (Mokwa et al. 1985). This holds also for the initial slope of the curves in Fig. 4.

The response of an oxidic gas sensor passes usually a maximum with increasing temperature (Fig. 5). By proper choice of the operation temperature at least some separation of different gases can be obtained. Gas sensors of the oxidic type are also applied to analyze the human breath for traces of alcohol. In the experiments of Fig. 6 the air was bubbled through water containing the desired fraction of alcohol. In this case a sudden admission of the test gas is possible and therefore the initial slope of the current at constant voltage is used. The addition of various metals as catalysts exerts a strong influence upon the sensitivity.

At high temperatures the bulk conductivity of several n-type oxides with a large band gap depends reversibly on the ambient oxygen pressure. For an oxygen sensor based upon this phenomenon a fast diffusion of donors is required to establish an equilibrium throughout the sample. Especially titania has been studied to be applied for combustion control in a similar way as zirconia, see Table 1. (Logothetis et al. 1983).

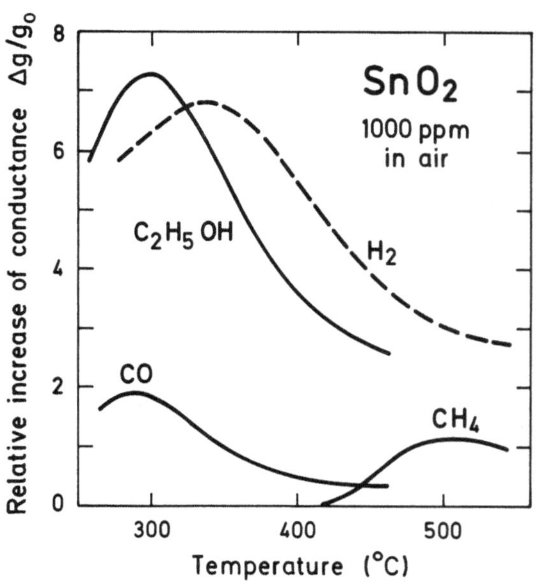

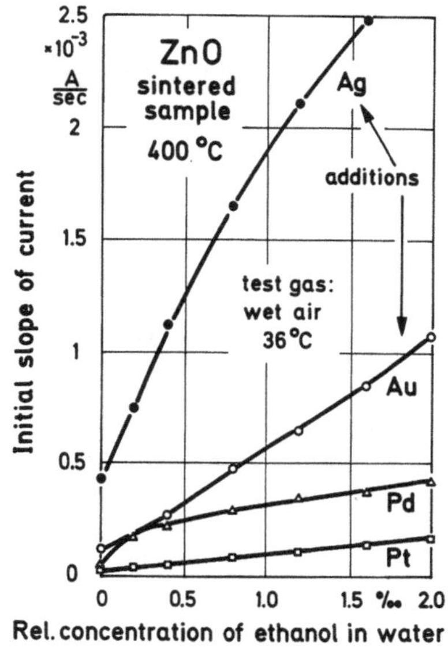

Fig. 5. Sintered layer with an addition of 0.05 wt.% Sb producing bulk donors. Relative increase of conductance g by various reducing gases in air as a function of temperature. After Komori et al. (1983).

Fig. 6. Sintered ZnO layers with different metal additions of about 1%. Initial slope of the current upon admission of the test gas as a function of the ethanol concentration. Voltage 1.35 V. After Schulz et al. (1979).

II. 1.2 Sensors employing organic semiconductors

The properties of a group of organic semiconductors, the phthalocyanines, make them especially apt for gas sensors detecting strongly oxidizing gases. Beeing p-type conductors they are in this respect complementary to the n-type oxides. The adsorbed oxidizing molecules inject holes increasing the conductivity. Fig. 7 shows the response of a thin film of Cu-phthalocyanine to bromine in air. The elevated temperature is again required for a sufficiently fast response.

Halogen containing molecules can be detected, although they do not decompose at the surface of the organic semiconductor. If the molecules are decomposed at a hot metal surface, the resulting radicals act upon the semiconducting film increasing the conductance (Fig. 8). This sensor is able to detect escaping anesthetic gases in the air of hospitals or vapors of solvents in factories or in dry cleaning shops, if the molecules contain halogens. For the studied examples the maximum allowed work place concentration is detectable. Programmed temperature modulation of the sensor and also of the catalyst with a cycle of 1 min allows to use the initial slope of the conductance for sampling of a signal (Thoren et al. 1987).

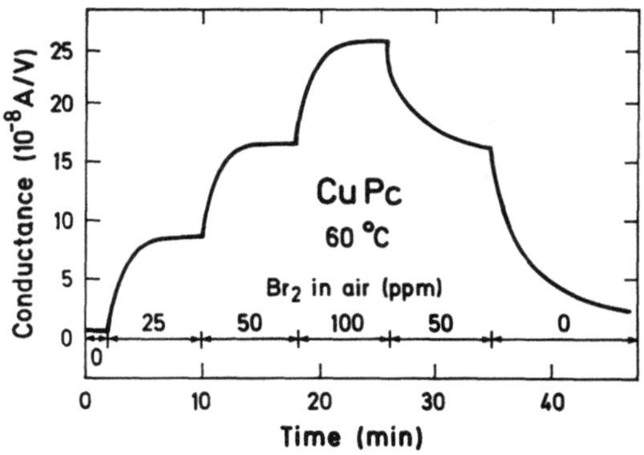

Fig. 7. Cu-phthalocyanine, thin film. Variation of conductance g during exposure to different concentrations of bromine in air. The sensor is shown by the inset of Fig. 8. For bromine no catalyst is required. After Laurs et at. (1987).

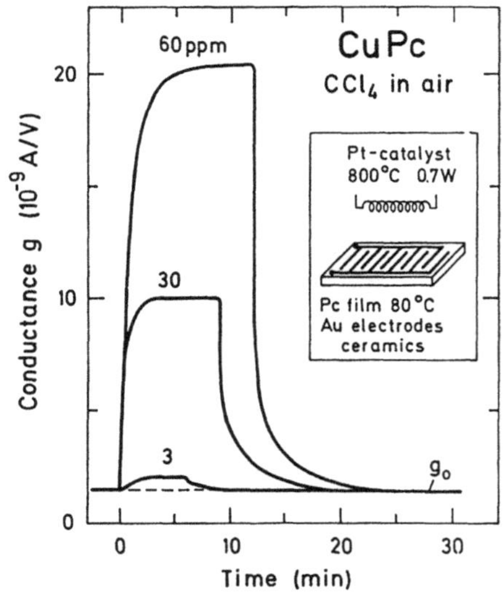

Fig. 8. Cu-phthalocyanine, thin film. Conductance g during transient exposure to various concentrations of carbon tetrachloride in air. No signal is observed, if the catalyst is not heated. The inset shows the arrangement of sensor and catalyst in a schematic way. Dimensions a few mm. After Laurs et al. (1987). See also Heiland and Kohl (1985).

II. 2 Structured semiconductor sensors

With exception of the ionsensitive field effect transistor (ISFET, Table 1) a thin film of a transition metal plays a decisive role for these devices (Lundström et al. 1987). The catalytic properties of the metals are important as for the homogeneous oxidic sensors with metal additions (Kohl 1989 Montreux). Fig. 9 shows schematic drawings of three different types of gas detecting devices comprising a thin metal film. Hydrogen molecules dissociate at the metal surface and penetrate the film by diffusion or through pores. At the interface between metal and insulator hydrogen atoms (protons) are adsorbed forming a dipol layer which causes a step of the electrostatic potential. There may be also a contribution from hydrogen atoms (protons) diffusing into the insulator and producing slow response and drift phenomena. To avoid the penetration of hydrogen atoms into the SiO_2 an additional layer of Si_3N_4 or Al_2O_3 is deposited (Fig. 10, inset).

For the recovery the oxygen of the air makes a contribution. Water molecules are faster formed at the suface than hydrogen molecules. The presence of oxygen also sets a limit in temperature and lifetime of the device by slow oxidation of the metal film. In the sketches (d)-(f) of Fig. 9 the influence of hydrogen in air can be seen as a shift or variation of the characteristic.

II. 2.1 Diodes

A Schottky depletion layer is induced, if an n-type semiconductor is contacted with a metal of higher work function. The space charge layer is depleted of electrons to a depth of about 10^{-5} cm. The potential drop across this depletion layer steers the resistance of the device and causes a rectifier characteristic. The scheme of such a diode and of a simple circuit for measurement of the characteristic are indicated in Fig. 9 (c).

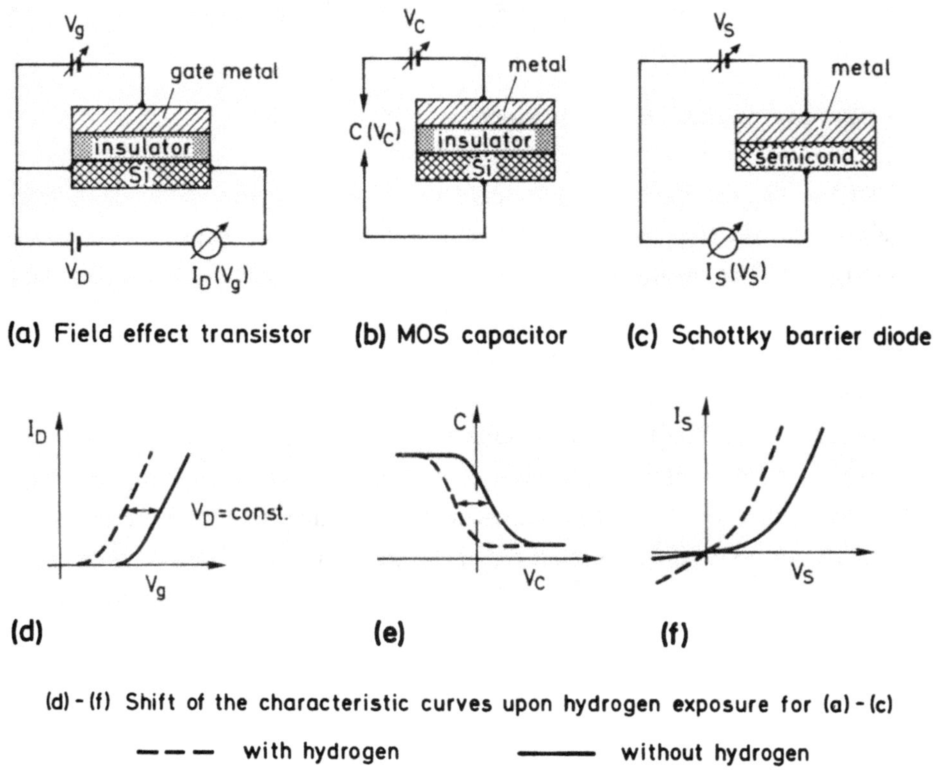

Fig. 9. Schemes of three devices using a catalytically active film of a transition metal. Below each drawing the corresponding characteric and its change by hydrogen exposure are sketched. The double arrow in (d) and (e) indicates the voltage shift dV. After Lundström et al. 1986).

A dipole layer formed by hydrogen atoms at the metal-semiconductor interface not only shifts the characteristic parallel to the voltage axis but also changes the shape of the curve (f). An interpretation is possible assuming a decrease of the metal work function and therewith of the barrier height. Studies with ZnO and TiO_2 have been published (Ito 1979 and Yamamoto et al. 1980).

II. 2.2 MOS capacitors

A transition metal film and a semiconductor separated by an insulating layer form an MOS (metal/oxide/ semiconductor) capacitor (Fig. 9, (b) and (e)). In the characteristic the maximum capacity appears if an accumulation space charge layer is induced in the semiconductor, whereas the low capacity corresponds to depletion. The hydrogen produces an electrical double layer between metal and insulator and changes thereby the electric field inducing the space charge layer on the semiconductor.

II. 2.3 Gas sensitive field effect transistors

The field effect transistors (FET's) for gas sensing have the same sequence of layers as the MOS capacitors (Fig. 9 (a) and (d)). They are MOSFET's whose gate consists of a transition metal. The phenomenon of the field effect is explained in Fig. 11 for the case of an n-type semiconductor and air or vacuum as the insulator. The induced charge is distributed between trapping states at the semiconductor surface or within the space charge layer and the conduction band. Only the latter fraction causes a change of the surface conductivity. Therefore trap densities at the interfaces of field effect devices are kept low by careful preparation. If the sign of V_g is reversed also a depletion layer can be induced decreasing the surface conductivity. Returning to Fig. 9 we see, that the hydrogen causes about the same shift dV of the transistor characteristic as of the MOS capacitor characteristic. An example is shown in Fig. 10. The sensitivity reaches down to about 1 ppm.

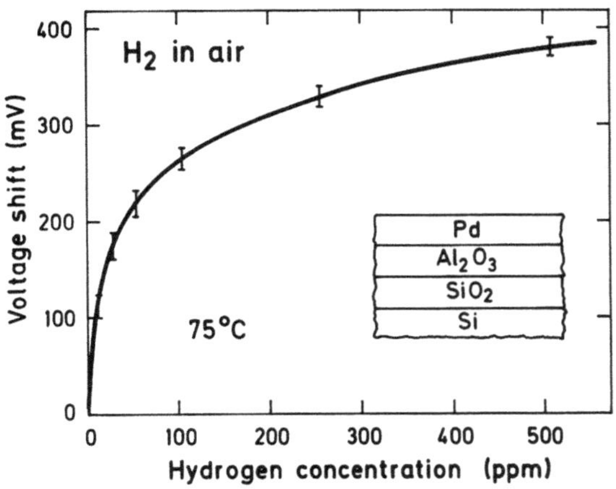

Fig. 10. Calibration curve for an MOS gas sensor. Voltage shift as a function of the hydrogen concentration in air. The bars show the maximum and minimum values during constant hydrogen cycling for 18 hours. The inset gives the sequence of the layers in the structure. After Lunström et al. (1986).

Also a sensitivity to some larger hydrogen containing molecules has been reported. If a catalytic decomposition takes place at the metal surface, the hydrogen penetrates and produces a signal. A high sensitivity has been found with very thin porous metal films, especially for ammonia in air. This fact led to the assumption, that the field of adsorbed dipols at the metal surface penetrates through holes in the film and induces the signal. Also CO can be detected with gates having holes or a fine slit (Dobos et al. 1983).

As an application of the hydrogen sensitive devices a leak detector has been built (Lundström et al. 1987). The ammonia sensor is recommended not only for industry and agriculture, but also for biological purposes. Biosensors can be composed by combination with enzymes producing ammonia as the intermediate product, see Fig. 1 (Winquist et al. 1985).

II. 2.4 Ion sensitive field effect transistors (ISFET's)

ISFET's are applied for the measurement of ion concentrations in electrolytes. For comparison the scheme of a glass electrode is shown in Fig. 12 (a). This electrode is widely used for pH-measurements (Bergveld et al. 1978). There are two solutions separated by a glass membrane. One solution serves as the reference with known pH value, the other one represents the test solution. The surface of the glass is hydrated and protons penetrate into the gel layer building up a diffusion potential. There is also a contribution from ion exchange at the glass surface. Some charge transport through the bulk of the glass is required. The carriers are sodium ions. The difference of the two double layer potentials is measurend by means of an electrometer.

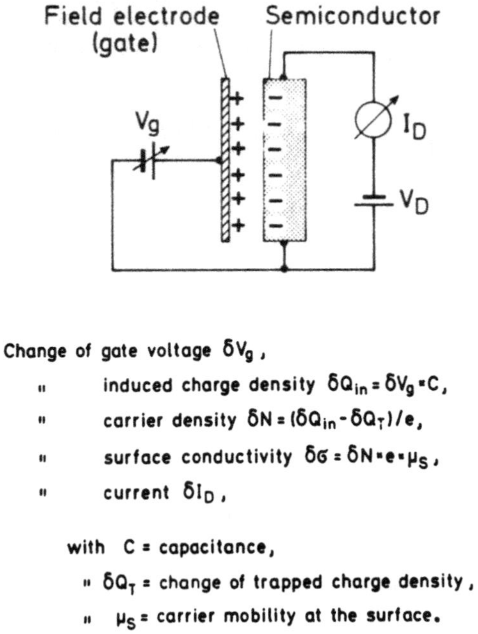

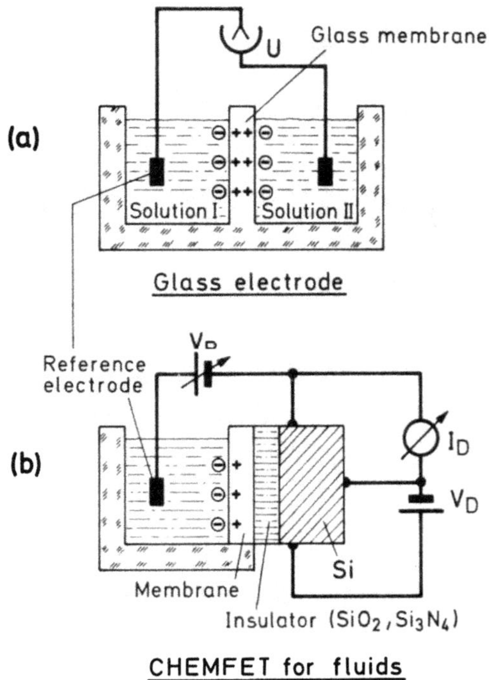

Fig. 11. Scheme for explanation of the field effect. The equations hold for an n-type semiconductor.

Fig. 12. Simplified schemes of (a) a glass electrode, (b) an ion sensitive field effect transistor. An ISFET belongs to the category of CHEMFET's, chemical sensitive FET's.

The principle of an ISFET is demonstrated by a simplified drawing in Fig. 12 (b). The metal gate of the MOSFET (Fig. 9 (a)) is replaced by the liquid. For pH-measurements the silicon oxide layer is usually protected by a layer of Si_3N_4, Al_2O_3 or Ta_2O_5. In many cases a membrane covers the insulating layer. As shown in Fig. 12 (b) there is no reference solution. If the electrical double layer at the surface or within the interior of the membrane or at the insulator surface changes with the ion concentration of the fluid, the resulting change of the potential appears at the insulator. This variation of the voltage across the insulator correspondes to a change of the field strength at the silicon surface and therewith of the induced charge, the surface conductivity and the current I_D. The field effect is explained in Fig. 11.

The possibilities of miniaturization are demonstrated by two medical applications. With pH-ISFET's studies were performed on the dental plaque. Within the oesophagus pressure and pH could be measured simultaneously and at different sites (Van den Vlekkert et al. 1988).

Besides the pH-determination the concentrations of many kinds of ions can be observed. For this purpose special membranes are prepared. Also many types of biosensors have been proposed (Turner et al. 1987). The main limitation to many applications remains the problem of stability especially for the invasive pH-measurement and for the implantation of glucose sensors. Also encapsulation without leakage represents a difficult task.

A detailed description and discussion of ISFET properties and also a comparison with ion selective electrodes are published elsewhere (Janata 1985).

II. 3. SENSORS EMPLOYING ION CONDUCTORS

The principles of solid-state electrochemical gas sensors have been presented elsewhere (Weppner 1987).

II. 3.1 Oxygen sensors employing zirconia

At sufficiently high temperature (500-700 °C) oxygen ions in zirconia ceramics are mobile. There are two arrangements for oxygen sensing, the potentiometric and the amperometric one.

The potentiometric sensor consists of a ceramics stabilized by additions and exposing two porous Pt-electrodes in different atmospheres. The voltage of this electrochemical concentration cell is proportional to the logarithm of the ratio of the oxygen pressures in the two atmospheres at the two electrodes and to the absolute temperature according to the Nernst equation. One atmosphere can serve as a reference. The most important application exists in the measurement of the oxygen concentration in combustion gases, especially in the exhaust of gasoline engines as lambda-sensor (Velasco et al. 1982). The stoichiometry of the air-fuel mixture can be observed and adjusted in a feedback circuit. By the catalytic action of the Pt-electrodes the thermodynamic equilibrium of the gas mixture is achieved, i.e. remaining traces of fuel and oxygen react at the electrodes. For other applications the high temperature of operation (500 to 700 °C) has to be considered.

The amperometric sensor makes use of mass transport control by diffusion. The oxygen diffuses through a diaphragm to the zirconia ceramic. With an applied voltage the ion conductor acts as an oxygen pump transporting the oxygen from the cathode to the anode where it is neutralized and released. The flow of the oxygen gas through the diffusion barrier is proportional to the pressure, and the current within the zirconia is proportional to the arriving oxygen flow. Thus a linear relation between electrical current and oxygen pressure is obtained within a certain range, if an appropriate voltage provides a saturation current.

II. 3.2 Other ion conductors for gas sensing

There are many ion conductors possibly apted for gas sensing, if at a sufficiently high temperature an interaction between defect concentration and ambient atmosphere takes place. Examples are given in Table 1 (Gauthier et al. 1977). A high temperature is required for oxygen sensing with zirconia. In contrast a film of LaF_3 has been reported to work in a potentiometric circuit at room temperature (Miura et al. 1989). However, for the detection of gaseous oxygen water vapour plays an important role. Also oxygen dissolved in liquid water can be observed.

III. CONCLUSION

The list of solid state chemical sensors looks promising and indeed there are many fields where such sensors are needed. However, at present the application is limited mainly by two problems. For some purposes the calibration is not stable enough. Recalibration provides not always a satisfying solution. Also the selectivity is not sufficient for a variety of applications. By combination of several sensors with different distributions of sensitivity an improvement for the recognition of certain gases has been achieved (Mueller this book). The surface chemistry of some important sensors is beeing studied by means of basic research (Kohl 1989 Montreux). Better understanding of the processes connecting the primary reaction and the electrical output signal will allow to improve the properties of the chemical sensors.

TABLE 1. The most important solid state chemical sensors. Material and properties.

SOLID STATE CHEMICAL SENSORS

HOMOGENEOUS SEMICONDUCTOR SENSORS

Changes of the electronic conductance by gases admixed to the air.

Sensors employing anorganic semiconductors

Preferentially n-type metal oxides with a large band gap, e.g. SnO_2, ZnO, TiO_2, WO_3, sintered layers, thin films, with small additions of noble metals as catalysts:

at 150 - 500 °C	detection of e.g. H_2, H_2S, CO, AsH_3, PH_3, hydrocarbons, in air,
at 800 - 1000 °C	measurement of the O_2 concentration, e.g in the exhaust of combustion processes.

Sensors employing organic semiconductors

e.g. phthalocyanines, thin films:

at 20 - 160 °C	detection of halogenes and NO_2 in air, of halogen containing molecules with a separated catalyst at about 700 °C.

STRUCTURED SEMICONDUCTOR SENSORS

Diodes

Transition metal films on an n-type semiconductor, e.g. Si, InP, ZnO or TiO_2:

at 20 - 200 °C	detection of H_2 and some hydrogen containing molecules in air by a change of the characteristics C/U or I/U.

MOS capacitors

Transition metal films separated by an insulating layer from a semiconductor, mostly Si. Detection properties similar to those of diodes. Shift of the C/U characteristic.

Field effect transistors, Si based FET's

With transition metal gate, MOSFET's,

applications similar to those of diodes, shift of the I_D/V_g characteristic,

with porous metal gate detection of other gases, e.g. CO and NH_3

biosensors with NH_3 as intermediate,

with insulating gate, ion sensitive FET's, ISFET's,

used in fluids in combination with a reference electrode, measurement of pH, and, with special membranes, of the concentration of various ions, biosensors.

SENSORS EMPLOYING ION CONDUCTORS

Zirconia, sintered ceramics

at 600 - 900 °C	potentiometric sensors (Nernst's law), monitoring of the oxygen concentration (log. scale) in the exhaust of combustion processes, application in automobiles as lambda sensor, amperometric sensors (diffusion limited gas flow), linear dependence on oxygen concentration.

Several other ion conductors

e.g. K_2SO_4, K_2CO_3, $Ba(NO_3)_2$ mostly

at high temperatures measurement of SO_x, CO_x and NO_x respectively,

LaF_3 for O_2 at room temperature, also in water.

REFERENCES

Aucuturier JL et al. (eds) (1986) Proceedings of the second international meeting on chemical sensors. Université Bordeaux

Bergveld P, De Rooij NF and Zemel JN (1978) Physical mechanisms for chemically sensitive semiconductor devices. Nature 273:438-443

De Rooij NF Chairman Transducers '89, 5th International Conference on Solid-State Sensors and Actuators, Montreux, Switzerland, June 1989

Dobos K, Strotman R and Zimmer G (1983) Performance of gas-sensitive Pd-gate MOSSFET's with SiO_2 and Si_3N_4 gate insulators. Sensors and Actuators 4:593-598

Gauthier M and Chamberland A (1977) Solid-state detectors for the potentiometric determination of gaseous oxides. J Electrochem Soc 124:1579-1583

Heiland G (1982) Homogeneous semoconducting gas sensors. Sensors and Actuators 2:343-361

Heiland G and Kohl D (1985) Problems and possibilities of oxidic and organic semiconductor gas sensors. Sensors and Actuators 8:227-233

Ito K (1979) Hydrogen sensitive Schottky barrier diodes. Surface Sci 86:345-352

Janata J (1985) Chemically sensitive field effect transistors in Janata J and Huber RJ (eds) Solid state chemical sensors. Academic press Orlando Florida p 65-118

Kohl D (1989) Surface processes in the detection of reducing gases with SnO_2-based devices. Sensors and Actuators 18:71-116

Kohl D (1989) Chemistry of solid state gas sensors employing noble metal additions. Transducers'89 International Conference Montreux. To be published in Sensors and Actuators

Komori N, Sakai S and Konatsu K (1983) Sintered SnO_2 sensor for methane in Seiyama T et al (eds) Proc. Int. Meeting on Chemical Sensors, Fukuoka, Japan. Kodansha Tokyo and Elsevier Amsterdam, p 57-61

Laurs H and Heiland G (1987) Electrical and optical properties of phthalocyanine films. Thin solid Films 149:129-142 and private communication.

Logothetis EM and Kaiser WJ (1983) TiO_2 film oxygen sensors made by chemical vapour deposition from organometallics. Sensors and Actuators 4:333-340.

Lundström I, Armgarth M, Spetz A and Winquist F (1987) Gas sensors based on catalytic metal-gate field-effect devices. Sensors and Actuators 10:399-421

Lowe CR (1985) An introduction to the concepts and technology of biosensors. Biosensors, an international journal, 1:3-16

Miura N, Hisamoto J, Yamazoe N, Kuwata S, Salardenne J (1989) Solid-state oxygen sensor using sputtered LaF_3 film. Sensors and Actuators 16:301-310.

Mokwa W, Kohl D and Heiland G (1985) An SnO_2 thin film for sensing arsine. Sensors and Actuators 8:101-108

Schmid RD Chairman Biosensor Workshop May 1989. Braunschweig, FRG.

Schulz M, Bohn E and Heiland G (1979) Messung von Fremdgasen in Luft mit Halbleitersensoren. Technisches Messen tm J 521-3:405-414

Stiewe H (1983) Sensors of biological organisms - biological transducers. Sensors and Actuators 4:689-704

Thoren W, Lagois W and Laurs H, Heiland G (1987) Gas sensor for low concentrations of anesthetic vapors. EUROSENSORS I, conference at Cambridge GB, programme and abstracts:p. 204

Turner AFP, Karube I and Wilson GS (1987) Biosensors, fundamentals and applications, Oxford University Press, Oxford

Van den Vlekkert HH, Kloeck B, Prongue D, Berthoud J, Hu B and de Rooij NF (1988) A pH-ISFET and an integrated pH-pressure sensor with back-side contacts. Sensors and Actuators 14:165-176

Velasco G, Schnell JPh and Croset M (1982) Thin solid state electrochemical gas sensors. Sensors and Actuators 2:371-384

Weppner W (1987) Solid-state electrochemical gas sensors. Sensors and Actuators 12:107-119

Winquist F, Spetz A, Armgarth M, Lundström I and B Danielson (1985) Biosensors based on metal-oxide-semiconductor structures. Sensors and Actuators 8:91-100

Yamamoto N, Tonomura S, Matsuoka T and Tsubomura (1980) A study on a palladium-titanium oxide Schottky diode as a detector for gaseous components. Surface Sci 92:400-406

PURIFICATION OF AN ODORANT BINDING PROTEIN FROM HUMAN NASAL MUCOSA

P. Pelosi[1], C. Maremmani[2] and A. Muratorio[2]

1) Istituto di Industrie Agrarie, via S. Michele,4 - 56100 Pisa, Italy
2) Istituto di Clinica Neurologica, via Roma,67 - 65100 Pisa, Italy

INTRODUCTION

The nasal mucosa of several mammals contains a protein that binds several odorants with dissociation constants in the micromolar range and is not found anywhere else in the body. This protein, that is very abundant, accounting for up to 1% of total soluble protein in the nasal mucosa, has been first identified and purified from the cow (Pelosi et al.,1982; Bignetti et al.,1985a; Pevsner et al.,1985) and fully characterized both in terms of structure and of binding specificity (Bignetti et al.,1985b, 1987; Cavaggioni et al.,1987; Pelosi and Tirindelli, 1988; Pevsner et al., 1986, 1988b; Topazzini et al., 1985).

Despite the great amount of data collected so far for this protein, its function is still unknown. While it seems clear that it is not an olfactory receptor, a role in olfaction is a reasonable hypothesis, but still to be unambiguosly proved. Support for such hypothesis is to be found in comparison between "in vitro" binding constants for a wide screen of odorants and "in vivo" olfactory data, such as odour quality and olfactory thresholds.

A comparative study on the presence of odorant binding proteins in different animals was aimed at identifying the most suitable species for studying both biochemistry and behaviour (Baldaccini et al., 1986). This investigation, far from solving the problem, raised more questions, as odorant binding proteins, similar to the bovine one, were identified only in some mammmal species, with no clear relation to their sense of smell.

In another type of study, the binding constants were measured with a great number of different odorants, in order to define the best ligands for the protein, with relation to chemical structure as well as to odour (Pelosi and Tirindelli, 1988; Topazzini et al.,

NATO ASI Series, Vol. H 39
Chemosensory Information Processing. Edited by D. Schild

1985). Although initial studies, performed with a limited number of odorants, indicated some trend between olfactory thresholds and binding constants, subsequent more detailed investigation has indicated that there is no correlation.

Another approach to understand the physiological function of odorant binding proteins has taken into account its histological localization. This study showed that in the cow the protein is mostly abundant, and secreted, at the level of tubulo-acinar glands, under the nasal respiratory tissue (Avanzini et al., 1987; Pevsner et al.,1986a), while in the rat it is found in the lateral nasal glands, from where it is delivered at the tip of the nose (Pevsner et al., 1986b). Once more, the results only made the picture more complex.

The main problem of assigning a function to this protein arises from the lack of data available on animal behaviour to correlate with binding constants.

It seemed reasonable at this point to undergo a study on the human odorant binding protein, that could provide the unique opportunity for comparing results from ligand binding experiments with the enormous wealth of psychophysical data available in the literature,as well as with known pathological disorders of the olfactory system.

IDENTIFICATION OF THE HUMAN OBP

We we were alerted to the presence of a human binding protein similar to bovine OBP by a private communication from J.Pevsner, who detected a low, but specific, binding of tritiated 2-methoxy-3-isobutylpyrazine to human mucus.

Following this observation, we measured binding of the same odorant and of tritiated 3,7-dimethyloctanolto a crude extract of nasal mucosa obtained from authopsy. In both cases the concentration of the ligand bound was very low: this may indicate that they are not the optimum ligands for this protein.

SDS-electrophoresis revealed the presence in both mucus and tissue extract of a protein of molecular weigth around 16 kdal. The results indicate that there is a slight difference in molecular weight between the protein present in the mucus and that isolated from the mucosa.

PURIFICATION AND PRELIMINARY CHARACTERIZATION OF THE PROTEIN

When a preincubated mixture of crude extract of nasal tissue and tritiated 3,7-dimethyloctanol was applied to a Mono-Q column and eluted with a 0-0.5 M NaCl gradient, radioactivity was present in fractions eluted at about 0.35 M NaCl, as reported in fig. 1.

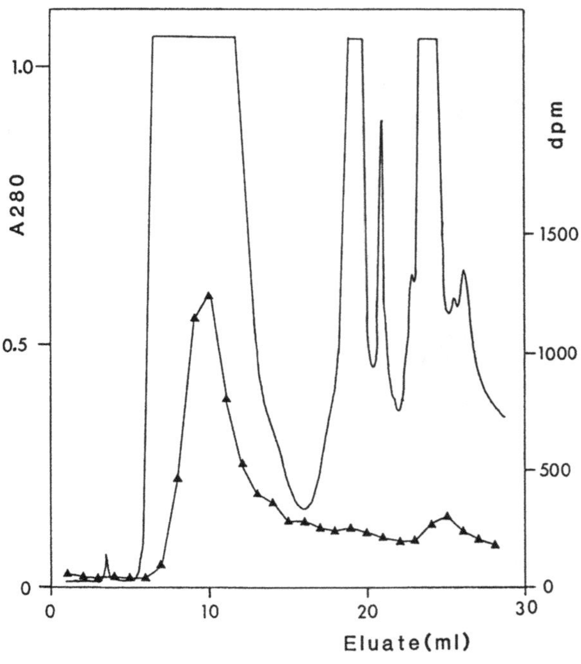

Fig. 1. Elution profile on Mono-Q column of a crude extract of human nasal mucosa, preincubated with tritiated 3,7-dimethyl-octanol. Continuous line represents absorbance at 280 nm; triangles indicate radioactivity counts on the single fractions. Elution was performed in Tris/HCl 20 mM, pH=7.4, using a NaCl gradient 0-0.5 M. The second peak of radioactivity, found around fraction 25, represents the bound ligand and was eluted at 0.35 M NaCl.

The protein profile of the single fractions, as analyzed by SDS electrophoresis indicated a 16 kdal protein as the best candidate for the 3,7-dimethyloctanol binding species. We therefore purified this species to homogeneity, by repetitive anion exchange chromatography on DE-52, using a salt gradient of 0.2-0.5 M NaCl. In this way we managed to obtain nearly one mg of pure protein from about 20 g of nasal mucosa, using three passes through the same column.

The purified protein had a molecular weight of 16 kdal under dissociating conditions and an isoelectric point of 4.1 as measured by isoelectric focusing in a 4.0 to 6.5 pH gradient of Ampholines. It is thus smaller and more acidic that the bovine OBP. Fig. 2 shows the purified protein taken from olfactory mucosa.

In comparison with the protein found in mucus, there appears to be some discrepancy in molecular weight: this aspect is under investigation.

The purified human OBP is not retained by an affinity column bearing antibodies to bovine OBP. This finding is in agreement with previous observations that bovine and rat OBPs do not cross react with their relative antibodies (Pevsner et al., 1988b).

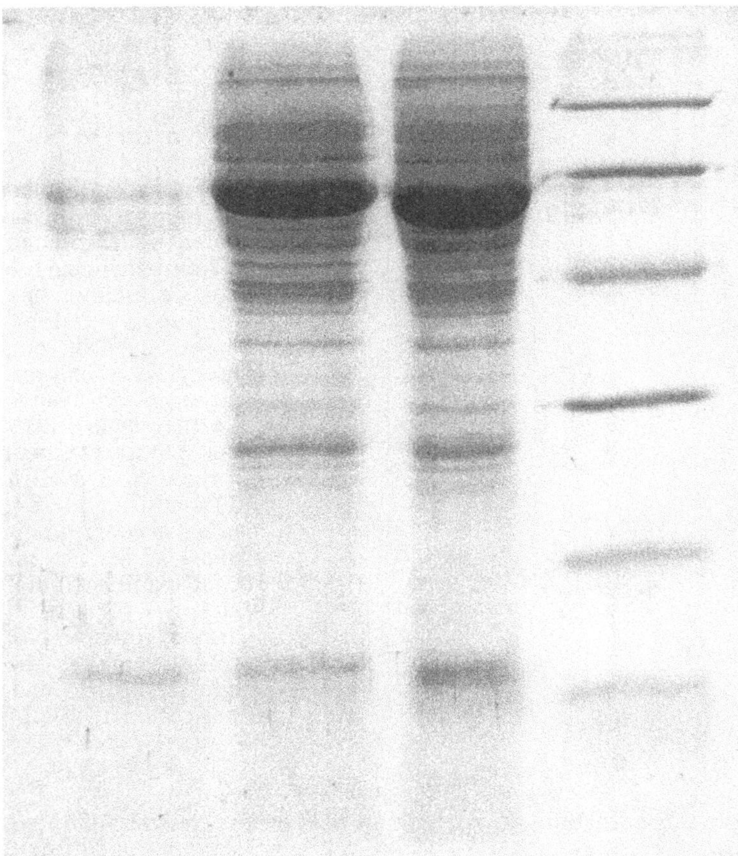

Fig 2. SDS-PAGE of the human OBP purified from authopsy (lane 1), total soluble proteins from human nasal mucus (lane 2) and a mixture of the two samples (lane 3). Standards (lane 4) were from top: beta-galactosidase (116 kdal), BSA (66 kdal), ovalbumin (45 kdal), carbonic anhydrase (29 kdal), trypsin inhibitor (20 kdal), alpha-lactalbumin (14 kdal).

Direct sequencing has failed, indicating that the N terminal is blocked, being this a common feature with the bovine OBP, where the terminal amino group is acetylated (Cavaggioni et al.,1987). Enzymatic digestion, following the same protocol successfully used for the bovine protein are currently being applied.

DISCUSSION AND FUTURE WORK

The physical properties of human OBP appeared to be very similar to those of bovine OBP and we were able to follow the same protocol for purification.

One major problem encountered with the study of this protein relates to the fact that the tissue can be obtained only one or two days after death, when it is likely that some degradation has occurred. The mucus contains the protein in higher relative amount; however, we have shown that the two proteins are slightly different. We are currently characterizing both forms of the protein in order to understand the relation between them.

It is our opinion that two types of study will give a partial answer to the question of the physiological function of OBPs.

The first is the histochemical localization of human OBP, using the specific antibodies already produced and purified. To understand where the protein is produced and to which part of the nasal cavity is then delivered would certainly add information towards a possible function in the detection and recognition of odours. Moreover, the potential discovery of subjects deficient in this protein could help in correlating the presence of OBP with possible misfunction of the olfactory system. On the other hand, if the complete absence of the protein does not modify olfactory function, then we could conclude that the OBPs have no role in olfaction. In this type of study, the presence of the protein could be easily detected by SDS electrophoresis of the mucus; the use of human subjects would be essential for detecting subtle malfunctions of olfaction, such as modified thresholds or adaptation, that is practically impossible to study in behaviour experiments with animals.

The second type of investigation is a very wide screening of odorants, where their binding constants to the human OBP will be measured and correlated with the corresponding olfactory thresholds. This kind of correlation has previously been attempted using the binding costants of the bovine protein and the human olfactory thresholds: no clear relationship has been found and therefore it is essential to compare data taken from the same animal. Antibodies to the three purified OBPs show no cross reaction. This indicates that correlation of any data between animals will be suspect.

The purification and characterization of human OBP will give a stronger basis to correlation with psychophysical data and physiological defects.

ACKNOWLEDGEMENTS - We thank dr. K.C. Persaud for helpful advice and discussion. We are also grateful to prof. G. Bevilacqua for his cooperation in obtaining authopsy material.

REFERENCES

Avanzini,F., Bignetti,E., Bordi,C., Carfagna,G., Cavaggioni,A., Ferrari,G., Sorbi,R.T. and Tirindelli,R. 1987. Immunocytochemical localization of pyrazine-binding protein in the cow nasal mucosa. Cell Tissue Res., 247: 461

Baldaccini,N.E., Gagliardo,A., Pelosi,P. and Topazzini,A. 1986. Occurrence of a pyrazine binding protein in the nasal mucosa of some vertebrates. Comp. Biochem. Physiol., 84B: 249

Bignetti,E., Cavaggioni,A., Pelosi,P., Persaud,K.C., Sorbi,R.T. and Tirindelli,R. 1985a. Purification and characterization of an odorant binding protein from cow nasal mucosa. Eur. J. Biochem. 149: 227

Bignetti,E., Tirindelli,R., Rossi,G.L., Bolognesi,M., Coda,A. and Gatti,G. 1985b. Crystallization of an odorant binding protein from cow nasal mucosa. J. Mol. Biol. 186: 211

Bignetti,E., Damiani,G., De Negri,P., Ramoni,R., Avanzini,F., Ferrari,G. and Rossi,G.L. 1987a. Specificity of an immunoaffinity column for odorant-binding protein from bovine nasal mucosa. Chem. Senses 12: 601

Cavaggioni,A., Sorbi,R.T., Keen,J.N., Pappin,D.J.C. and Findlay,J.B.C. 1987.Homology between the pyrazine binding protein from nasal mucosa and major urinary proteins. FEBS Lett. 212: 225

Pelosi,P., Baldaccini,N.E. and Pisanelli,A.M. 1982. Identification of a specific olfactory receptor for 2-isobutyl-3-methoxypyrazine. Biochem. J. 201: 245

Pelosi P. and Tirindelli R. 1988. Structure/activity studies and characterization of the pyrazine binding protein. Int. Conf. on Chemosensory Transduction: "Receptor and transduction mechanisms in taste and olfaction", Philadelphia (USA), March 1988

Pevsner,J., Trifiletti,R.R., Strittmatter,S.M. and Snyder,S.H. 1985. Isolation and characterization of an olfactory receptor protein for odorant pyrazines. Proc. Natl. Acad. Sci. USA 82: 3050

Pevsner,J., Sklar,P.B. and Snyder,S.H. 1986. Odorant-binding protein: localization to nasal glands and secretion. Proc. Natl. Acad. Sci. USA 83: 4942

Pevsner,J., Reed,R.R., Feinstein,P.G. and Snyder,S.H. 1988a. Molecular cloning of odorant-binding protein: member of a ligand carrier family. Science 241: 336

Pevsner,J., Hwang,P.M., Sklar,P.B.,Venable,J.C. and Snyder,S.H. 1988b. Odorant-binding protein and its mRNA are localized to lateral nasal gland implying a carrier function. Proc. Natl. Acad. Sci. USA 85: 1

Topazzini,A., Pelosi,P., Pasqualetto,P.L. and Baldaccini,N.E. 1985. Specificity of a pyrazine binding protein from cow olfactory mucosa. Chem. Senses 10: 45

THE DESIGN OF AN ARTIFICIAL OLFACTORY SYSTEM

J.W. Gardner[1], P.N. Bartlett[2],
G.H. Dodd[2] and H.V. Shurmer[1]

Departments of Engineering[1] and Chemistry[2]
University of Warwick
COVENTRY CV4 7AL, U.K.

1. BACKGROUND AND INTRODUCTION TO AN ARTIFICIAL OLFACTORY SYSTEM

1.1 Towards an Electronic Nose

The aim of our research is to design an instrument which will mimic the human sense of smell. This ambition distinguishes our approach from that used in the related area of gas sensing. In the latter area, the object is usually to detect a single gas or vapour (frequently a toxic material) which has exceeded some predetermined threshold value. We use our sense of smell not only for detecting noxious vapours but also in a much more sophisticated way for classifying and grading smells.

The human sense of smell is still the primary 'instrument' used for quality evaluation of a broad range of industrial products including the following:

beverages ....wine, beer, whisky, fruit juices, tea,coffee,milk;
foodstuffs............cheese, butter, fish, and a broad range of natural and fabricated foodstuffs;
perfumes ....extrait perfumes, soaps, cosmetics

In the industries associated with these products, the physicochemical variables of the finished goods are usually measured using conventional analytical instrumentation, but the crucial property of flavour or odour, often the attribute of the product which will be

NATO ASI Series, Vol. H 39
Chemosensory Information Processing. Edited by D. Schild

most easily noticed by the consumer, is measured subjectively by sensory analysis. The sense of smell is the key element of this sensory analysis.

Our design philosophy is based on the belief that Nature has evolved a sophisticated, sensitive, and reliable system for detecting and evaluating odours - the sense of smell. In the first instance, our objective is to understand how our sense of smell works and to incorporate the design principles of the natural system into an electronic analogue. Further on, we will explore alternative designs for an olfactory system which may utilize features of electronic systems (such as ultra-fast switching) which are not available in the biological domain. The remainder of Section 1 summarises our current understanding of olfactory mechanisms.

1.2 The chemical senses

The ability to respond to chemicals in the environment is a fundamental property of living cells. In the case of a free-living cell such as a bacterium, a protozoan, or a leucocyte from blood plasma, this chemoreception usually results in the cell moving up the concentration gradient towards the source of the chemical stimulus. The various types of free-living cells show a selective response to stimuli and this selectivity is usually due to specific receptor proteins often located in or on the plasma membrane of the cell.

All multicelluar organisms also respond to environmental chemicals. With the increasing complexity of organisms as we move up the evolutionary tree, there is a regular progression towards expression of distinct physiological funtions in specialized tissues. Thus in mammals, which are regarded as valid models for human physiological systems, we find that the task of detecting environmental chemicals is shared by at least four chemical senses; olfaction, gustation, the trigeminal sense, and the vomeronasal system. This latter sense is not used by humans and will not be considered further here.

The different chemical senses are located in distinct tissues; respectively, the olfactory epithelium, the tongue, and in the case of trigeminal sense, receptor cells are found in all mucous membranes and also in the skin. All of these tissues send nerves to the central nervous system. Thus, when we refer to a chemical sense we mean both the peripheral tissue which senses the stimulus and also the parts of the brain which further analyze the signal. So, in the case of higher organisms, including humans, we find a complexity in the chemoreception system which far exceeds that in the unicellular organisms.

In humans, the sense of taste is used for detecting non-volatile chemicals. The stimulus must be introduced into the mouth before a sensation can be initiated. Volatile chemicals can stimulate all three main human chemoreceptor systems. A small number of volatile chemicals can dissolve in the liquid on the tongue and can stimulate the sense of taste. Some off-odours, e.g. those resulting from peroxidation of unsaturated fatty acids in rancid dairy products, are sometimes described as having a 'metallic taste' due to this kind of effect. The trigeminal system responds to many volatile chemicals and is thought to be particularly important for the detection of irritant and chemically reactive vapours. Virtually all organic vapours stimulate the sense of smell. Under normal conditions, most non-volatile chemicals cannot reach the human olfactory epithelium and so are unable to stimulate olfaction. However, such molecules, when presented as aerosols, can reach the sensory tissue and can then stimulate a response; this is a relatively unexplored aspect of olfaction (Mather & Dodd 1988).

Electronic noses will, in the future, find a role as flavour quality control devices. It should be noted that the sensation of flavour is due to the simultaneous stimulation of all the chemical senses together with integration of the signals from the components senses by the higher brain centres. Unfortunately, the word 'taste' is often used loosely to refer to the complex sensation of flavour. In fact, the olfactory component is frequently the main contribution to the flavour of beverages e.g., whisky, and comestibles.

Anthropomorphic bias leads us to underestimate the importance of the role of olfaction in the lives of mammals. The sense of smell is used for the location of and the evaluation of food. It is also a primary sense used in the detection of and the evaluation of potential sexual partners (this involves the biologically important sub-class of odorants - the pheromones). Olfaction, again using pheromones, is used by many mammals for a variety of social communication which we usually associate with either vision or audition. The communication includes recognition of the following; individuals, families, groups, social status, home, and the home range (Shorey 1976).

The outstanding features of olfaction are extremely high sensitivity and extremely great discrimination. Interestingly, these properties are found in the same device.

1.3 Odorants

The sensation of a smell arises from stimulation of olfactory sensory cells by molecules of the smell - the odorants. Until about 20 years ago a number of hypotheses on olfaction attributed the sensation to various kinds of vibrations emanating from the vapour. Since the transduction mechanism can be explained in terms of the simpler hypothesis of molecules rather than the more complex hypothesis of vibrations, the latter type of hypothesis has fallen into disuse.

We can distinguish between simple and complex odours (the point of distinction being the nature of the stimulus and not the quality of the sensation). A simple odour consists of one kind of odorant molecule. A complex odour has a mixture of odorant molecules. Simple odours, as defined here, are essentially laboratory curiosities, since virtually all natural odours are mixtures. Indeed, many industrially important smells which are of interest to us as stimuli for our electronic nose are exceedingly complex mixtures and contain several tens of different types of odorant molecules; e.g., flavours and perfumes. The subtle differences between, for example, the odours of different grades of coffee, are attributed to variations in the relative ratios of many flavour molecules in this complex odour.

Table 1.1 **Typical Odorants**

Odorant	Odorant Type	Threshold in H_2O (μg/l)
	Sweaty Cheesy	250
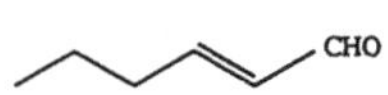	Green Leaves	316
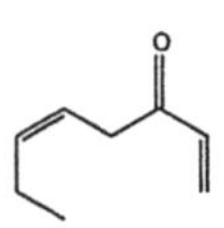	Metallic-Engine oil -Fusty-Fungal	0.001
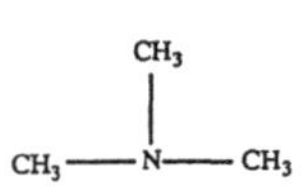	Fishy	0.47
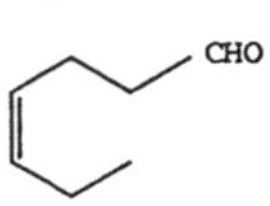	Off-Flavour of White Fish	0.04
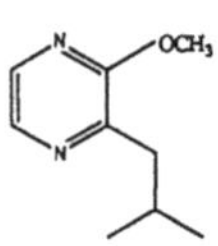	Green Peppers	0.002
	Lemons	10

Odorants are typically small, hydrophobic, organic molecules with a mass range of 34-300 Da (Arctander 1969). Most odorants contain a single polar group. Some representative odorants from the several thousand known are shown in Table 1.1. A majority of odorants contain oxygen. This atom will usually be in one of the typical polar groups shown in the table. Nitrogen occurs less frequently in odorants. An important class of natural odorants containing nitrogen are the pyrazines and other heterocyclic odorants which are found in the odours of roasted products. Sulphur occurs in many important natural odorants from animals. These

Table 1.1 The properties of some typical odorants. All the odours shown are important in flavour chemistry.

odours, which often originate from microbial metabolism of animal precursor molecules are often important signalling odorants (pheromones).

The key question about odorants which we need for the rational design of our electronic nose is this; how do the physico-chemical properties of the odorant molecule correlate with the properties of the odour?[1]

The three important properties are:

* the odour type
* the threshold value
* the intensity curve

The properties of the molecules which determine these properties are discussed in detail elsewhere (Beets 1978; Moulton 1975). We can summarize our present understanding of this key question in olfaction by saying that the size, shape, and polar properties of the odorant molecule determine its odour properties. However, the rules which apply are poorly understood and most new synthetic odorants are designed semi-empirically (and even then most studies focus on a related family of odorants).

There are two difficulties in understanding the rules. First, whether there are fundamental dimensions of odour quality has yet to be determined. Most classifications of odour type are empirical and use common names to signify an odour; e.g., fruity, flowery, musk. Even a single, highly pure odorant[2] may need several descriptors to describe its odour. Further, because of subjective associations to the odour and also because of the many types of specific smell deficiencies (specific anosmias) which are known, the description of an odour by one individual is of limited value. An odour description should be the consensus opinion of a panel of trained smellers. The panel should take account of the specific anosmias of its members. Curiously, most odour evaluation panels span a narrow age range and fail to take this factor into account. It is

1) There are, at present, two separate strands to this question. One aspect of our work concentrates on the responses of our instrument to pure, single odorants. This study will lead to a set of predictive rules for classifying the simple odours. In a separate study we are also investigating the responses of our nose to complex odours such as flavours. Ultimately, the responses to the complex odorants must be explained in terms of the responses to the individual odorants in the complex odour (and it is likely that this account will involve interactions between the responses of the individual odorants). It is probable that the classification of the complex odours (the ones of practical importance to industry) will be achieved empirically long before the theoretical explanation (the academic goal) can be given.

2) Highly pure odorants are a rarity in laboratories; especially in those laboratories studying human responses to odours. This deficiency of technique is responsible for much misleading information in the literature.

now known that there are significant differences in responses to odour with increasing age (Van Toller et al. 1985).

The second difficulty is the large number of parameters which are currently used in quantitative structure-activity relationships for describing the physicochemical properties of molecules. The large number of terms in the regression equations demand that the number of odorants studied in odour-structure experiments should be prohibitively large (Dodd 1976). The special difficulty here is the huge amount of work which is required for studying even a small number of odorants, see, e.g., Van Toller et al. 1985.

Two examples will illustrate the complex inter-relationship between the shape and size of an odorant, and its odour. Suppose we keep the polar group constant in a series of molecules and also the size of the molecule (i.e., the number of atoms in the molecule remains constant). Now let us examine how the odour type changes as we alter the shape of the molecule (i.e., we have a series of isomers).

Table 1.2

	R2									
R1	Methyl	Ethyl	Propyl	Butyl	Pentyl	i-Pentyl	i-Butyl	i-Propyl	s-Butyl	t-Butyl
Formate	E	E/F	E/F	E/F	E/F	F/E	F/E	F/E	F/E	M/E
Acetate	F/E	F/E	F	F	F	F	F/E	F/E/M	F/M/E	M
Propionate	F/E	F/E	F	F	F	F	F/E	F/M/E	F/M	M
Butyrate	F/E	F/E	F	F	F	F	F	F	F	M/F
Pentanoate	F/E	F/E	F/E	F/E	F/E	F	F	F	F/M	M
i-Pentanoate	F/E	F	F	F	F	F	F	M/F	M	M
i-Butyrate	F/E	F	F	F	F	F/M	M/F	M	M	M
tri-Methyl Acetate	F/M	F/M	F/M	F/M	F/M	F/M	F/M	F/M	F/M	M

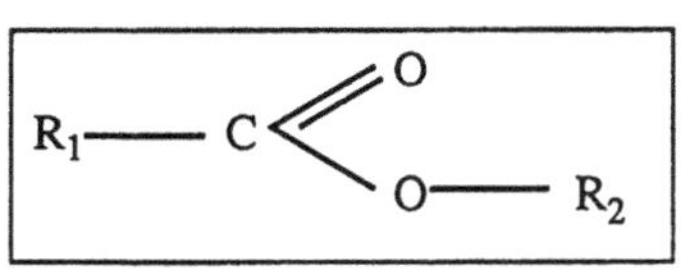

Table 1.2 Table showing how the odour quality of a series of molecules with the same mass and the same polar group varies as the shape of the molecules are varied.

F - fruity odour.
E - ethereal odour
M - minty odour

The minty esters have compact spherical molecules. The fruity esters have flexible molecules which can adopt a non-spherical, linear conformation.
The data are taken from an unpublished study by Squirrel and Dodd.

The results of one such experiment which we carried out in our laboratory are shown in Table 1.2. There is a graduation of odour quality which is correlated with the shape of the molecules. The linear molecules, such as that in the top left compartment, have fruity odours with no hint of a minty note, whereas the spherical molecules, such as that in the lower right compartment, have a pure mint note with no hint of fruitiness. Many of the molecules with intermediate shapes have both fruity and minty notes.

In the second example, shown in Table 1.3, we will also keep the polar group constant but will vary both the size and shape of the molecules (though all will be compact rings with a basic similarity in shape). The odour descriptions are taken from Arctander and from Moncrieff. There is a regular gradation in odour type with increasing ring size.

Table 1.3 Table showing the odours of simple cyclic ketones. Both the mass and shape of the molecules alter throughout the series.

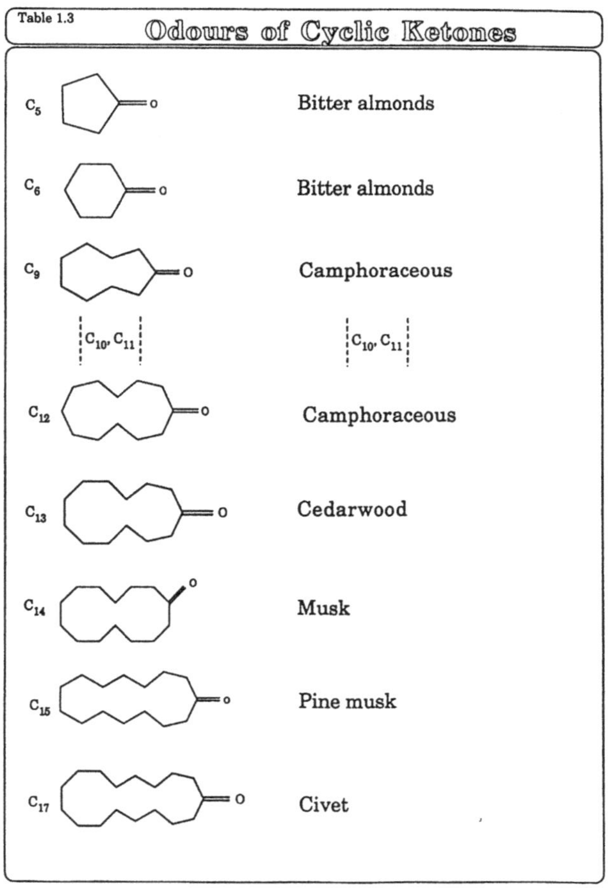

Studies of the odour quality with molecular shape led Amoore to propose a classification of odours (Amoore 1970). The Amoore scheme is now known to be oversimple but the data in his book emphasize the key role of the shape/size parameter in determining the quality of an odour.

We do not have space here to consider the other odour properties in detail. Some examples of olfactory thresholds are given in Figure 1.1. The very low values for some of the odorants serve to warn us of potential difficulties in olfactory work. For example, a minor impurity (<0.1%) with a very low threshold in an odorant (with a purity of >99.9%, this would normally be regarded as a pure substance in a chemical laboratory) with a high threshold, can dominate the odour profile. In other words, in such an instance it is the odour of the impurity you smell and not the odour of the supposed odorant.

Odorants with exceedingly low threshold values can also dominate the odour profile of complex natural flavours and flower smells. For example, the pyrazine odorant shown in Table 1.1 is a minor component of the multicomponent oil obtained from green peppers. Yet, this pyrazine gives the oil its characteristic smell.

The intensity of an odour is measured by the slope of the line obtained in a graph such as that shown in figure 1.1. Odorants with a high intensity, such as dimethylether, have a high vapour pressure and so can achieve a high concentration in the air; thereby giving a strong smell sensation. In contrast, low intensity odorants, such as the musk cyclopentadecanone, have the converse behaviour. They have a low vapour pressure and so the concentration of molecules in the air is not high. Never-the-less, such molecules, including the musk shown, frequently have low thresholds and so can be smelt.

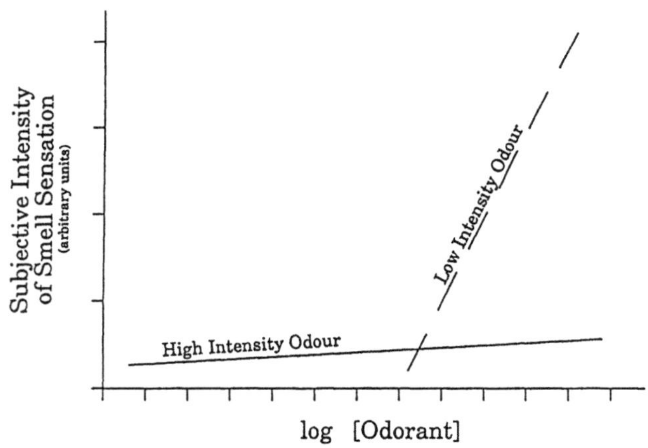

Figure 1.1 Typical intensity curves for odorants. Diethylether is a typical example of an odorant with a low intensity curve. Odorants with a high molecular mass, such as musks, give high intensity curves: the odour can be greatly diluted with little alteration in the odour strength.

The figure also provides the answer to the question of the lasting power of different odorants. As odorants evaporate, the low intensity ones will quickly reach a sub-threshold level whereas the high intensity odorants can still be detected when a tiny fraction only of the original amount of material remains. This phenomenon is a key factor in commercial perfumery. The strategies which our electronic nose will use when dealing with odours of differing intensities has yet to be fully worked-out.

1.4 The Sensory Tissue

The sensory tissue of the olfactory system has a structure which is basically similar for all higher animals. The tissue is a patch of thin epithelium which lies high up in the nose, beneath the front of the brain cavity. The tissue lies on a number of cartilaginous structures, the turbinates which are often arranged as a series of whorl-like structures when viewed up the nasal cavity, as shown in figure 1.2. The turbinates are a device for increasing the surface area of the epithelium within the nasal cavity.

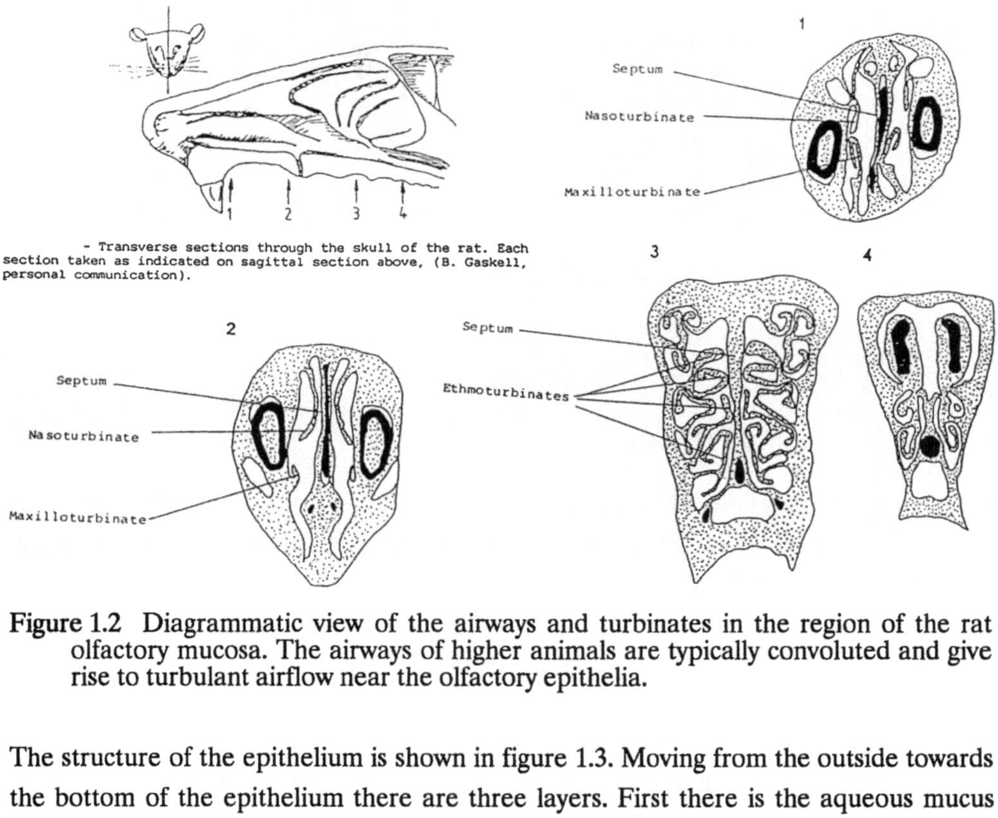

Figure 1.2 Diagrammatic view of the airways and turbinates in the region of the rat olfactory mucosa. The airways of higher animals are typically convoluted and give rise to turbulant airflow near the olfactory epithelia.

The structure of the epithelium is shown in figure 1.3. Moving from the outside towards the bottom of the epithelium there are three layers. First there is the aqueous mucus layer which can vary between 10-100 μ in depth. Then there is the layer containing the sensing cells, the olfactory neurones. There would typically be between 50-100 million neurones in a single epithelium. These neurones have a conspicuous dendrite, of typical dimensions 40-80 μ by 1 which protrudes into the mucus layer as a terminal swelling. From the terminal swelling a number of cilia, typically between 5-20, depending on the organism, radiate out into the mucus layer. The neurones are separated by large columnar epithelial cells, the sustentacular cells.

The bottom layer, the basal layer, has a number of cells including the important undifferentiated basal cells. These cells mature and differentiate into functional neurones throughout the life of an animal (this might indicate that the sensing cells are damaged in the act of "sniffing").

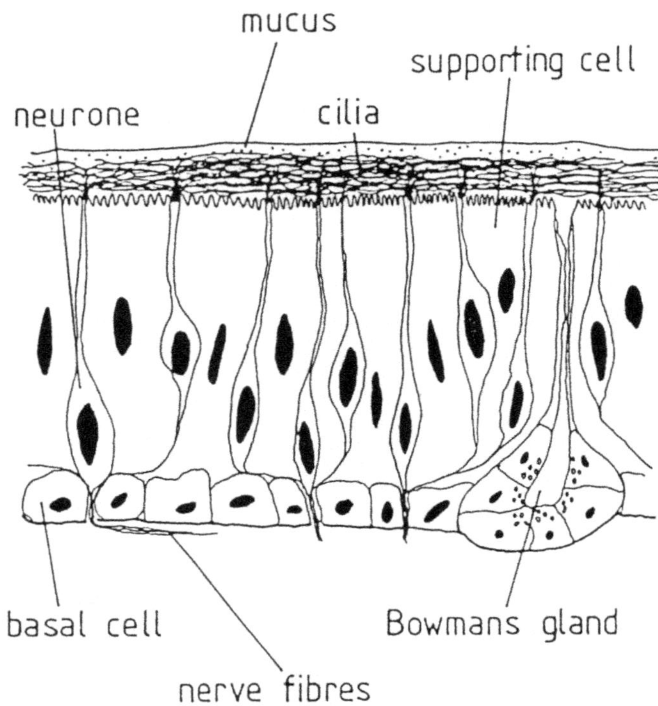

Figure 1.3 Diagram of the cells and structures found in a vertebrate olfactory epithelium. The diameter of the dendrite of the olfactory neurone is about 1 micron.

Starting in this layer, there are a number of glands which feed into the mucus layer. The axons of the sensory neurones are collected into bundles in this layer and run along the bottom of the layer into the brain.

In comparison with other nerve tissues and other sensory tissues, there are three unique points about olfactory neurones. First, they have a limited lifespan, about 28 days in the mouse; and then they die and are replaced by basal cells which grow into neurones. This is an interesting point for the design of the sensors in the electronic nose: perhaps they too will be damaged in normal use and will have to be replaced.

Second, they are the only ciliated neurones known. Interestingly, modified cilia are frequently used as a sensory structure in a cell, e.g., the rod of the rod cells in the retina.

Third, though the neurones lie outside the brain (and indeed grow into the brain) the axons pass through fine holes in the skull and synapse with neurones in the olfactory bulb. Expressing this idea from another viewpoint, we could say that they are the only neurones in the central nervous system which have direct access to the outside world.

The epithelium contains the receptor elements which must be simulated in the electronic analogue of the nose. Interestingly, an odorant must traverse a thin layer of aqueous medium, the mucus, before it encounters the sensory membrane. Whether it is essential to mimic this feature in the electronic analogue is not yet clear.

The second cells in the olfactory system occur in the olfactory bulb. There are two points of interest about the organization of the bulb which are of interest for our purposes. First, there is a marked convergence as we pass from the primary cells in the system, the olfactory sensing neurones, to the secondary cells, the mitral cells, in the bulb. There are between 1-20 thousand primary cells connected to a single secondary cell. This suggests that a secondary cell is processing and integrating information from its many input cells. Consistent with this, is the observation that secondary cells respond to distinct categories of odorants whereas the primary cells appear to show only a weak selectivity (Holley et al. 1974). The convergence phenomenon is, we think, a key point in the design of an artificial olfactory system.

The second feature of interest in the bulb is the complex interactions between the secondary and higher cells, in a way reminiscent of the organization of the cells in the retina. Presumably, higher order analysis of the signals takes place at this level; a feature of interest for our design.

At a higher level still, the olfactory bulb makes connections with components of the limbic system, the area of the brain from which moods and emotions arise; and also with the cortex. These connections are treated elsewhere (Van Toller & Dodd 1988). The aesthetics of odours (the hedonic response) are poorly understood (Van Toller & Dodd 1988). We ignore this crucial facet of our responses to odorants in the design of our electronic nose. Our instrument will, for the forseeable future, be calibrated by the response of a sensory panel of trained noses.

1.5 Molecular Aspects of Olfactory Receptor Mechanisms

The molecular mechanisms of olfaction have been relatively neglected by biochemists in comparison with the many studies on visual mechanisms. Until twenty years ago there were virtually no biochemical investigations on olfactory mechanisms. The absence of experimental data did not, however, deter several investigators proposing exotic mechanisms for olfactory transduction. Most of these speculative mechanisms (Wright 1964; Moncrieff 1951) violated well-established principles of cellular mechanisms and did not lead to useful hypotheses which could be tested experimentally.

During the past two decades serious biochemical studies have been carried out on the molecular mechanisms of olfactory transduction. Two key themes, also shared with the many receptors for chemicals (e.g. hormones, neurotransmitters etc), have been, first, the nature of the olfactory plasma membrane (and including the possibility that the liquid crystalline phospholipid phase might be the receptor site for odorants) and, second, the possibility of the odorants acting as regulatory ligands for an enzyme system. Both of these themes are still under active investigation.

Prior to the modern understanding of the phospholipid phases in membranes, Davies in a series of pioneering experimental studies (Davies 1971) in olfaction, showed that the action of odorants could be explained in terms of the odorants puncturing a hole in the lipid region of the membrane, and thereby initiating a receptor potential. Later studies, used an understanding of the liquid crystalline properties of the membrane phospholipids and showed the odorants could alter the conductance of phospholipid bilayers (Cherry et al. 1970).

These and subsequent studies have shown that odorants, which are typically small hydrophobic ligands, could trigger off the initial events in olfactory neurones by binding to the phospholipids. The events subsequent to binding might involve either a phase change in the phospholipid alone, or a propagation of this effect by inducing a conformational change in a membrane enzyme. Studies on the phospholipid receptor hypothesis have suffered through the lack of a suitable preparation of the ciliary plasma membrane. A recent study has provided information on the lipid composition (including the fatty acid profile) of rat olfactory epithelium (Russell et al 1989). The next crucial development must be a similar analysis of the ciliary membrane, when the methodology allows this to be carried out.

The hypothesis of odorants acting as regulatory ligands for enzymes is an old and well-established concept for which there is increasing evidence. The original proposal was made by Ruzicka in his doctoral thesis (Ruzicka 1920), long before either enzymes or their regulatory mechanisms were understood. The first experimental studies were made by Koch who proposed that odorants regulated the activity of the Na^{+} K^{+} Mg^{2+} ATPase in the olfactory plasma membranes (Koch and Norring 1969). Only weak regulatory effects were found by other workers (Dodd 1970). Further work is required to establish whether the effects of odorants on this enzyme are artifacts or whether they constitute a transduction mechanism.

The discovery of an exceptionally high activity of adenylate cyclase in olfactory mucosa (Kurihara and Koyama 1972) was made in 1972, a time when this enzyme was established as the enzyme whose activity was regulated by several hormones binding to receptor proteins in the appropriate plasma membranes (Robinson et al. 1968). The key role of this enzyme in producing a second messenger, cAMP, together with the recognition that the cooperative interactions between ligands in allosteric enzymes (Monod et al. 1965) provided a basis for understanding the initial events of olfactory coding (Dodd 1974, Dodd 1976) and stimulated work to establish this enzyme as a candidate for an olfactory transduction mechanism (Persaud & Dodd 1982).

It was shown that the summated receptor potential in the tissue, the electro-olfactogram, could be regulated by altering the intracellular concentration of the second messenger, cAMP (Minor and Sakina 1973, Menevse et al. 1977). The first demonstration of odorant regulation of the olfactory adenylate cyclase was made with a thiol odorant, hexane thiol (Menevse et al. 1974). However, the membrane preparations available then were unsatisfactory for more detailed studies.

Sonicated preparations of olfactory mucosa gave cleaner (but still impure) ciliary membrane vesicles which have been the basis of subsequent studies. The activation of the olfactory adenylate cyclase by odorants has been established in both frog and rat (Pace et al. 1985, Sklar et al. 1986, Shirley et al. 1986) and involves the classical mechanism of a G-protein. As the ciliary preparations are improved, the activation by odorants is being given at odorant concentrations which approach the physiological range. Indeed, it is now possible to study the effects of simple odorant mixtures on the response of the enzyme (Shirley et al. 1989).

An alternative second messenger system for hormones and neurotransmitters, the phosphoinositide system has come into prominence in the past decade. Phosphoinositides are found in rat olfactory mucosa and there is a high level of phosphoinositide metabolism in the tissue (Wood et al 1988) but to date the activity has not been shown to be modulated by odorants.

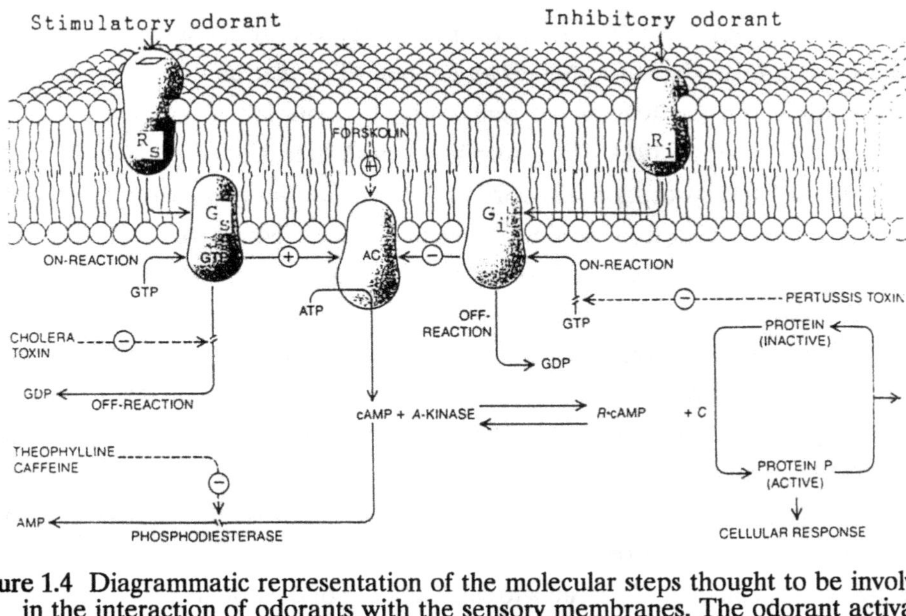

Figure 1.4 Diagrammatic representation of the molecular steps thought to be involved in the interaction of odorants with the sensory membranes. The odorant activates an enzyme system which leads ultimately to the production of a receptor potential.

R - olfactory receptor protein.
AC - adenylate cyclase.
G - GTP-binding proteins.

Receptor mechanisms involving adenylate cyclase usually include receptor proteins. The search for olfactory receptor proteins is now a major issue in olfactory research. Distinctive glycoproteins have been found in olfactory cilia (Chen et al. 1986) and some of them are labelled by the protein concanavalin A. This protein selectively inhibits the odorant response of rat olfactory mucosa (Shirley et al. 1987). Odorant binding proteins are known (Pelosi et al. 1981, Pevsner et al. 1985, Persaud et al. 1988). A multiple profile-multiple receptor site model has been a useful guide in explaining discrimination between odours (Polak, 1973). Current models of olfactory transduction are outlined in figure 1.4.

1.6 The History and Evolution of Electronic Noses

A number of studies during the past 40 years have contributed to this general area. Most of this work has been on the topic of transducers and the object of these studies, in contrast to our electronic nose project, has been to discover materials which would respond with high specificity to a single gas or vapour, such as carbon monoxide. Much of this work, in fact, has been for sensors for toxic gases. The seminal work in this area is reviewed below.

The Turkish scientist, Tanyolac, starting in 1950, made a number of studies on the effects of odours on a number of surfaces, including the measurement of changes in surface tension following adsorption of an odorant (Tanyolac and Eaton 1950). One of the first studies on the reaction of semiconductor materials to gases was made by Brattan and Bardeen in 1953. Then, starting in 1957, Moncrieff, a pioneer in olfactory studies, studied the adsorption of odorants to both the intact olfactory epithelium in severed heads and also to a variety of coated surfaces (Moncrieff 1961). Starting in 1962, Dravnieks, made a series of comprehensive studies on the effect of odorants on the contact potentials of a number of coated materials (Dravnieks and Trotter 1965). Wilkens, Hartman and Rosano used a complex system involving both the adsorption and the oxidation or reduction of odorants at an electrode (Wilkens et al. 1964). The conductivity changes following the interaction of odorants with charge carriers in materials was studied in 1965 (Buck et al. 1965).

In the late 1960s, as the first biochemical studies on olfactory mechanisms were underway, there were two reports of natural materials as possible odour transducers. One involved the binding of odorants to single crystals of β-carotene (Misra et al. 1968). These crystals exhibit semiconductivity and the conductivity alters when odorants adsorb to the surface. It was suggested that carotenoids might be the natural transducers in the receptor cells, but little work has been done to substantiate this claim. The other study showed that the conductivity of phospholipid bilayers, formed from phosphatidylcholine, increased when they were exposed to odorants (Cherry, et al. 1970). Interestingly, these membranes showed a differential effect, with some odorants exhibiting much larger conductivity changes than others. Since the ciliary membranes of the olfactory neurones contain phospholipids, it is possible that these lipids may be the natural transducers for at least some classes of odorants.

Studies on models for the processing of the information from olfactory transducers lagged behind the work on artificial transducers. The seminal work in this area was the book 'Models of the Nervous System' by Deutsch (Deutsch 1967). Chapter 10, entitled - The Instinctive Recognition of Chemical Patterns - gives a detailed formal model for an olfactory system and was one of the inspirations for our current project.

The first experimental electronic nose based on the likely discrimination principles used by the natural olfactory system, was described in 1982 (Persaud and Dodd 1982). This device, though crude by the standards of our present instrument, could solve elementary problems of interest to both the flavour and fragrance industries and was also of use in the analysis of discrimination mechanisms in the mammalian olfactory system. The results obtained with this device suggested that it may not be necessary to postulate specific olfactory receptor proteins. This is a crucial point for the strategy of the biochemical approaches to olfactory mechanisms.

2. ARTIFICIAL ODOUR TRANSDUCERS

2.1 Inorganic Semiconductors

There are many forms of inorganic gas sensors/transducers (Moseley and Tofield 1987), including MOSFET transistors, MIS capacitors, Schottky barrier diodes, hot wire resistors, SAW devices and metal oxide resistors (for reviews see Wohltjen 1984 and Fouletier 1982/83). Within the latter category falls the tin oxide Taguchi gas sensor (TGS) which is at present the most widely used semiconductor gas sensor of all (Taguchi 1971). Variants of it have been produced for a great variety of applications and it can be utilized for an artificial olfactory system. At the heart of the TGS sensor (Fig. 2.1) is a small ceramic tube former upon whose outer surface is deposited a sintered bulk semiconductor composed mainly of tin oxide with catalytic additives (e.g. palladium), which are varied according to the desired detection function. The layer of semiconductor is used as a varistor element between two deposited electrodes in the form of an interdigitated metallization pattern on the outer cylindrical surface of the ceramic tube. Within the tube runs a healing element required to maintain an optimum semiconductor operating temperature of typically 350 °C, which consumes up to about 1W in heating power. This form of construction is conducive to high gas sensitivity by virtue of a detecting element of large surface-to-volume ratio. The symmetry of the arrangement minimizes the energy required and facilitates a substantially uniform surface temperature.

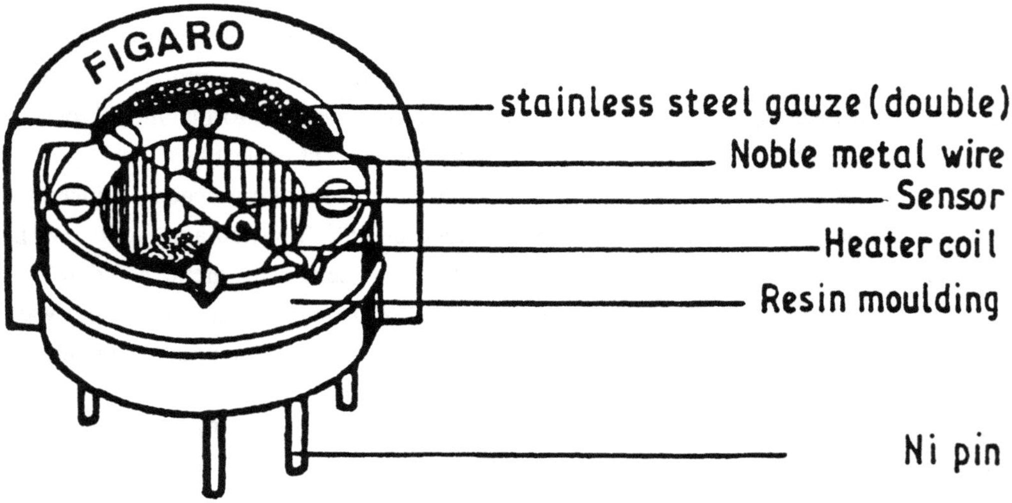

Figure 2.1 Configuration of a Taguchi gas sensor.

The mounting may incorporate a resin or ceramic base and units are designed to plug into a 7-pin tube socket. The upper and lower openings in the sensor case are covered with a flameproof double layer of fine mesh stainless steel gauze to guard against ignition problems in explosive atmospheres.

The manufacturer recommends that the TGS sensors are operated for an initial stabilization period of up to 1 minute after storage. There is also a time dependency in sensitivity for sensors which have been stored for more than a week or two. The sensitivity which is recorded immediately following initial stabilization may decrease by some 10-20% over a period of several days and then increase again over a few weeks to a fairly stable intermediate value. Further variations in sensitivity may arise if the heater supply voltage is not stabilized. As TGS sensors are sensitive to temperature and humidity, compensation is required where high accuracy and reliability are demanded. Response times to particular gases are typically ten seconds with recovery in about 1 minute, for gas concentrations of a few p.p.m.

The mechanisms in gas adsorption processes in SnO_2 sensors are not fully understood but it appears that the changes in sensitivity involve gas-solid and gas-gas interactions. The former is the more important and is discussed below.

Atmospheric air comprises mainly nitrogen and oxygen but there are also traces of argon and other noble gases. Tin oxide is not responsive to nitrogen or argon (Clifford and Tuma 1982) but the oxygen, which has high reactivity and is electronegative, does interact with the semiconductor surface. There are various ways in which gas-solid interactions can occur (Green 1981) but the principal ones are that of physi- or chemi-adsorption. The former is the attraction between gaseous molecules and surface atomic species arising from dipole-dipole attractions. Chemical adsorption occurs between oxygen molecules or atoms and the semiconductor in which a reversible chemical bond is formed at higher temperatures. The electrical charge between the oxygen molecule and the surface becomes redistributed and the effect propagates into the bulk material (McAleer et al. 1987).

It is speculated that oxygen acts as an acceptor for electrons, since the resistivity of SnO_2 increases with oxygen partial pressure. This leads to a decrease in conductivity as the electrons near to the surface become progressively depleted. The distribution of the gas as it diffuses through the porous material determines the speed of sensor response. The diffusion rate and chemical reaction are temperature dependent and so a high operating temperature is often conducive to a fast response.

When a reducing or combustible gas is present, it can become adsorbed on to the surface and react with the oxygen. When multiple reactive gases are present, mutual reactions also occur and the competing rates for each reaction lead to complicated steady state conditions. The semiconductor surface, is also vulnerable to "poisoning" by ambient conditions which include particular gases and selective protection against these is a desirable aim in manufacture.

The principal manufacturer of gas sensors, Figaro Inc., offers ranges of devices either for domestic or industrial usage. The domestic range includes detectors for combustible gases, intended for use in caravans, boats, etc., as well as detectors covering CO, H_2 and cigarette smoke, intended for ventilation control. There are also types aimed at the control of volatile matters from food, alcohol detection and combustion exhaust monitoring. The industrial range includes devices intended for use in detecting various additional gases, including ammonia, hydrogen sulphide, toluene, xylene, freons, NOx and SOx. Fig. 2.2 shows typical responses for three different types of Taguchi sensor to alcohols.

Figure 2.2 Typical responses of three Taguchi sensors to methanol, ethanol, propanol (Gardner 1989b).

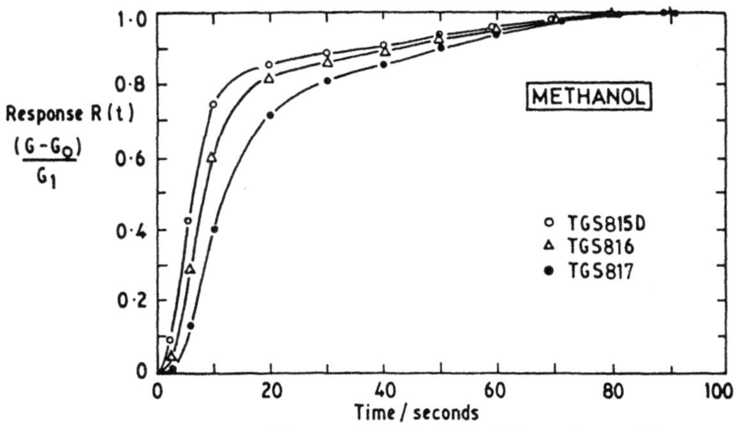

Observed response of three Tagauchi gas sensors to 15ppm of methanol in air.

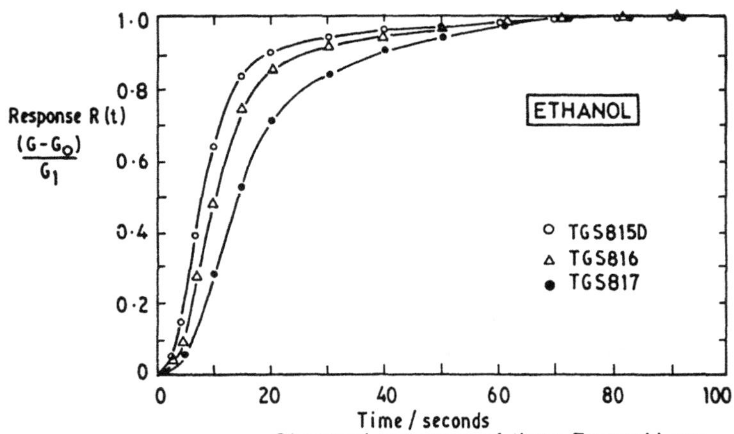

Observed response of three Tagauchi gas sensors to 15ppm of ethanol in air.

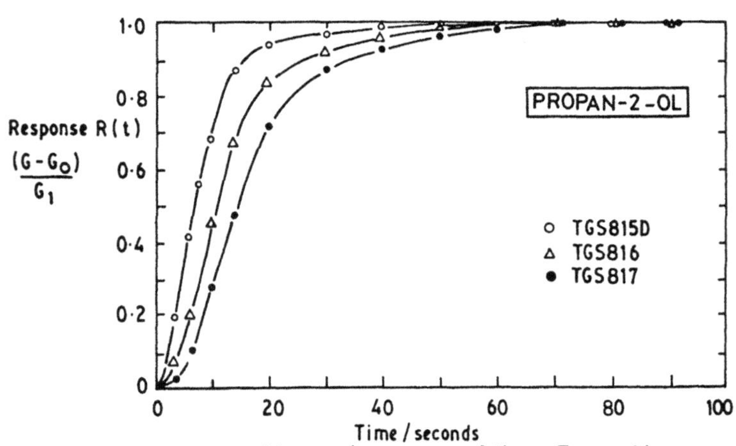

Observed response of three Tagauchi gas sensors to 15ppm of propanol in air.

Integrated Sensors

Discrete commercial devices currently available are enabling excellent progress to be made towards the development of artificial olfactory systems, through the use of sensing heads incorporating up to twelve different devices, although such composite units are somewhat bulky. However, it is desirable for high discrimination to have sensing units with still more detecting elements, where overall size would become an embarrassment if discrete devices were used. The solution lies in the development of integrated devices in chip form and work along these lines is under way, as indicated by one example quoted below.

Demarne and Grisel (1988) at Neuchâtel (Switzerland) have described a CO sensor of integrated form on a silicon base. In their structure, SiO_2 is first deposited on the silicon surface in two stages, involving thermal growth and chemical vapour deposition. A film of gold is then added over a molybdenum adhesion layer, patterned and etched to realize a resistance heater. A further SiO_2 layer is then deposited by r.f. sputtering, patterned and etched to open holes to the gold surface. Finally a SnO_2 thin sensitive film is sputtered and patterned, using a lift-off process. Operation up to 400 °C was reported with less than 0.2W heating power.

2.2 Conducting Polymers

It has been known for many years that the chemical oxidation of pyrrole produces an insoluble polymeric material, pyrrole black. More recently Diaz et al. (1979) reported the formation of coherent films of poly(pyrrole) at electrode surfaces by the controlled electrochemical oxidation of the monomer in acetonitrile. This publication marked the start of a large body of work on the study of conducting polymers formed by electrochemical polymerisation. These materials have attracted wide interest because of their great potential for application in many areas; including their use as battery electrodes (MacInnes et al. 1981), as electrochromic display materials (Garnier et al. 1983), as ion gates (Burgmeyer and Murray 1984), in microelectronic devices (Thackeray et al. 1985), for controlled drug release (Zinger and Miller 1984), for electrocatalysis (Haimerl and Merz 1987), and in chemical sensors (O'Riordan and Wallace 1986).

This diversity of applications stems from the unique combination of properties possessed by this class of materials. First and foremost because these are electronic conductors built up from monomers joined together into conducting chains it is easy to envisage ways to adapt the chemical structure of the polymers (Bidan et al. 1984, Diaz et al. 1980, Eaves et al. 1985). This is a very significant advantage since it holds the potential for an infinite variety of tailor-made organic, electronically conducting, materials. In principle it should be possible to adapt the material to the particular need. Secondly these materials possess the interesting property of being switchable between conducting and insulating states. Frequently this change is associated with changes in the chemical properties of the polymer, notably the colour and the amount of incorporated counter ion. Thirdly these materials may be prepared as coherent films by electrochemical polymerisation at electrode surfaces. This is advantageous because it means that their preparation is generally simple, that the film thickness is readily controlled, and that the spatial distribution of polymer is well controlled. It is disadvantageous because the films are difficult to characterise and the materials are not amenable to polymer processing technology. These problems are gradually being overcome as processing technologies are developed and more is known of the materials (Street 1986, Bocchi and Gardini 1986, Nazzal and Street 1985, Wang et al. 1985, Jen et al. 1987).

Our prime present concern is the application of conducting polymers to the development of intelligent gas sensors. In this context the combination of properties possessed by these materials, most particularly the ability to control their deposition and the large variety of polymers available, make them highly suitable transducer materials. In the next section we will describe the electrochemical properties and preparation of such polymers before moving on to describe their application to gas sensors.

Electrochemistry of Conducting Polymers

A wide range of aromatic and heteroaromatic monomers undergo electrochemical oxidation to yield adherent films of conducting polymer under suitable conditions (see Diaz and Bargon 1986, Chandler and Pletcher 1985 for reviews). The oxidation reactions are most often carried out in non-aqueous solvents, such as acetonitrile, for which the anodic range is much greater than for water. It is possible, however, to grow conducting polymer films from aqueous solutions provided the monomer is sufficiently readily oxidised (Asavapiriyanont et al. 1984a, 1984b, Bartlett and Whitaker 1987). Table 2-1 lists a number of monomers which have been used to grow conducting polymer films; this list is representative but by no means exhaustive.

TABLE 2-1 : Some heteroaromatic monomers which yield conducting polymers

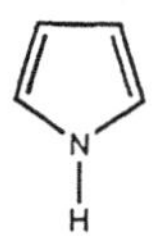

pyrrole[a,b]

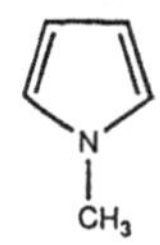

N-methylpyrrole[c]

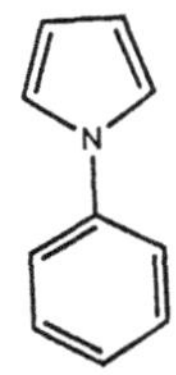

N-phenylpyrrole

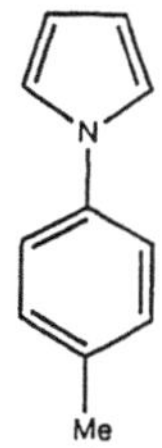

N-*p*-tolypyrrole[d]

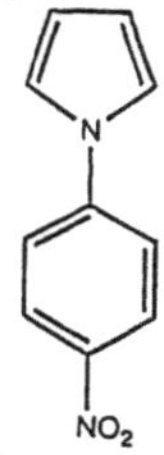

N-*p*-nitrophenylpyrrole[d]

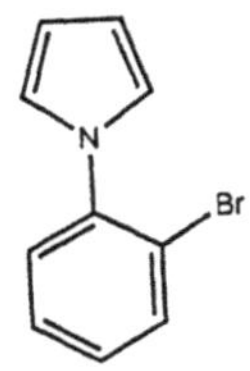

N-*o*-bromophenylpyrrole[e]

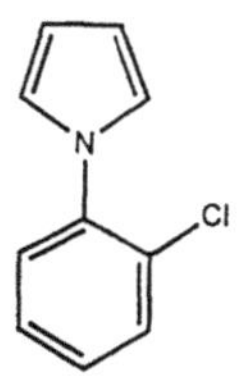

N-*o*-chlorophenylpyrrole[e]

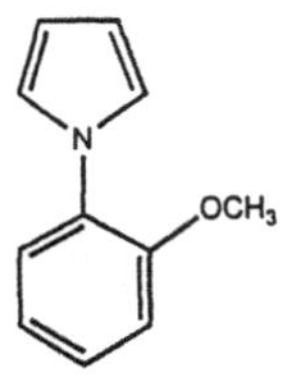

N-*o*-methoxyphenylpyrrole[e]

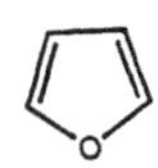

furan[f]

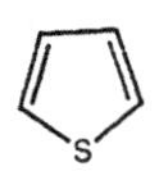

thiophene[g]

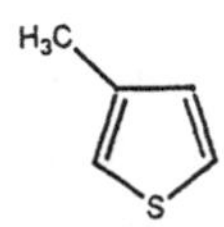

3-methylthiophene[h]

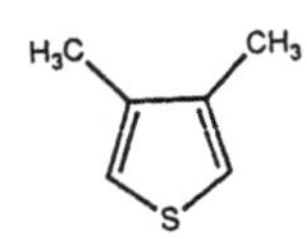

3,4-dimethylthiophene[h]

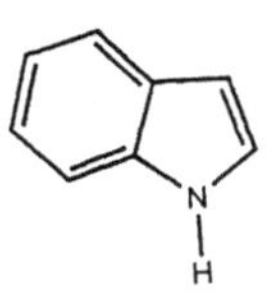

indole[i]

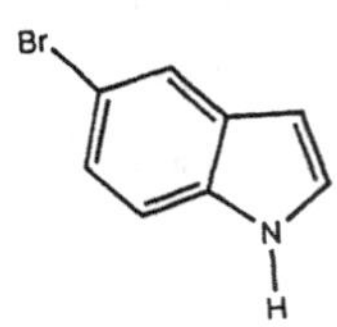

5-bromoindole[i]

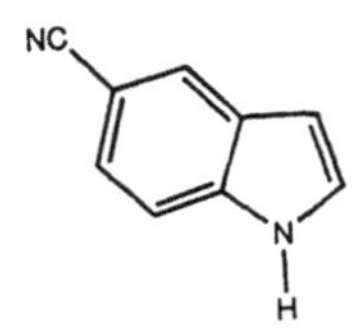

5-cyanoindole[i]

---

a	Diaz *et al.* 1979	b	Genies *et al.* 1983	c	Asavapiriyanont *et al.* 1984b
d	Salmón *et al.* 1982	e	Salmón *et al.* 1983	f	Tourillon and Garnier 1982
g	Waltman *et al.* 1983	h	Garnier *et al.* 1983	i	Waltman *et al.* 1984

We shall consider the electrochemical polymerisation of pyrrole as a fairly typical example. The mechanism of the polymerisation has been discussed by a number of authors (Inoue and Yamase 1983, Genies et al. 1983, Takakubo 1989, Baker and Reynolds 1988). Fig. 2.3 shows the general mechanism.

Figure 2.3 The general mechanism for the electrochemical polymerisation of pyrrole (adapted from Chandler and Pletcher 1985).

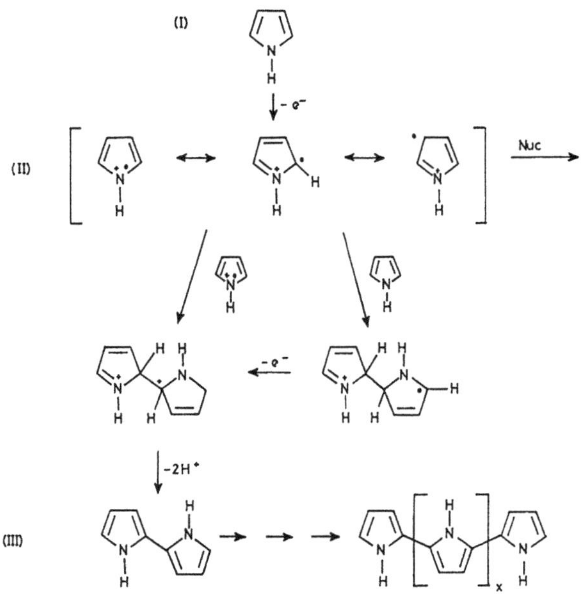

Electrochemical oxidation of the pyrrole monomer (I) gives a radical cation (II). Subsequent dimerisation and loss of protons gives the dimeric species (III). Chain growth then occurs by oxidation of the dimer and reaction with further monomer to produce oligomers and eventually polymeric material. Support for this mechanism comes from a number of sources. Firstly attempts to polymerise α-substituted pyrroles are generally unsuccessful whilst both N- and β-substituted pyrroles can be polymerised. These observations are consistent with a predominantly α-linked structure. Secondly, films of the polymer cannot be grown successfully in the presence of nucleophilic anions (NUC), or solvents which would react with the radical cation (II). Finally the film only continues to grow at potentials at which the monomer is oxidised. Fig. 2.4 shows the course of a typical polymer deposition experiment carried by cycling the potential of the electrode in a solution of monomer. With each cycle the quantity of polymer deposited at the electrode increases as witnessed by the increase in the current for the oxidation and reduction of the polymer.

Once grown the polypyrrole film, in common with other electrically conducting polymer films, can be switched between conducting and insulating forms (Feldman et al. 1985). In the as-grown, oxidised state, the polymer is conducting. Table 2-2 summarises data for the conductivities of a number of conducting polymer films.

TABLE 2-2 : The conductivities of some oxidised conducting polymers

Polymer	S cm^{-1}	Anion	Reference
Poly(pyrrole)	2-8	chemically polymerised	a
	29-100	BF_4^-	b
Poly(3-methylpyrrole)	4	BF_4^-	c
Poly(*N*-methylpyrrole)	0.001	BF_4^-	b
Poly(thiophene)	10-100	BF_4^-	d
	0.02	BF_4^- or PF_6^-	c
Poly(3-methylthiophene)	1	PF_6^-	c
Poly(3-octylthiophene)	78	PF_6^-	e
Poly(3-hexylthiophene)	95	PF_6^-	e
Poly(furan)	10-80	BF_4^-	d
Poly(indole)	0.005-0.01	ClO_4^-	d
Poly(azulene)	0.01-0.1	ClO_4^-	d

a Bocchi and Gardini 1986 b Diaz *et al.* 1982 c Waltman *et al.* 1983

d Tourillon and Garnier 1982 e Sato *et al.* 1986

For polypyrrole the fully oxidised film carries a charge of about +1/4 (the precise value depends on the choice of counter ion) for every pyrrole unit in the film. In other words, on the average the charge is delocalised over four pyrrole units. This charge is balanced by the charge on anions incorporated in the film from the growth solution. In very simple terms the polypyrrole can be thought of as an extended π-system, Fig. 2.5. When the polymer is oxidised there is a gap, or hole, in this π-system which can migrate along the polymer chain carrying the current. On reduction all of these gaps are filled so that there are no longer any mobile charge carriers and consequently the film becomes insulating (see Brédas et al. 1983, Patil et al. 1988 for a more detailed discussion). Also upon electrochemical reduction the polymer becomes neutral and the incorporated counter ions are expelled into solution. In this fully reduced, neutral form the polymer is insulating. This change in oxidation state and conductivity of the film is accompanied by a change in colour, from black for the oxidised form to almost colourless for the reduced form. These changes are summarised in Fig. 2.6 which shows a typical cyclic voltammogram for polypyrrole in acetonitrile.

The behaviour found for polypyrrole is typical of that for other conducting polymers. In most cases the electrochemically polymerised conducting polymers studied to date have only been conducting when oxidised, in other words the mobile charge carriers have been holes. Two notable exceptions are poly(bipyridine) (Schiavon et al. 1985) and poly(thiophene) (Aizawa et al. 1986) which both exhibit conductivity when reduced.

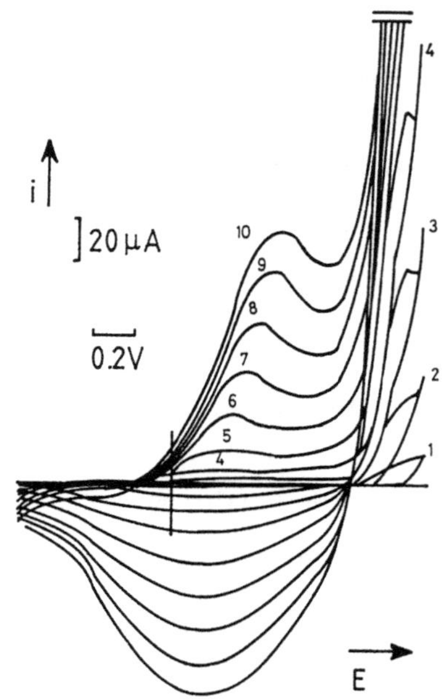

Figure 2.4 Typical cyclic voltammogram for the growth of a poly(pyrrole) film across the gap between two microband electrodes from a solution of 0.05 mol dm^{-3} pyrrole in acetonitrile containing 0.1 mol dm^{-3} tetraethylammonium tetrafluoroborate.

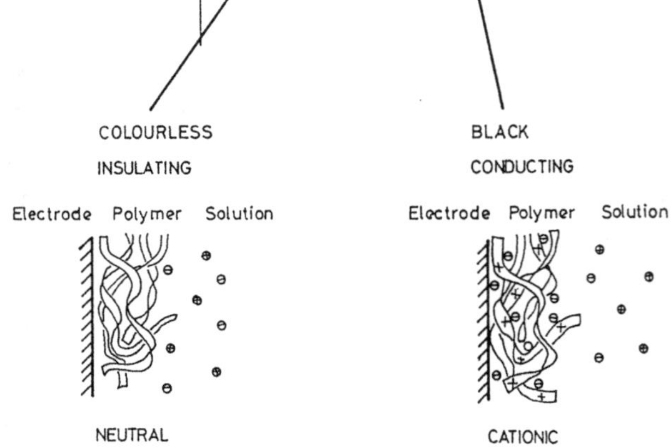

Figure 2.5 The structure of poly(pyrrole).
Figure 2.6 Cartoon showing a typically cyclic voltammogram for poly(pyrrole) in acetonitrile and the structure of the polymer film in the oxidised and reduced forms.

Conducting polymers as Gas Sensors

One type of gas sensor based on conducting polymers makes use of the fact that the conductivity of the polymer film is altered on exposure to different gases. Devices which operate in this way are called chemiresistors. This effect has been observed for chemically prepared samples of polypyrrole impregnated on filter paper (Nylander et al. 1983), for free standing films (Gustafsson and Lundström 1987) and using electrochemically deposited material on interdigitated electrodes (Hanawa et al. 1988, Miasik et al. 1986) or across pairs of microband electrodes (Bartlett et al. 1989, Bartlett and Ling-Chung 1989a, 1989b).

A general view of such a chemiresistor is shown in Fig. 2.7. The current/voltage characteristics of such a device are often close to ohmic. Under these circumstances it is sufficient to measure either the applied voltage at constant current, or the current at constant voltage.

Figure 2.7 General view of a chemiresistor based on an electrodeposited conducting polymer film.

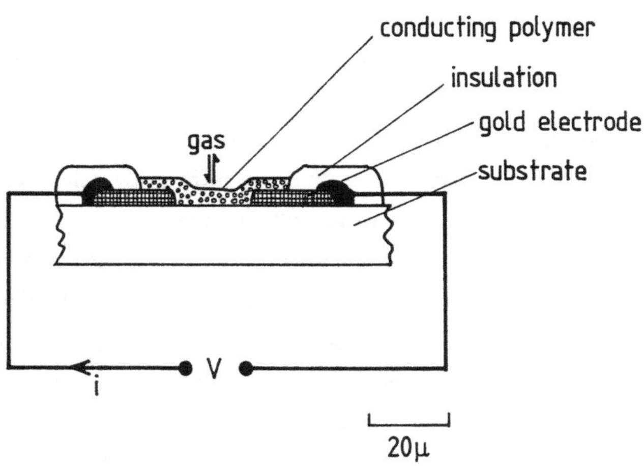

The devices are fabricated by controlled electrochemical deposition of the polymer across the gap between two electrodes (Bartlett et al. 1989). Typically the gap between these electrodes will be of the order of 10 microns wide.

Poly(pyrrole) has also been used as a gas sensitive material in suspended gate field effect transistors, Fig. 2.8 (Cassidy et al. 1986, Josowicz and Janata 1986, 1988, Josowicz et al. 1987). In this case the change in the work function of the polymer on exposure to the gas or vapour is measured. The response characteristics are similar to those for the much simpler chemiresistor structure.

At present our knowledge of the physical mechanisms giving rise to the changes in conductivity or work function observed on exposure of the devices to gases or vapours is incomplete. Indeed there is considerable interest in the mechanism of conduction of the polymers themselves. In order to be able to design gas sensors for commercial application based on this technology it is important to improve our understanding of these effects.

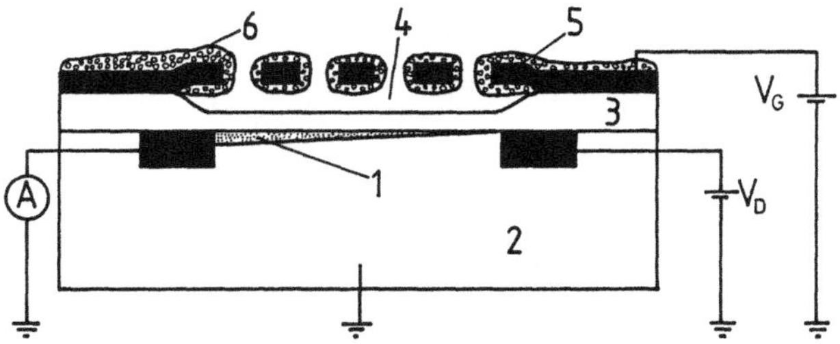

Figure 2.8 Schematic of a suspended gate field effect transistor. 1: the inversion channel, 2: Si substrate, 3: Si_3N_4/SiO_2 insulator, 4: air gap, 5: metal mesh, 6: polymer coating (adapted from Cassidy et al. 1986).

The most common application of conducting polymers to the area of gas sensors has been in the development of sensors specific for particular gases, for example ammonia (Nylander et al. 1983, Miasik et al. 1986), and nitric oxide (Hanawa et al. 1988). The concept of the artificial olfactory system is essentially different because the goal is not to make one sensor for each type of gas or vapour but rather to make an array of sensing elements and to use pattern recognition techniques to identify the gases or vapours presented to the device.

To devise such an array it is important that a number of criteria be met. Firstly each element of the array should behave differently. Clearly, it is no good having ten sensor elements in the array which all behave in exactly the same manner for all vapour samples. Secondly a significant number of sensor elements will be required so that it is essential that a variety of transducer materials are available. Thirdly it must be feasible to construct and then manufacture the devices in quantity. Fourthly the power consumption and physical size of the device should not be excessive. Finally the devices should exhibit a reasonable lifetime and reproductibility between devices; calibration over a large set of vapours is not feasible. In addition the normal criteria of reasonable response time, sensitivity, and ease of operation should apply.

A consideration of these criteria reveals that conducting polymers are really rather suitable transducers for such applications. Clearly there is a practically limitless variety of monomers which may be polymerised. Fabrication using electrochemical polymerisation is well suited to the deposition of the materials on microelectronic substrates with the required spatial definition. Furthermore the materials operate at ambient temperatures so their power consumption is low.

We have carried out a preliminary investigation of the responses of a number of polymers with a variety of different organic vapours. Fig. 2.9 shows some of the results for typical responses obtained upon exposure to different saturated vapours. Note that the response times are all of the order of seconds and that the polymers recover the baseline value very rapidly.

Figure 2.9 Typical responses of conducting polymer films to exposure to saturated: methanol and ethanol: PP polypyrrole; PMeP poly(N-methylpyrrole); PCI poly(S-carboxyindole; PAN polyaniline. The polymers are exposed to vapour at (a) and then to air againat (b).

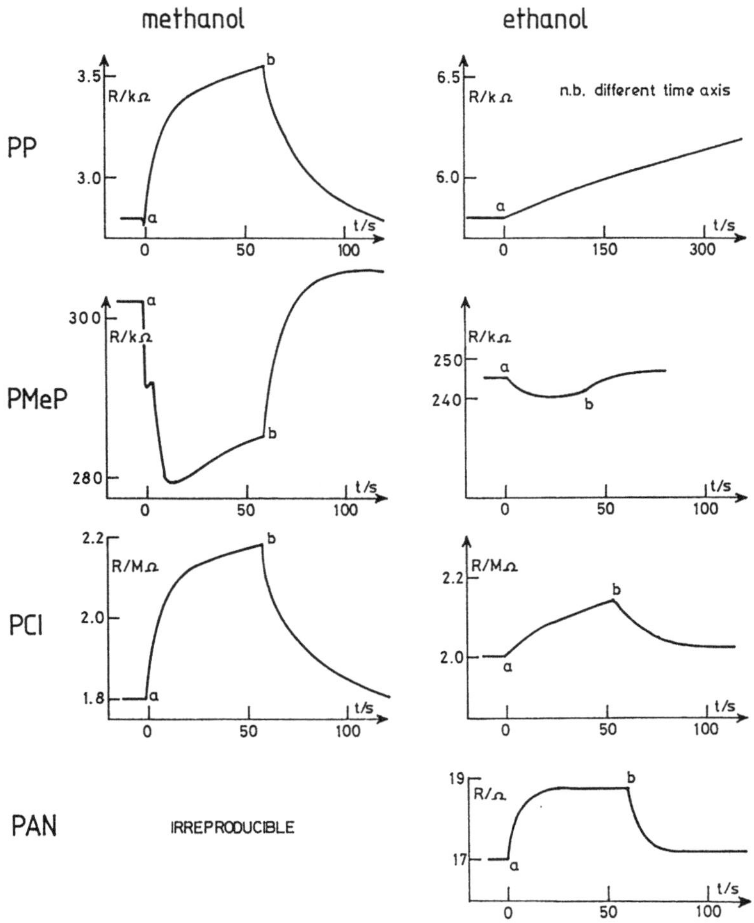

Table 2-3 summarises results for four different polymers and five different vapours (Bartlett and Ling-Chung 1989b). The results were obtained using one particular set of deposition conditions and choice of counter anion. It should be possible to improve the gas sensing properties of these materials by systematic optimisation of these parameters and we are attempting to do this in our laboratory at present.

Table 2-3 : Responses of four different polymers to five different vapours.

VAPOUR	POLYMER pyrrole	N-methylpyrrole	5-carboxyindole	aniline
methanol	-	?	-	?
ethanol	-[a]	?	-	-
acetone	?	?	-	-
ether	0	?	0	0
toluene	?	?	+[b]	?

-	conductivity decreases	a	slow response at room temperature
0	conductivity unchanged	b	small response
+	conductivity increases		
?	irreproducible/poor stability		

2.3 Other chemical sensing materials

Inorganic semiconductors and conducting polymers are by no means the only sensing materials available for use in gas sensors and we will briefly discuss some of the others here.

One class of organic materials which have been investigated in some detail in recent years is the metal phthalocyanines, Fig. 2.10.

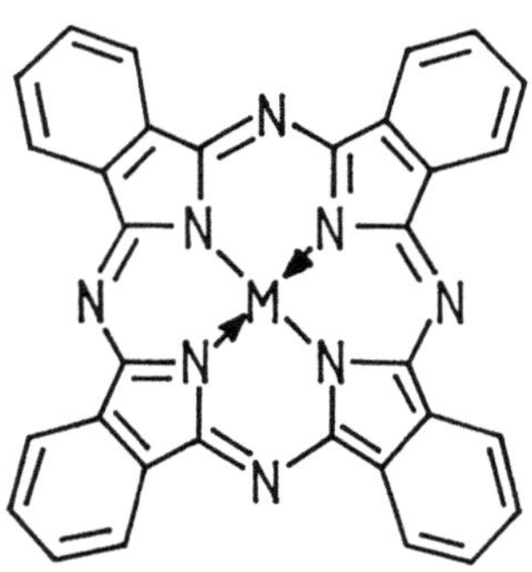

Figure 2.10 Metal phtalocyanine

Gas sensing studies of these materials have concentrated mainly on the unpolymerised metal phthalocyanines (Pc) deposited onto suitable sensor substrates by vacuum sublimation. The resulting thick film devices show significant conductivity changes on exposure to NO_2 (Miasik et al. 1986, Chadwick et al. 1986). The precise details of the interaction between the gas and the solid are not well understood but it is believed that the NO_2 molecules partially replace O_2 adsorbed on the phthalocyanine surface from the air. The phthalocyanines are also responsive to other high electrophilic gases such as HCl, SO_2 or H_2S. They have the disadvantage that their recovery times are slow, even at 170 °C and there are frequently memory effects. Improved performance has been observed using Pc films laid down by Langmuir-Blodgett techniques (Baker et al. 1983, Bott and Jones 1984, Jones and Bott 1986).

The Langmuir-Blodgett (LB) technique is named after Irving Langmuir and Katharine Blodgett who devised a means to control and deposit monomolecular films. The technique is limited to materials that have amphiphilic long-chain molecules terminated by a hydrophilic and hydrophobic molecule at each end (Pitt 1983). Using a Langmuir-Blodgett trough, such as that shown in Fig. 2.11, monolayers of an LB film are transferred from the water onto a substrate that is periodically dipped at constant film surface tension. At Warwick University LB films have been prepared from a NIMA Technology trough from stearic acid. It was found that films consisting of 10 layers were sensitive to a variety of substances such as methanol, ethanol, methylamine, acetone, ammonia and tobacco smoke (Imanpour 1986). There appears to be good scope for further exploitation of LB films in gas sensing as recently reported by Moriizumi (1988).

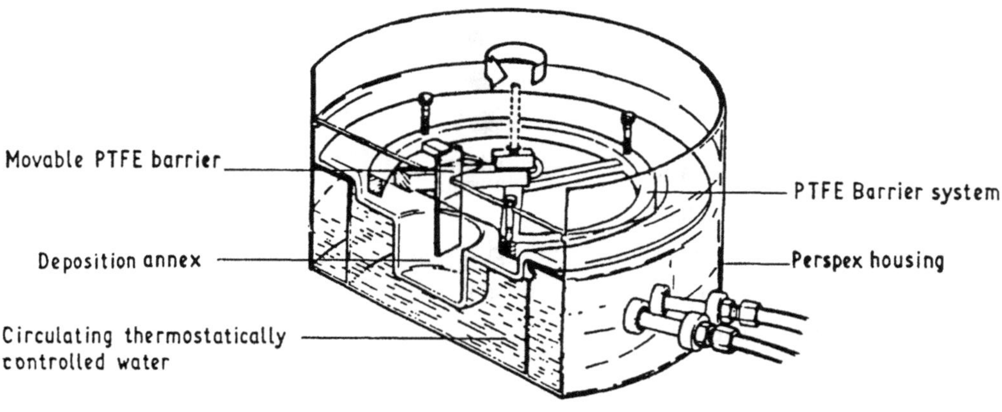

Figure 2.11 General view of a Langmuir-Blodgett trough (NIMA Technology).

Finally we should mention the use of adsorbant materials such as stationary-phase material for gas chromatography (Carey et al. 1986, 1987) and zeolites (Mu ller and Lange 1986). The basic idea in this case is to use materials known to adsorb small organic molecules and to use them as part of the sensor devices. Carey et al. 1987 have used arrays of gas-chromatographic stationary-phases coated onto piezoelectric sensors to construct arrays for use with pattern recognition techniques. This system basically detects mass changes brought about by adsorption of the gas or vapours. In the system of Müller and Lange the zeolites are employed as filters over MOS sensors. Again they used pattern recognition techniques to process their results.

3. DESCRIPTION OF THE WARWICK ARTIFICIAL OLFACTORY SYSTEM

3.1 Introduction

Our basic philosophy is to develop an electronic analogue of the mammalian olfactory system. It should be capable of classifying and discriminating between complex mixtures. Our interest arises firstly from a desire to further our understanding of mechanisms underlying olfaction and secondly because such a system would have enormous potential for application.

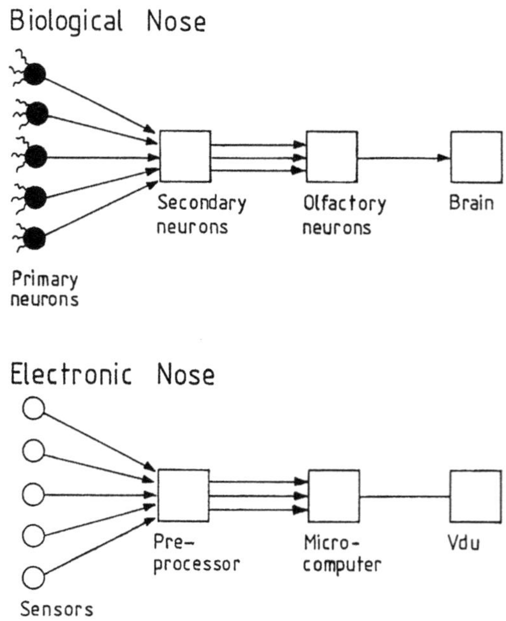

Figure 3.1 Basic arrangements of mammalian and electronic noses.

The basic arrangements for the mammalian and electronic noses are compared in Fig. 3.1. As an example of the former, a human nose has about 2 x 10^7 primary receptors in the form of neurons terminated in mucus. Some ten thousand primary neurons may synapse into a single secondary neuron, which feeds *via* the cribriform plate into the olfactory cortex of the brain. In an electronic nose gas detection is via transducers which each incorporate a layer of material on which gases are adsorbed, resulting in either a change of conductivity or electric charge (see Section 2). By incorporating into electrical circuitry transducers whose response characteristics differ from each other, output voltages may be obtained which can be used to identify the types and concentrations of the gases present in the environment to which the gas-sensitive material is exposed. The procedure by which such identification is achieved involves conversion of analogue signals, derived from the sensor circuitry, into digital inputs for a microcomputer which processes the information.

3.2 Experimental apparatus

A considerable effort at Warwick University involves the use of commercial SnO_2 transducers. The manufacturer's (Figaro Inc.) recommended test circuit is shown in Fig. 3.2, the sensor being part of a potential divider network with the load resistor R_L of 4 kΩ. When the potential V_s dropped across the load resistor R_L changes as a result of exposure to a reactive gas, the device resistance R_s changes proportionately in accordance with the relationship

$$R_s = (V_s\text{-}V_o)\,R_L/V_o \tag{3.1}$$

Changes in the potential difference across R_L are monitored as a function of odorant concentration for gaseous input stimuli.

All of the early work in establishing the feasibility of an artificial olfactory system employed three sensors, mounted together on a perspex panel which formed the top of a 5 litre flask (Fig. 3.3). The panel was fitted with a rubber septum, through which the odorants to be tested could be injected with a Hamilton syringe. A small electric motor was used to operate a stirring paddle within the flask. The whole system was kept at a constant temperature by immersing the flask in a water bath at 25 °C.

The load resistors, together with the manufacturer's recommended heat purging circuitry for the sensors were contained in a box and the resulting three outputs were

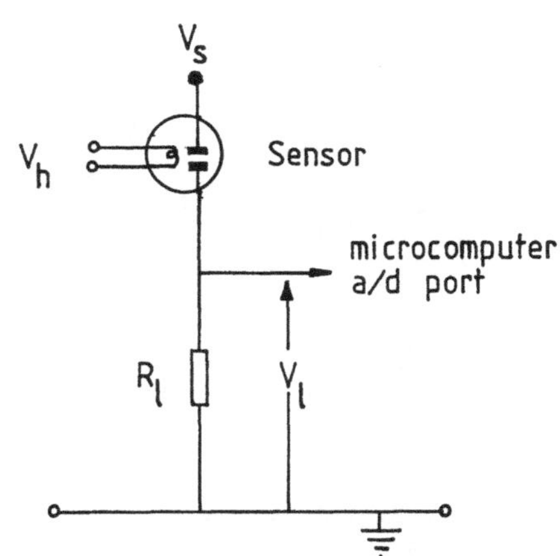

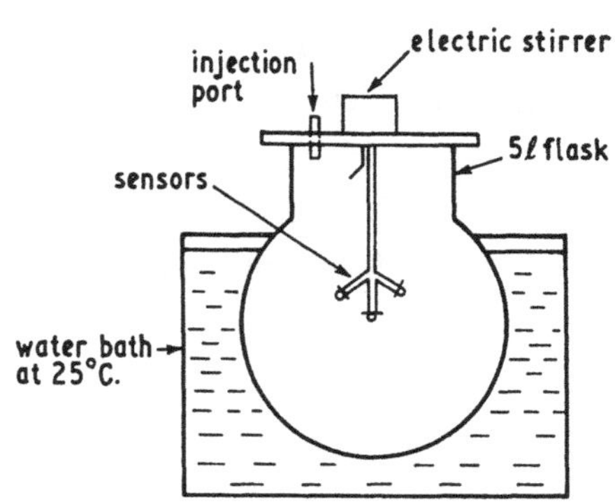

Figure 3.2 Recommended test circuit of Figaro gas sensors.

Figure 3.3 Experimental arrangement of 3-element electronic nose.

connected to the analogue inputs of a BBC microcomputer A/D convertor. Power supplied to both sensor, V_s, and heater, V_H, were connected to 5V stabilized power supplies. The sensors were burnt in for a period of one week prior to the commencement of tests.

Software written for the BBC microcomputer enabled experimental work to be controlled, measurements taken, results processed and either output to a printer or kept on file for future use.

Subsequent developments have led to the replacement of the 3-device sensing unit with one incorporating 12 sensors within a correspondingly larger flask of 20 litres. The outputs of the sensors are mulitplexed and fed into a linearising circuit (Fig. 3.4). This gives an output voltage which is, under certain circumstances, proportional to gas concentration, a feature which greatly facilitates subsequent information processing procedures (Fig. 3.5). These are currently being developed with the aid of an Opus IV PC-AT microcomputer, which has replaced the BBC microcomputer and can cope effectively with the requirements for implementing pattern recognition techniques.

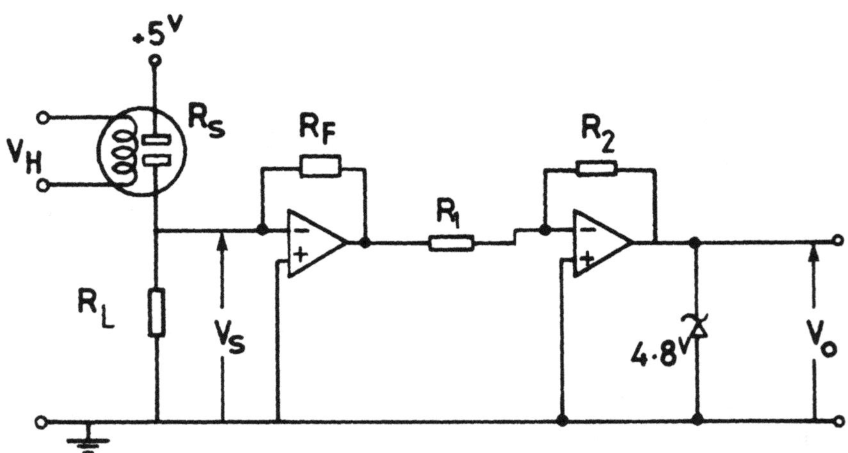

Figure 3.4 Linearising circuit of SnO_2 sensors.

The test schedule will vary according to the objective. As an example, we will consider a procedure for identifying a single gas, subject to the following constraints:

1) Only one gas or gas mixture is present at any one time.
2) Only gases for which the system has been calibrated occur in the environment under test.
3) The concentration of gas is constant during the time taken for the sensors to respond.

Such a schedule would be appropriate in an industrial situation for leak detection, for instance, or in the food industry for quality control involving flavours or taints. A system block diagram for an instrument utilising this schedule is illustrated in Fig. 3.6.

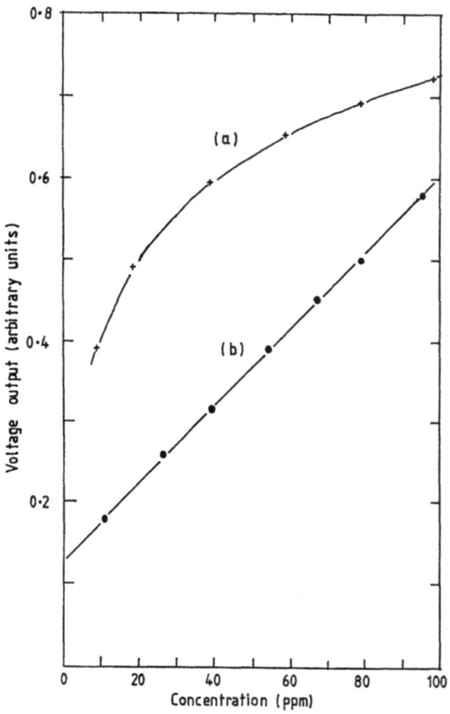

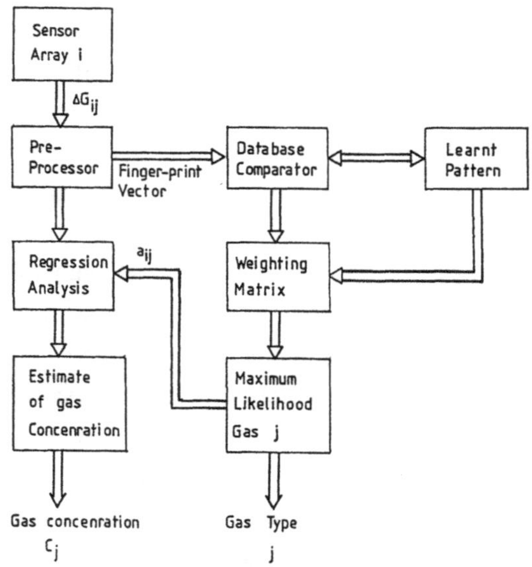

Figure 3.5 Improvement of response of SnO_2 sensor to ethanol for op-amp circuit (a) compared to potential divider (b) (see Shurmer et al. 1989).
Figure 3.6 System block diagram for the identification of a gas and determination of its concentration.

In this system, the array of sensors is sampled at regular time intervals and the data supplied to a pre-processor. This produces a gas finger-print by deriving a set of coefficients independent of gas concentration. By comparing the finger-print vector with data obtained during calibration, the most probable gas type is found and its concentration calculated from a polynomial equation, using a least squares multiple linear regression algorithm. Such an approach was initially developed for the 3-sensor array, written in BASIC to be user-friendly, performing all calculations automatically and advising the user when to re-calibrate. Pattern recognition techniques are discussed in detail in the next section.

4. PATTERN RECOGNITION IN ARTIFICIAL OLFACTORY SYSTEMS

4.1 Background

Several developments during the last fifty years have provided a tremendous impetus for the design of algorithms for processing signals from distributed sensor arrays. These include the discovery of RADAR and SONAR which generated the need for complex processing algorithms; the development of semiconductor technology for the fabrication of 1,024 x 1,024 element CCD arrays for optical imaging; the study of image analysis in such diverse fields as medical tomography, and non-destructive testing. These have produced a plethora of processing algorithms based upon statistical techniques such as Correlation, Maximum Likelihood, Minimum Entropy etc. However, these sophisticated algorithms are not generally applied to chemical sensor arrays, presumably due to the relatively slow response time of a chemical sensor (1-100s) compared to an optical sensor (1ns-1 μs).

The use of integrated sensor arrays in olfaction was postulated by Persaud and Dodd in the early 1980's, and designed to overcome problems encountered by selectivity. Its arrival acknowledged the partial failure to identify specific odours or vapours by a single sensing material. The earliest work involved the investigation of different types of semiconducting material in the array, such as metal oxides (Ikegami and Kaneyasu 1984, 1985, Kaneyasu et al 1987) and Langmuir-Blodgett films (Fard 1985). Simultaneously, statistical methods were applied to the detection of flammable and toxic vapours by others (Zarcomb and Stretter 1984, Stretter et al 1986, Müller and Horner 1986). At Warwick we have applied a variety of algorithms, some of which we will now discuss.

4.2 Analysis of Multi-sensor Arrays

The detailed analysis of a multi-sensor array depends upon the physical nature of the sensing materials (e.g. inorganic or organic semiconductor). However, the majority of devices operate on the principle of a change in conductance following the introduction of a single odour/vapour or a mixture of them.

Consider the general case of a sensor i whose conductance $G_{ij}(t)$ changes as a function of time following the introduction of a measurand j of concentration C_j where

$$G_{ij}(t) = G_i° + G_{ij}(C_j) \qquad (4.1)$$

and $G_i°$ represents the sensor conductance in the absence of any test species and is concentration independent, it may be related to physical properties such as device geometry and temperature and often suffers from long term variation. The term $G_{ij}(C_j)$ represents the time depended concentration response of sensor i to measurand j. Typically, the response time is between 10-100s for a wide range of sensing materials (e.g. SnO_2, ZnO, MgPc, CuPc, conducting polymers) and tends to be dominated by either diffusion (Gardner 1989 a,b) or reaction (Bartlett et al. 1989 c).

So far, the analysis of multi-sensor arrays has been simplified by neglecting the time-dependent terms and treating only the asympotic step- response ΔG_{ij}, where

$$\Delta G_{ij} = G_{ij}(t \rightarrow \infty) - G°_i \qquad (4.2)$$

There is substantial evidence (e.g. McAleer et al. 1987) that the step-response of a semiconductor sensor may be related to the measurand concentration C_j by

$$\Delta G_{ij} = k_{ij}(C_j)^r \qquad (4.3)$$

where the power index r lies between 0.2 - 1.0 for a range of devices and chemicals sensed (Yamazoa and Seiyama 1985).

4.3 Identification of a single chemical species

The identification of a single chemical species j by an array of n-sensors is relatively straight-forward when a sensor i' has high specificity since

$$\Delta G_{ij} = k_{ij}(C_j)^r \quad \text{when } i = i' \qquad (4.4)$$
$$= 0 \quad \text{when } i \neq i'$$

Although metal oxide sensing elements may be sensitised to a particular species by metallic doping they tend to be non-specific and to show differing degrees of response to different chemicals; this behaviour underpins the need for an array processing system.

Using a tin oxide multi-sensor array (n = 3) we have been able to identify various chemical species using a ratiometric method (Shurmer et al. 1989). The technique removes the concentration-dependence of the conductance by calculating the normalised parameter Δg_{ij} where

$$\Delta g_{ij} = \Delta G_{ij} / \sum_{i=1}^{n} |\Delta G_{ij}| = k_{ij} / \sum_{i=1}^{n} k_{ij} \qquad (4.5)$$

The resulting array of n-values is compared to the known pattern Δg^{s}_{ij} and the root-mean-square deviation gives a confidence level. The technique may be used to identify simple odours with a small number of sensing elements and relies upon the principles of vector space analysis. For example, the identification technique may be developed by plotting the outputs from pairs of sensors on a rectilinear grid. This was initially used successfully with an array of three sensors for the identification of alcohols at differing concentration levels and is illustrated in Fig. 4.1.

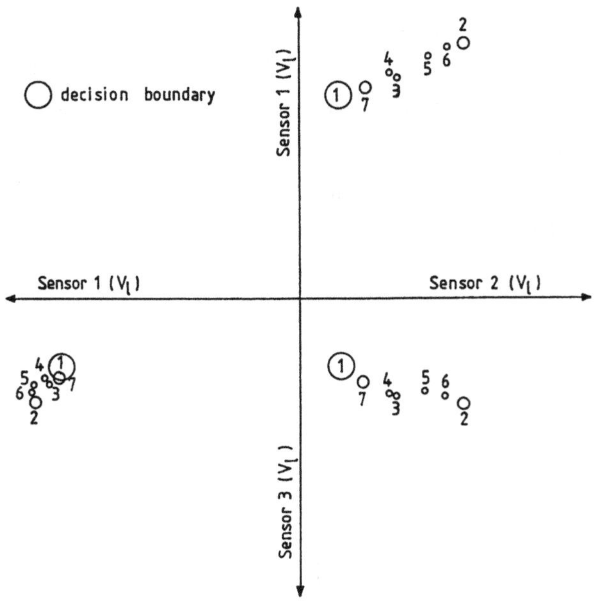

Figure 4.1 Discriminating response of the electronic nose to (1) air (2) 20 ppb ethanol (3) 10 ppb ethanol (4) 10 ppb methanol (5) 20 ppb methanol (6) 20 ppb ethanol + 20 ppb methanol (7) 20 ppb acetone.

Random variations in digitised sample voltages fall within circles of tolerance, constructed by monitoring random variations which occur with the system operated in a calibration mode. The circles are used as decision boundaries in testing unknown materials. If, for all three plots, the 2-dimensional vectors generated by an unknown gas each lie within the circles corresponding to a particular gas used in calibration, then a positive identification is made. Failing this, the most likely possibility for the unknown species is established (Shurmer et al. 1987, Gardner 1988).

4.4 Identification of mixtures of species

Consider the case of an array of n-sensors with an environment containing m-odours. Then, provided the principle of superposition holds then

$$\Delta G_i = \sum_{j=1}^{m} k_{ij} (C_j)^r \tag{4.6}$$

For dilute mixtures, the principle of linear superposition has been shown to hold for tin oxide sensors (Shurmer et al. 1989). Thus, for binary mixtures the concentrations of each constituent can be determined from the equations - provided there are at least two sensors. In the case where a limited set of chemical species are introduced, the output from the sensor array may also be analysed when the interdependency is determined *a priori.* For example, Mueller and Horner (1986) have used multi-variable linear regression analysis to calculate the conditional regression coefficient, i.e. $k_{ij.j'}$, where sensor response to chemical species j depends upon the concentration j'. The application of such a correlation technique is limited by sensor drift and the need for exhaustive testing. Furthermore, the sensor array cannot test for any new species because its concentration dependence is unknown. However, unlike some other techniques it not only identifies the species present but also gives their concentration.

Cluster separation procedures to partition apparently random results into clearly separated groups has been dramatically successful when applied to diluted smoke from different brands of tobacco, mainly Burleigh and Virginian blends (Shurmer et al. 1989). Fig. 4.2(a) shows a scatter diagram formed by plotting the raw digitised output of two sensors on a rectilinear grid for smoke from four brands of tobacco. The results for each brand overlap with those from at least one of the other brands. When each of the coordinates is divided by the standard deviation of the set to which it belongs and the results are re-plotted we obtain, the modified scatter diagram shown in (b), where each brand has become clearly identifiable.

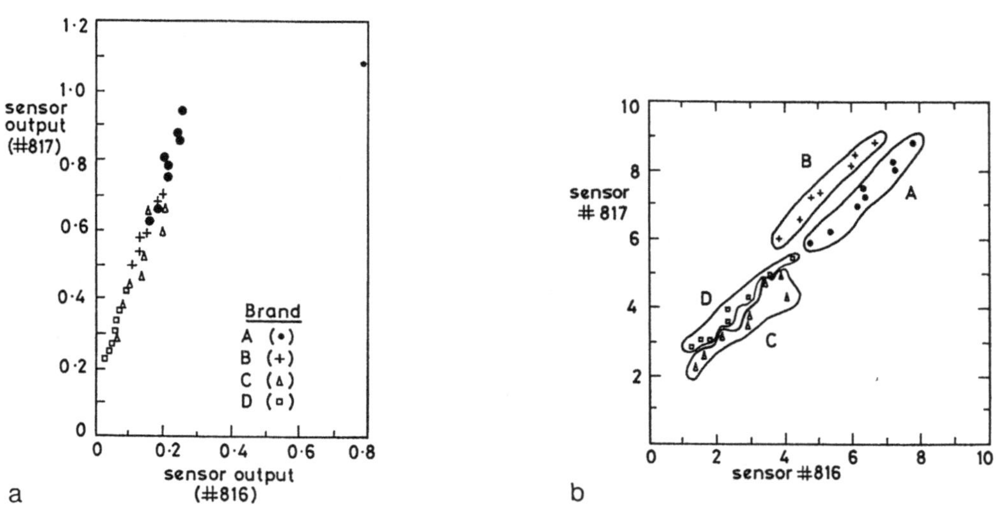

Figure 4.2 Scatter diagram for four brands of tobacco (a) raw data (b) processed data (Shurmer et al. 1989).

More recently, the performance of a 12-element chemosensor array has been tested against various beverages and spirits. Preliminary results show that identification is possible using a maximum likelihood method (as described earlier) and we intend to publish our results in the near future.

4.5 Other techniques

By far the majority of pattern recognition techniques that have been applied to chemical sensor arrays are based upon standard statistical techniques. However, more powerful techniques are also being applied such as Principal Component Analysis and Factor Analysis. Furthermore, Schiffman et al. (1981) have developed a mathematical procedure called Multidimensional Scaling which can transform objects into a two-dimensional spatial configuration. The technique has been applied to beverages and compares taste images. Alternatively, a novel approach has been proposed by Gardner and Hines (1988), namely to analyse the electronic nose by an Artificial Neural Network technique (ANN). ANNs offer several advantages over classical algorithms such as fault-tolerance, flexibility and adaptability. The inherent parallelism leads to a high speed of operation which could be an advantage when processing large arrays. The method is non-parametric and heuristic, this makes it difficult to interpret results but is closer to the real olfactory system and may yet prove to be the most beneficial approach.

5. REFERENCES

Aizawa M, Yamada T, Shinohara H, Akagi K, Shirakawa H (1986) Electrochemical fabrication of a polypyrrole-polythiophene p-n junction diode. J Chem Soc Chem Commun 1315-1317.

Amoore JE (1970) Molecular Basis of Odor. Charles C. Thomas, Springfield, Illinois, USA.

Arctander S (1969) Perfume and Flavour Chemicals. Published by the author, USA.

Asavapiriyanont S, Chandler GK, Gunawardena GA, Pletcher D (1984a) The electrodeposition of polypyrrole films from aqueous solutions. J Electroanal Chem 177:229-244.

Asavapiriyanont S, Chandler GK, Gunawardena GA, Pletcher D (1984b) The electrodeposition of poly-N-methylpyrrole films from aqueous solutions. J Electroanal Chem 177:245-251.

Baker CK, Reynolds JR (1988) A quartz microbalance study of the electrosynthesis of polypyrrole. J Electroanal Chem 251:307-322.

Baker S, Roberts GG and Petty MC (1983) Phthalocyanine Langmuir-Blodgett film gas detector. Proc IEE Part I: Solid State and Electron Devices 5 : 260-263.

Bartlett PN, Archer PBM, Ling-Chung SK (1989) Conducting polymer gas sensors. Part I: fabrication and characterisation. Sensors and Actuators 19 125-140.
Bartlett PN, Ling-Chung SK (1989a) Conducting polymer gas sensors. Part II: response of polypyrrole to methanol vapour. Sensors and Actuators 19 141-150.
Bartlett PN, Ling-Chung SK (1989b) Conducting polymer gas sensors. Part III: results for four different polymers and five different vapours. Sensors and Actuators in press.
Bartlett PN, Whitaker RG (1987) Electrochemical immobilisation of enzymes. Part II. Glucose oxidase immobilised in poly-N-methylpyrrole. J Electroanal Chem 224:37-48.
Beets MGJ (1978) Structure-Activity Relationships in Human Chemoreception. Applied Science Publishers, London.
Bidan G, Deronzier A, Moutet J-C (1984) Electrochemical coating of an electrode by a poly(pyrrole) film containing the viologen (4, 4'-bipyridinium) system. J Chem Soc Chem Commun 1185-1186.
Bocchi V, Gardini GP (1986) Chemical synthesis of conducting polypyrrole and some composites. J Chem Soc Chem Commun 148.
Bott B and Jones TA (1984) A highly sensitive NO_2 sensor based on electrical conductivity changes in phthyalocyanine films. Sensors and Actuators 5 43 - 53.
Brédas JL, Silbey R, Boudreaux DS, Chance RR (1983) Chain-length dependence of electronic and electrochemical properties of conjugated systems: polyacetylene, polyphenylene, polythiophene, and polypyrrole. J Amer Chem Soc 105:6555-6559.
Buck TM, Allen FG and Dalton M (1965) Detection of chemical species by surface effects on metals and semiconductors. in Surface Effects in Detection (eds. T Bregman and A Dravnieks) Spartan Books Inc, USA.
Burgmeyer P, Murray RW (1984) Ion gate electrodes. Polypyrrole as a switchable ion conductor membrane. J Phys Chem 88:2515-1521.
Carey WP, Beebe KR and Kowalski BR (1987) Multicomponent Analysis using an array of piezoelectric crystal sensors. Anal Chem 59 1529-1534.
Carey WP, Beebe KR, Kowalski BR, Illman DL and Hirschfield T (1986) Selection of adsorbates for chemical sensor arrays by pattern recognition. Anal Chem 58 149-153.
Cassidy J, Foley J, Pons S, Janata J (1986) Polymer coatings on suspended metal mesh field effect transistors. Anal Chem Symp Ser 25:309-314.
Chadwick AV, Dunning PBM and Wright JD (1986) Applications of organic solids to chemical sensing. Mol Cryst Liq Cryst 134 137-153.
Chandler GK, Pletcher D (1985) The electrochemistry of conducting polymers. In: Specialist periodical reports, electrochemistry Vol. 10, Royal society of Chemistry, London p117-150.
Chen Z, Pace U, Ronen D and LAncet D (1986) Polypeptide p95: a unique glycoprotein of olfactory cilia with transmembrane properties. J Biol Chem 261 : 1299-1305.
Cherry RJ, Dodd GH and Chapman D (1970) Small molecule-lipid membrane interactions and the puncturing theory of olfaction. Biochem Biophys Acta 211 : 409-416.
Clifford PK and Tuma DT (1982) Characteristics of semiconductor gas sensors. Sensors and Actuators 3 233-254.
Davies JT (1971) Recent developments in the 'penetration and puncturing' theory of odour. in Taste and Smell in Vertebrates (eds. G Wolstenholme and J Knight) p265-291.
Demarne V and Grisel A (1988) An integrated low-power thin film CO gas sensor on silicon. Sensors and Actuators 13 301-313.
Deutsch S (1967) Models of the Nervous System, John Wiley and Sons.
Diaz AF, Bargon J (1986) Electrochemical synthesis of conducting polymers. In: Skotheim TA (ed.) Handbook of conducting polymers Vol. 1, Marcel Dekker, New York p81-115.

Diaz AF, Castillo J, Kanazawa KK, Logan JA, Salmon M, Fajardo O (1982) Conducting poly-N-alkylpyrrole polymer films. J Electroanal Chem 133:233-239.

Diaz AF, Kanazawa KK, Gardinin GP (1979) Electrochemical polymerisation of pyrrole. J Chem Soc Chem Commun 635-636.

Diaz AF, Lee W-Y, Logan A, Green DC (1980) Chemical modification of a polypyrrole electrode surface. J Electroanal Chem 108:377-380.

Dodd GH (1970) The Adenosine Triphosphatase activities of the olfactory mucosa of some common animals. Comp Biochem Physiol 36 : 633-637.

Dodd GH (1974) Structure and function of chemoreceptor membranes. in Transduction Mechanisms in Chemoreception (ed. T M Poynder) p103-111, IRL Press, London.

Dodd GH (1976) Ligand-binding phenomena in chemoreception. in Structure-Activity Relationships in Chemoreception (ed. G Benz), p 55-61, IRL Press, London.

Dodd GH, Menevse A and Poynder TM (1977) An allosteric-membrane-enzyme hypothesis for olfaction. in Olfaction and Taste VI (eds. J Le Magnen and P MacLeod) p59. IRL Press, London.

Dodd GH and Persaud K (1981) Biochemical mechanisms in vertebrate primary olfactory neurons. in Biochemistry of Taste and Olfaction (eds. RH Cagan and MR Kare), Academic Press, p333-358.

Dravnieks A and Trotter PJ (1965) Polar vapour detection based on thermal modulation of contact potentials. J Sci Instruments 42 : 624.

Eaves JG, Munro HS, Parker D (1985) Electroactive thin films from the anodic electropolymerisation of iron (II) and ruthenium (II) tris(N-bipyridylpyrrole) complexes. J Chem Soc Chem Commun 684-685.

Fard M (1985) A study of semiconductor sensors for an electronic olfactory system. MSc thesis. University of Warwick.

Feldman BJ, Burgmeyer P, Murray RW (1985) The potential dependence of electrical conductivity and chemical charge storage of poly(pyrrole) films on electrodes. J Amer Chem Soc 107:872-878.

Fouletier J (1982/3) Gas analysis with potentiometric sensors: A review. Sensors and Actuators 3 295-314.

Gardner JW (1988) Pattern recognition in the Warwick electronic nose. 8th Int. Biennial Congress of the European Chemoreception Research organisation.

Gardner JW and Hines EL (1988) Integrated sensor array processing in an electronic nose. IEE Coll. 130:7/1-4.

Gardner JW (1989a) Electrical conduction in solid-state gas sensors. Sensors and Actuators 18:371-385.

Gardner JW (1989b) A diffusion-reaction model in tin oxide gas sensors. Semicond Sci Technol 4 : 345-350.

Garnier F, Tourillon G, Gazad M, DuBois JC (1983) Organic conducting polymers derived from substituted thiophenes as electrochromic material. J Electroanal Chem 148:299-303.

Genies EM, Bidan G, Diaz AF (1983) Spectroelectrochemical study of polypyrrole films. J Electroanal Chem 149:101-113.

Green M (1981) Lectures on gas-solid interactions. Sensors and Actuators 1 379-381.

Gustafsson G, Lundström I (1987) The effect of ammonia on the physical properties of polypyrrole, Synth Metals 21:203-208.

Haimerl A, Merz A (1987) Catalysis of quinone-hydroquinone redox reactions at polypyrrole benzenesulphonate-coated platinum electrodes. J Electroanal Chem 220:55-65.

Hanawa T, Susumu K, Yoneyama H (1988) Gas sensitivity of polypyrrole films to NO_2. J Chem Soc Faraday Trans 1 84:1587-1592.

Holley A, Duchamp A, Revial MF, Juge A and MacLeod P (1974) Qualitative and quantitative discrimination in the frog olfactory receptors: analysis from electrophysiological data. Ann NY Acad Sci 237 : 102-114

Ikegami A and Kaneyasu M (1985) Olfactory detection using integrated sensor. IEEE 9:136-139.

Ikegami A and Kaneyasu M (1984) Smell identification by using pattern recognition algorithm. Proc. SICE. Jap. 673-674.
Imanpour M (1986) Development of processing conditions for production of Langmuir-Blodgett films for an electronic nose. MSc dissertation, University of Warwick.
Inoue T, Yamase T (1983) Electrochemical properties of electro-chemically polymerised polypyrrole film in the dark and under light irradiation. Bull Chem Soc Jpn 56:985-990.
Jen K-Y, Maxfield M, Shacklette LW, Elsenbaumer (1987) Highly-conducting, poly (2,5-thienylene vinylene) prepared via a soluble precursor polymer. J Chem Soc Chem Commun 309-311.
Jones TA and Bott B (1986) Gas induced electrical conductivity changes in metal phthyalocyanines. Sensors and Actuators 9 27-37.
Josowicz M, Janata J (1988) Suspended gate field effect transistor. In: Seiyama T (ed.) Chemical sensor technology Vol. 1. Elsevier, New York p153-177.
Josowicz M, Janata J, Ashley K, Pons S (1987) Electrochemical and ultraviolet-visible spectroelectrochemical investigations of selectivity of potentiometric gas sensors based on polypyrrole. Anal Chem 59:253-258.
Josowicz M, Janata J (1986) Suspended gate field effect transistor modified with polypyrrole as alcohol sensor. Anal Chem 58:514-517.
Kaneyasu M, Ikegami A, Arima A and Iwanaga S (1987) Smell Identification using a thick-film hybrid gas sensor. Trans. Comp. Hybrids. Manu. Technol. 10:267-273.
Koch RB and Norring RL (1969) Fractionation of olfactory tissue homogenates. Isolation of a concentrated plasma membrane fraction. J Neurochem 16 : 145-157.
Kurihara K and Koyama N (1972) High activity of adenyl cyclase in olfactory and gustatory organs. Biochem Biophys Res Comm 48 : 30-34
MacInnes D, Druy MA, Nigrey PJ, Nairns DP, MacDiarmid AG, Heeger AJ (1981) Organic batteries: reversible n- and p-type electrochemical doping of polyacetylene, $(CH)_x$. J Chem Soc Chem Commun 317-319.
Mather R and Dodd GH (1988) The delivery of an aqueous aerosol containing non-volatile odorants to the olfactory epithelium of the rat. Chemical Senses 13 : 247-255.
McAleer JF, Moseley PT, Norris JOW, Williams DE (1987) Tin dioxide gas sensors. J Chem Soc Faraday Trans 1 83 1323-1346.
Menevse A and Dodd GH and Poynden TM (1977) Evidence for the specific involvement of cAMP in the olfactory transduction mechanism. Biochem Biophys Res Comm 77 : 671-677.
Menevse A, Cheng L, Menco B and Dodd GH (1974) Adenyl Cyclase activities of olfactory and gustatory plasma membranes. 1st ECRO Congress, Paris, p 32.
Miasik JJ, Hooper A, Tofield BC (1986) Conducting polymer gas sensors. J Chem Soc Faraday Trans 1 82:1117-1126.
Minor AV and Sakina NL (1973) Role of cAMP in olfactory reception (in Russian) Neirofiziologia 5 : 414-422.
Misra TN, Rosenberg B and Switzer R (1968) Effect of adsorption of gases on the semiconductive properties of all-trans β-carotene. J Chem Phys 48 : 2096-2102.
Moncrieff RW (1951) The Chemical Senses Leonard Hill Ltd, London.
Moncrieff RW (1961) An instrument for measuring and classifying odors. J App Physiol 16 : 742.
Monod J, Wyman J and Changeux JP (1965) On the nature of allosteric transitions : a plausible model J Mol Biol 12 : 88-118.
Moriizumi T (1988) Langmuir-Blodgett films as chemical sensors. Thin Solid Films 160 413-429.
Moseley PT, Tofield BC (eds) (1987), Solid state gas sensors. Adam Hilger, Bristol.
Moulton DG, Turk A and Johnston JW (1975) Methods in Olfactory Research, Academic Press.
Müller R and Horner G (1986) Chemosensors with pattern recognition. Siemens Forsch-u. Entwickl-Ber. Bd. 13 Nr. 3, Springer-Verlag.

Müller R and Lange E (1986) Multi-dimensional sensor for gas analysis. Sensors and Actuators, 9 39-48.
Nazzal AI, Street GB (1985) Pyrrole-styrene graft copolymers. J Chem Soc Chem Commun 375-376.
Nylander C, Armgarth M, Lundström I (1983) An ammonia detector based on a conducting polymer. Anal Chem Symp Series 17:203-207.
O'Riordan DMT, Wallace GG (1986) Poly(pyrrole-N-carbodithioate) electrode for electro-analysis. Anal Chem 58:128-131.
Pace U, Hanski E, Saloman Y and Lancet D (1985) Odorant-sensitive adenylate cyclase may mediate olfactory reception. Nature 316 : 255-258.
Patil AO, Heeger AJ, Wudl F (1988) Optical properties of conducting polymers. Chem Rev 88:183-200.
Pelosi P, Pisanelli A, Baldacinni N and Gagliardo A (1981) Binding of isobutylmethoxypyrazine to cow olfactory mucosa. Chemical Senses 6 : 77-85.
Persaud K and Dodd GH (1982) Analysis of discrimination mechanisms in the mammalian olfactory system using a model nose. Nature 299:352-355.
Persaud K, Pelosi P and Dodd GH (1988) Binding and metabolism of the urinous odorant 5α-androstan-3-one in sheep olfactory mucosa. Chemical Senses 13 : 231-245.
Pevsner J, Trifiletti RR, Strittmatter SM and Snyder SH (1985) Isolation and characterization of an olfactory receptor protein for odorant pyrazines. Proc Natl Acad Sci USA 82 : 3050-3054.
Pitt CW (1983) Langmuir-Blodgett films - the ultrathin barrier. Electronics and Power 29 226-229.
Polak E (1973) Multiple Profile - Multiple receptor site model for Vertebrate Olfaction. J Theor Biol 40 : 469-484.
Robinson GA, Butcher RW and Sutherland EW (1968) Cyclic AMP Ann Rev Biochem 37 : 149-174.
Russell Y, Evans P and Dodd GH (1989) Lipid analysis of rat olfactory mucosa J Lipid Res at press.
Ruzicka L (1920) Fundamentals of odour chemistry Chemikerstg, No 14.
Salmón M, Carbajal ME, Aguilar M, Saloma M, Juárez JC (1983) Polyheterocyclic conductors from N-(ortho-substituted phenyl) pyrroles. J Chem Soc Chem Commun 1532-1533.
Salmón M, Diaz A, Goitia J (1982) Chemically modified conducting polypyrrole film electrodes. In: Miller JS (ed.) Chemically modified surfaces in catalysis and electrocatalysis. ACS symposium series 192:65-69.
Sato M, Tanaka S, Kaeriyama K (1986) Soluble conducting polythiophenes. J Chem Soc Chem Commun 873-874.
Schiavon G, Zotti G, Bontempelli G (1985) Cathodically deposited polypyridine films. A novel electroactive polyconjugated polymer. J Electroanal Chem 194:327-338.
Schiffman S, Reynolds ML and Young FW (1981) Introduction to Multidimensional Scaling. Academic Press New York.
Shirley SG, Polak E, Mather R and Dodd GH (1987) Effect of concanavalin A on the rat electro-olfactogram : differential inhibition of odorant responses. Biochem J 245 : 175-184.
Shirley SG, Robinson CJ, Dickinson K, Aujla R and Dodd GH (1986) Olfactory adenylate cyclase of the rat : stimulation by odorants and inhibition by Ca^{2+}. Biochem J 240 : 605-607.
Shirley SG, Robinson CJ and Dodd GH (1989) Olfactory adenylate cyclase of the rat : stimulation by odorant mixtures. Biochem J at press.
Shorey H (1976) Animal Communication by Pheromones, Academic Press.
Shurmer HV, Gardner JW and Chan TS (1989) The application of discrimination techniques to alcohols and tobaccos using tin oxide sensors. Sensors and Actuators 18:359-369.

Sklar PB, Anholt RH and Snyden SH (1986) The odorant-sensitive andenylate cyclase of olfactory receptor cells. J Biol Chem 261 : 15538-15543

Street GB (1986) Polypyrrole. From powders to plastics. In: Skotheim TA (ed.) Handbook of conducting polymers Vol. 1, Marcel Dekker, New York p265-291.

Stretter JR, Jurs PC and Rose SL (1986) Detection of hazardous gases and vapours: pattern recognition analysis of data from an electrochemical sensor array. Anal Chem 58:860-866.

Taguchi N (1971) British Patent No 1 257 155.

Takakubo M (1989) Molecular orbital study of the initial processes in the electrochemical polymerisations of pyrrole, thiophene, and furan. J Electroanal Chem 258:303-311.

Tanyolac NN and Eaton JR (1950) Study of odors. J Am Pharm Assoc 39 : 10.

Thackeray JW, White HS, Wrighton MS (1985) Poly(3-mehtylthiophene)-coated electrodes: optical and electrical properties as a function of redox potential and amplification of electrical and chemical signals using poly(3-methylthiophene)-based microelectrochemical transistors. J Phys Chem 89:5133-5140.

Tourillon G, Garnier F (1982) New electrochemically generated organic conducting polymers. J Electroanal Chem 135:173-178.

Van Toller C, Dodd GH and Billing A (1985) Ageing and the Sense of Smell, Charles C. Thomas, Illinois, USA.

Van Toller C and Dodd GH Perfumery - the psychology and biology of fragrance. Chapman and Hall, Ltd. London, 1988.

Waltman RJ, Diaz AF, Bargon J (1984) Substituent effects in the electropolymerisation of aromatic heterocyclic compounds. J Phys Chem 88:4343-4346.

Waltman RJ, Diaz AF, Bargon J (1983) Electrochemical studies of some conducting polythiophene films. J Phys Chem 87:1459-1463.

Wang TT, Tasaka S, Hutton RS, Lu PY (1985) Conductive composites prepared by electrochemically polymerizing pyrrole in poly(vinyl chloride) blended with electrolyte. J Chem Soc Chem Commun 1343-1344.

Wilkens WF and Hatman JD (1964) An electronic analog for the olfactory processes. Ann NY Acad Sci 116 : 608.

Wohltjen H (1984) Chemical microsensors and microinstrumentation. Anal Chem 56 87-103.

Wood M, Russell Y and Dodd GH (1988) Phosphoinositide Metabolism in the rat olfactory mucosa. Proc 8th Int ECRO Congress, Coventry, p140.

Wright RH (1964) The Science of Smell, Allen and Unwin.

Yamazoa N and Seiyama T (1985) Sensing mechanism of oxide semiconductor gas sensors. IEEE CH2127-9/85:376-379.

Zarcomb S and Stretter JR (1984) Theoretical basis for identification and measurement of air contaminants using an array of sensors having partially overlapping sensitivities. Sensors and Actuators 6:225-243.

Zinger B, Miller LL (1984) Timed release of chemicals from poly-pyrrole films. J Amer Chem Soc 106:6861-6863.

MORPHOLOGICAL BASIS OF INFORMATION PROCESSING IN THE OLFACTORY BULB[1]

N. Halasz

Institute of Biophyics
Biological Research Center of the
Hungarian Academy of Sciences
6701 Szeged, P.O.Box 521
Hungary

1. INTRODUCTION

A high diversity of odorous stimulants exists in our environment, some ofwhich can reach the olfactory organ where they may act simultaneously. Although a large variety of molecular receptors are beleived to be localizedon the receptive sites of the primary olfactory neurons, only a few may be essential for the detection of certain odorants. As regards morphological techniques used at the cellular level, subtypes of receptor cells are not sufficient to account for detection of the number of odorants which can have biological significance. According to our present knowledge, higher stations of the olfactory pathway must contribute to a large extent in the precise discrimination of the odorants. It seems very likely that brain regions closer to the sensory receptors may contribute to signal processing with a higher degree of specificity since more central areas have an increasing number/density of other afferents from other systems. Consequently, it is very probable that the main olfactory bulb (OB) plays a significant role in providing a structural and neurochemical basis for the discriminatory processing of olfactory stimuli.

This review emphasizes the structural characteristics of the OB neurons, their arrangement, connections and major neurotransmitters. Most of the studies considered

1) Abbrevitions used in text: ACh, acetylcholine; DA, dopamine; EM, electron microscope; EPL, external plexiform layer; GABA, gamma-aminobutyric acid; GCL, granule cell layer; GL, glomerular layer; 5-HT, serotonin; IPL, internal plexiform layer; MCL, mitral cell layer; ONL, olfactory nerve layer.

NATO ASI Series, Vol. H 39
Chemosensory Information Processing. Edited by D. Schild

in this review were conducted on mammalian tissue. Consequently, mammals will be the primary focus of the following discussions.

2. PRINCIPAL LAYERS AND NEURONS

A distinct arrangement in the OB of neuronal somata and components of the neuropil regions was an attractive feature for early investigators. It has remained so for all of us currently engaged with the morphology of this part of the forebrain. Although the refinement of tools for morphological studies has increased recognition of neuronal subtypes in the OB, the principal neuronal classes and main histological layers were characterized early in the classic works of Golgi (1875) and Cajal (1911).

Arrangement of the best known projection neurons, the mitral cells, into a distinct layer, their radial orientation and assemblies of local interneurons deeper or more superficially is characteristic for most of thevertebrates (see Nieuwenhuys, 1967). This arrangement largely defines the most appropriate localization and available space and position for all local elements.

The nasal epithelial afferents, axons of the olfactory receptor neurons, approach the main olfactory bulb from its anterior or ventral pole and spread on its surface forming the olfactory nerve layer. Moving deeper into the tissue, the following main layers can be distinguished: glomerular, external plexiform, mitral cell body, internal plexiform, granule cell layer and a periventricular (germinal) zone.

2.1 Olfactory nerve layer (ONL)

On the surface of the bulb, several million small axons run, bundled into discrete fascicles, to their targets within the next layer. The layer is thicker on the bulbar surface closely apposed to the epithelium. The unmyelinated axons contain several microtubules and become enlarged only at their terminal segment.

2.2 Glomerular layer (GL)

The olfactory nerve axons enter the glomeruli, which are the most complex structural (see Andres, 1965; Pinching and Powell, 1971b), and very probably physiological (see

Shepherd, 1972, 1979) units of the superficial OB. The glomeruli are ensheated by glial processes and also delineated by the juxtaglomerular cells (neurons and glia). Among these cells, local circuit neurons may predominate but somata of certain output neurons are also situated here, no matter which class of vertebrates the respective subject belongs to (see Andres, 1970).

The most numerous, although the smallest type of juxtaglomerular cells is the periglomerular cell. It has an intraglomerular dendritic arborization and an axon, although it never travels more distantly than 4-5 glomeruli. Apart from these, other superficial interneurons, short-axon cells, may also be distinguished. They have somewhat larger somata; their processes are primarily limited to the periglomerular space (see Pinching and Powell, 1971a-c), although some may extend into the EPL (Schneider and Macrides, 1978).

The third and largest neuron type of the juxtaglomerular cells belongs to the output neurons in the tufted cell category. According to their position, they have been called external or superficial tufted cells (see Cajal, 1911; Pinching and Powell, 1971a). However, since they may or may not have basal dendrites, in addition to their primary dendritic branch within the glomeruli (Pinching and Powell, 1971a; Macrides and Schneider, 1982), it seems justified to distinguish between subcategories. Those tufted cells, lying in the superficial two-thirds of the GL (in hamster) seem likely to have no output (see Schoenfeld and Macrides, 1984) beyond the intrinsic OB.

Surrounded by the juxtaglomerular cells, the glomerular neuropil is a complex of intermingled fibers of the primary olfactory nerve and from intrinsic bulbar (mitral, tufted and periglomerular) neurons as well as from central sources (see Andres, 1965; Pinching and Powell, 1971b).

2.3 External plexiform layer (EPL)

In most vertebrates, apart from lower fish, there is a distinct neuropil region between glomeruli and the mitral cells (see Cajal, 1911; Allison, 1953; Andres, 1970). Main components of the layer are the basal dendrites of the output (mitral and tufted) neurons and peripheral processes of those interneurons (granule cells) lying deeper than or alongside the output neurons. The layer is relatively free of myelinated fibers, although central pathways travel through the layer and also terminate here. The EPL includes several neuron types, belonging primarily to the output cell category (tufted cells) although local neurons were also observed in the EPL. Tufted cells send their primary (apical) dendrites to the glomeruli, while their basal dendrites are distributed more or less superficially in the EPL, according to the depth and size of somata of their parent cells. They differ systematically from each other, which justifies their classification into subcategories as in the GL (see Macrides and Schneider, 1982; Orona et al., 1984). Axon collaterals of the middle and deep tufted cells are distributed just deep to the mitral cells, within the IPL or GCL (see Kishi et al., 1984; Orona et al.,1984).

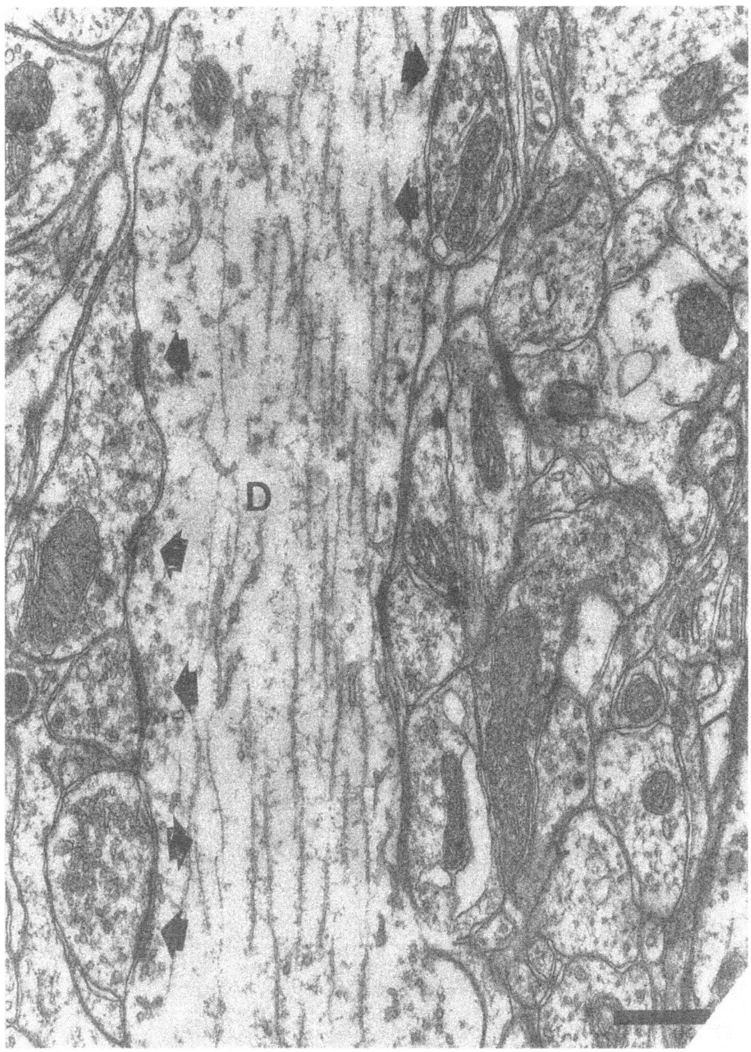

Fig. 1 A large dendrite (D) of an output (mitral or tufted) cell traverses through this picture forming several synapses with likely granule cell dendritic terminals. Supposed polarity of synapses is shown by arrows. Bar, 0.5 μm.

2.4 Mitral cell layer (MCL)

Mitral cells form a distinct, shell-like layer in the bulb of most vertebrates. They have their apical (primary) dendrites directed towards the glomeruli, while axons leave the somata from the opposite side. The basal (secondary) dendrites run obliquely in the EPL, where they contact the peripheral processes of the granule cells (see e.g. Andres, 1965; Rall et al., 1966; Price and Powell, 1970b; see also Fig.1). Mitral cells may also be classified as differing in the distribution of their secondary dendrites: those dendrites of the type I mitral cells arborize in the deep EPL, while type II cells distribute their secondary dendrites more superficially in the rat (Orona et al., 1984). Similar differences have also been described for the rabbit (Mori et al., 1983). An additional feature of the type I mitral cells of the rat, recognized by Orona et al. (1984), was the absence of local axon collaterals, while rat type II mitral cells, and also the regular (non-displaced) mitral cells of the rabbit had collaterals in the GCL (Mori et al., 1983; Kishi et al., 1984). Superficial granule cells are also present in the MCL.

2.5 Internal plexiform layer (IPL)

This layer is not always apparent but in many adult vertebrates a thin and relatively cell-free zone between the mitral cells and the granule cell layer is present (see Allison, 1953; Andres, 1970). It is traversed by the centrally directed axons of mitral and tufted cells, the vertically directed granule cell peripheral processes, axons of certain short-axon cells and central afferent pathways, in addition to the horizontally distributed local axon collaterals and varicose, terminal segments of central fibers. Somata of certain short-axon cells may also be situated here (see Schneider and Macrides, 1978).

2.6 Granule cell layer (GCL)

The highest number of local interneurons is confined to the deep aspect of the OB in most vertebrates. They appear in tightly packed stacks, arranged parallel with the layer, between narrow neuropil areas. The granule cells are regularly small, with little cytoplasm; spines may appear in all parts of the cell (see Price and Powell, 1970a). The cells have a shorter, less spiny central process and a longer peripheral process, arborizing richly in the EPL, giving rise to many large spines (gemmules, see Rall et al., 1966). Granule cells have no axons (Cajal, 1911). Although these neurons share many common characteristics in structure and neurochemistry, their classification into subtypes, according to the situation of the cell somata and branching patterns of the peripheral processes, seems reasonable, as suggested by Mori and Kishi (1982) and Mori et al.(1983) for the rabbit and by others for the rat and mouse (see e.g. Orona et al., 1983; Greer, 1987).

Another, though very heterogeneous class of interneurons are the deep short-axon cells (see Price and Powell, 1970d; Schneider and Macrides, 1978). They differ largely in their localization of somata, orientation and distribution of dendritic processes, and axonal arbors. The axons may stay in the GCL (or IPL) (e.g., those of the Blanes and Golgi cells in the hamster OB, see Schneider and Macrides, 1978), but may also penetrate the EPL (vertical and horizontal cells of Cajal, same reference as above).

The GCL is the layer in which most of the (myelinated) axons of the output neurons are gathering as they form the olfactory tracts which are directed towards the higher olfactory centers. Most of the central afferents are also associated with these tracts; they may terminate in the GCL or travel more superficially.

2.7 Subependymal (periventricular) zone

Just deep to the granule cells, an accumulation of small rounded cells can be observed. They are thought to be supplied by the rostral migratory stream (see Altman, 1969), and they may be the source of new outwardly migrating neurons, as seen even in the late adulthood (see Kaplan et al., 1985).

3. SYNAPTIC ORGANIZATION

Information about an odorant, detected by olfactory receptor cells is supplied by the peripheral afferents, while the output is conveyed to olfactory cortical regions by axons of the mitral (and certain tufted) cells. Microscopy has revealed that local interneurons also make contacts with the output neurons, suggesting a potent interaction with the bulbar output. These contacts are segregated into two main levels in the OB, according to the arrangement of the two types of the output cell dendrites. Primary (apical) dendrites may receive such influences through their terminal arbors within the glomeruli, while the second receptive level is represented by the secondary (basal) dendrites (and, at a lower level, by the output cell somata). The second level of interaction involves primarily the EPL. Synaptic contacts can, of course, be found in other layers; these may be involved in activation or inhibition of interneurons, and so, indirectly affect the output neurons.

A simple scheme of the connections is given in Fig. 2, stressing relations between principal bulbar cells and disregarding subcategories and synapses of the output cell axon collaterals. The scheme does, however, account for results, achieved with physiological and pharmacological approaches.

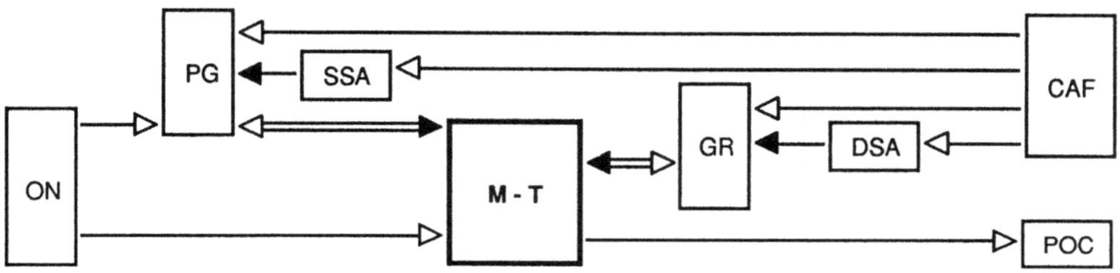

Fig. 2. Block scheme of the main synaptic contacts in the vertebrate OB. Arrows show polarity of synapses and indicate the most likely postsynaptic effects (open arrows, activation; closed arrows, inhibition). Double arrows: reciprocal synapses. CAF, central afferents; DSA, deep short-axon cells; GR, granule cells; M-T, mitral or tufted cells; PG, periglomerular cells; ON, olfactory nerve; POC, primary olfactory cortex; SSA, superficial short-axon cells. Slightly modified from Halasz, in press.

In the following Sections, a brief description is provided of the synaptic contacts, observed principally in mammals, and their segregation within the OB layers.

3.1 Glomerular layer

The most characteristic synapses of the GL are formed within the glomeruli between the olfactory nerve terminals and (primary) dendritic branches of the mitral and tufted cells. The morphology of this synapse (round synaptic vesicles and asymmetrical postsynaptic thickenings) and the available physiological results suggest that this synapse is excitatory. In addition, the olfactory nerve may also synapse on dendrites of local circuit neurons, as exemplified in mammals (see Pinching and Powell, 1971b; White, 1973), although they are rarely observed in goldfish (see Kosaka and Hama, 1982-83). Between intraglomerular dendrites of the mitral/tufted cells and periglomerular neurons, a reciprocal contact can frequently be observed. Axodendritic synapses (not involving the olfactory nerve) within the glomerular neuropil were also observed on periglomerular cell dendrites.

The juxtaglomerular neuropil lacks olfactory nerve axon terminals but is nevertheless rich in axodendritic and axosomatic synaptic contacts; the source of these presynaptic profiles is most likely other juxtaglomerular cells as well as central afferent fibers and local axon collaterals (see Pinching and Powell, 1971c; 1972a,b).

Besides histochemistry, ultrastructural characteristics of the synaptic contacts also help morphologists to make suggestions for the role of different contacts. Accordingly, symmetric synaptic thickenings and flat vesicles within the terminals of short-axon cells and periglomerular neurons suggests their inhibition upon postsynaptic counterparts (see Pinching and Powell, 1971a,b).

3.2 External plexiform layer

The dendrodendritic synapses between the peripheral processes of the granule cells and secondary dendrites of the output neurons, as determined first by Rall et al. (1966) in rats, are the most extensively studied contacts of the EPL. The reciprocal nature of this synapse was evidenced by serial sectioning. No morphological distinctions have yet been made between those synapses established by deeper or more superficially located granule cells, or according to different subtypes of the mitral or tufted cells. Other synapses in this region include axodendritic contacts (mostly on granule cell dendrites) made by terminals assigned to centrifugal afferent fibers and short-axon cells (see Price and Powell, 1970b,c). In addition, somata of the tufted and displaced mitral cells seem to have contacts similar to the reciprocal contacts made by their dendrites, so they may serve similar function.

3.3 Deeper layers

Mitral cells have numerous reciprocal contacts with endings resembling structurally the granule cell dendritic terminals. There are many synapses in the IPL and GCL too, but only a few have been characterized with EM in their relation to granule or short-axon cells. These include synapses on the somata as well as on central and peripheral dendritic shaft and spines of the granule cells (see Price and Powell, 1970a,b). These contain either spherical vesicles (suggesting terminals of fibers from olfactory cortical areas, diagonal band, brainstem, or recurrent collateral axons of the bulbar output neurons) or flattened vesicles (likely endings of the deep short-axon cells). Some of these terminals were more precisely identified, as to their possible origin, by lesion experiments, tracer injection and transmitter histochemical studies. A segregation of central inputs is seen according to their preferential termination in sublamina of the GCL; e.g., anterior prepiriform and periamygdaloid cortices and nucleus of the lateral olfactory tract send axons to the deep half of the GCL, while pars externa of the contralateral nucleus olfactorius anterior terminates mostly in the superficial GCL (see Luskin and Price, 1983; Macrides and Davis, 1983). Fibers from retrobulbar sources appeared to terminate on both types of interneurons in the GCL, but they may not have direct contact to the output neurons.

4. NEUROTRANSMITTER LOCALIZATION

Although a wide variety of neurotransmitter candidates have been mapped in various bulbar layers and neurons, we shall refer here only briefly to those substances which raised the most attention and were also histochemically localized to identified cellular constituents of the OB. Namely, histochemical indications will be summarized for the possible involvement in local signal processing of gamma-aminobutyric acid (GABA), excitatory amino acids (glutamate and aspartate), dopamine (DA), serotonin (5-HT), noradrenaline and acetylcholine (ACh). For more details, see reviews by Margolis (1981), Halasz and Shepherd (1983) and Macrides and Davis (1983).

Output neurons activate the primary olfactory cortical areas as well as groups of local circuit neurons in the OB. According to the high content of excitatory amino acids (glutamate and aspartate) and presence of specific receptors in areas of termination, they (or a dipeptide, N-acetylaspartyl glutamate) are strong candidates for the neurotransmitter of these cells. Aspartate (Saito et al., 1986), glutamate (Liu et al., 1989) and this dipeptide (Blakely et al., 1987) have also been localized to mitral/tufted cells by immunohistochemistry.

Suggestions for the synthesis of dopamine, mostly within a subpopulation of periglomerular and superficial tufted cells, were offered by tyrosine hydroxylase immunohistochemistry (see e.g. Halasz et al., 1977; Macrides and Davis, 1983; Baker, 1986; McLean and Shipley, 1988; see also Figs. 3 and 4).

In many periglomerular neurons, certain short-axon cells and in a vast majority of the granule cells strong evidence for the synthesis and content of GABA was obtained by immunohistochemistry (see Ribak et al., 1977; Mugnaini et al., 1984). A high uptake of radiolabeled GABA into presynaptic spines of the granule cells was observed (see Halasz et al., 1979, 1981). Label was also seen in axon terminals and other processes; the cellular labeling reflected predominantly glial uptake (see Fig.5). Coexistence of DA with GABA in the periglomerular cells was also reported (see Kosaka et al., 1985).

In addition to the evidence for intrinsic sources of glutamate and/or aspartate, DA and GABA, a minor supply of these substances by central pathways was also suggested (see e.g., aspartate: Watanabe and Kawana, 1984; dopamine: Fallon et al., 1978; GABA: Zaborszky et al., 1986). In contrast, ACh, NA and 5-HT was proven to be fed into the bulb exclusively by central pathways.

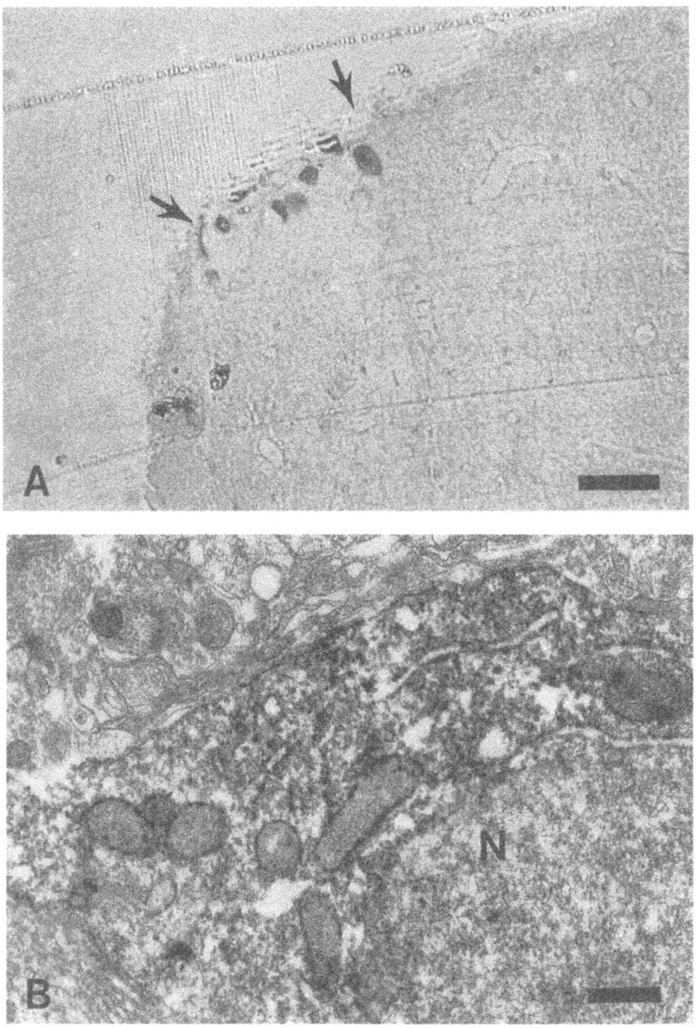

Fig. 3 A,B. Tyrosine hydroxylase immunoreactive (indirect PAP technique) juxtaglomerular cell somata. A: semi-thin section (of a re-embedded Vibratome section) reaching the depth of the periglomerular region (between arrows). Bar, 50 μm. B: Detail of a likely periglomerular cell, featuring immunoreactive cytoplasm around the nucleus (N). Bar, 0.5 μm.

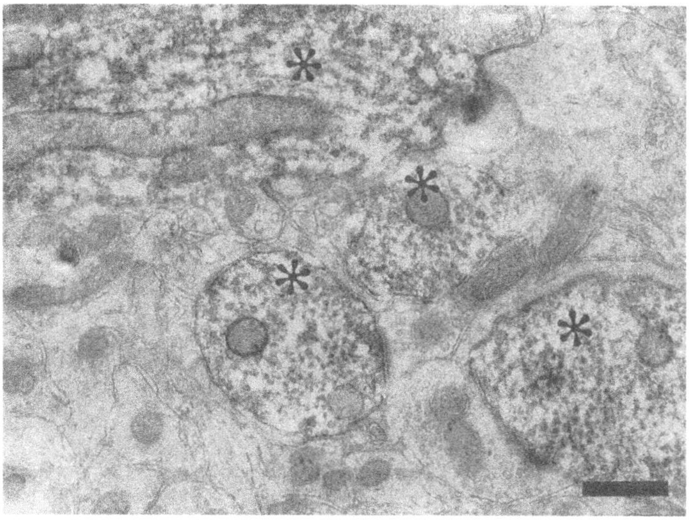

Fig. 4 Large, tyrosine hydroxylase immunoreactive (indirect PAP technique) dendritic profiles (asterisks) from the glomerular neuropil. The labeled fibers may be of tufted or periglomerular cell origin. Bar, 0.5 μm.

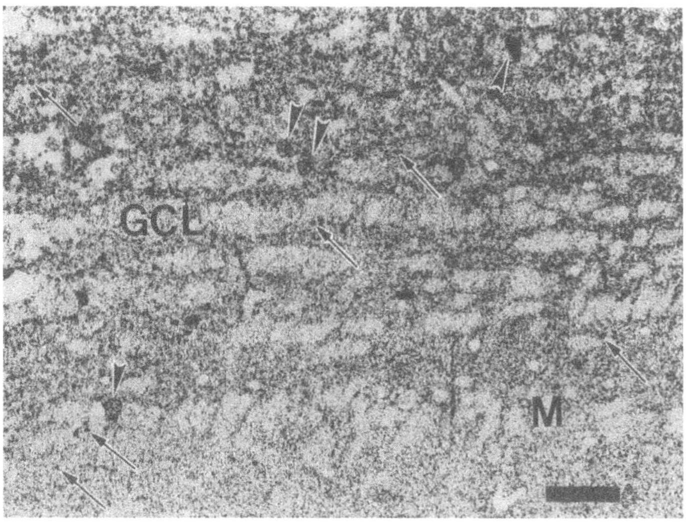

Fig. 5 Accumulation of silver grains over numerous varicosities (arrows) and a few cells (arrowheads), representing uptake of the radiolabeled GABA. This was typical of the granule cell layer (GCL) and also more superficially. Mitral (M) cells and most of the other cell somata were, however, devoid of the label as seen by this autoradiography. Bar, 50 μm.

Cholinergic fibers arising in the reticular and limbic, especially the diagonal band area, may be present in all layers of the OB (excluding ONL) (Shute and Lewis, 1967). However, they may synapse preferentially in the GL and close to mitral cells, where cholinoceptive neurons have recently been described (Nickell and Shipley, 1988). An unusually dense plexus of likely cholinergic fibers was reported in the posterior OB (Zheng et al., 1987).

Noradrenergic afferent fibers of the OB, originating in the locus coeruleus, were known since 1965 (see Dahlstrom et al.); they arborize primarily in the IPL and superficial GCL. The OB also has a rich serotoninergic innervation; these fibers can be distinctly visualized by immunohistochemistry (see e.g. Steinbusch, 1981) or autoradiography (see Halasz et al., 1978). Highest density of the 5-HT fibers and terminals appears in the intraglomerular neuropil (McLean and Shipley, 1987).

5. CONCLUSIONS

According to the scope of the present paper, characteristics of the histological, cellular and histochemical organization of neurons in the OB were briefly reviewed. In the following summary, a correlation of these findings will be attempted, related primarily to the output neuron - local circuit neuron interactions.

Odorants activate a specific population of olfactory receptor cells which transmit this signal onto the primary dendrites of topographically defined bulbar output neurons in the glomeruli. Also within the glomeruli, synapses for either a direct olfactory nerve input, or for dendrodendritic activation (by the output cell dendrites), are available on the dendritic arbor of periglomerular cells. A substantial proportion of these neurons were shown to contain GABA and/or dopamine. It is not known, whether the dendrodendritic or axodendritic activation (through the olfactory nerve axon) of the periglomerular cells would be more effective, but GABA release within a distance of a few glomeruli (according to the extension of the periglomerular cell axons) seems appropriate to spread inhibition around an activated glomerular area. As a result, that population of output neurons, having their primary dendrites in the activated zone, would have the chance to process further the signal from the primary olfactory nerve. This seems to be the first step in maintenance and amplification of specificity of an odorant signal in the OB.

The shaping of the pattern of activation across the bulb, of course, may depend on a number of further synaptic interactions. These may include connections with other neurons and pathways within the OB, in the GL and below. Activation of the GABA-containing granule cells through the secondary dendrites of output neurons seems likely the second, perhaps most important, level for intrabulbar inhibition. Both sets of GABA-containing small interneurons (periglomerular and granule cells) were shown to have inputs from central (mainly cholinergic and serotoninergic in the glomerular layer, and noradrenergic in the IPL and GCL) afferents. Through these central inputs limbic, reticular and brain stem regions may control local bulbar conditions.

A direct effect of central pathways upon the output cell activity seems unlikely. Similarly, synapses from axon collaterals of the output cells onto the same cell category are not a likely alternative. Instead, dissimilar distribution of axon collaterals suggests that they may activate different groups of interneurons (granule cells). In this way, subcategories of output neurons may activate a vertically segregated subpopulation of granule cells, which, through their dendrodendritic contacts onto spatially segregated secondary dendrites of sets of output neurons, may provide again some degree of specificity. A further opportunity for modulation of the output cell activity through the granule (or periglomerular) interneurons is offered by the likely involvement of disinhibitory short-axon cells activated by central pathways, output cell collaterals or the other local interneurons.

As discussed, numerous synaptic contacts have been observed in the bulb and characterized according to their fine structural details. Estimation of the significance of these connections is, however, complicated by the possibility of multiple interactions. These may result from the structural and neurochemical heterogeneity of the local neurons, their diverse arborization patterns and heterogeneous innervation by central afferents. This clearly indicates that there is a great need for further physiological and pharmacological experiments correlated with morphological studies at the level of fine resolution. In parallel, modeling of the conditions for local bulbar circuits and their correlation with those suggested by anatomical findings would give another opportunity for understanding more precisely the role of one single element in this signal processing system. Finally, theories about the likely interactions within and between local circuits would greatly benefit morphologists by directing their interest towards the existence of possible, but still undiscovered structural and neurochemical links.

ACKNOWLEDGEMENTS - I would like to express my special gratitude to colleagues, who participated in the original studies with immunohistochemistry and autoradiography, as discussed here, especially to Prof. T. Hokfelt, Dr.A. Ljungdahl and Dr. O. Johansson; to Dr. C.A. Greer for valuable suggestions and reading the manuscript; to Etelka Mitrov and Heather Cameron for assistance with illustrations.

REFERENCES

Allison AC (1953) The morphology of the olfactory system in the vertebrates. Biol Rev 28:195-244

Altman J (1969) Autoradiographic and histological studies of postnatal neurogenesis. IV.Cell proliferation and migration in the anterior forebrain, with special reference to persisting neurogenesis in the olfactory bulb. J Comp Neurol 137:433-458

Andres KH (1965) Der Feinbau des Bulbus olfactorius der Ratte unter besonderer Beruecksichtigung der synaptischen Verbindungen. Z Zellforsch 65:530-561

Andres KH (1970) Anatomy and ultrastructure of the olfactory bulb in fish, amphibia, reptiles, birds and mammals. In: Wolstenholme GEW, Knight J (eds) Ciba Found.Symp. Taste and Smell in Vertebrates. Churchill, London, p177

Baker H (1988) Neurotransmitter plasticity in the juxtaglomerular cells of the olfactory bulb. In Margolis FL, Getchell TV (eds) Molecular Neurobiology of the Olfactory System. Plenum, New York London, p185

Blakely RD, Ory-Lavollee L, Grzanna R, Koller KJ, Coyle JT (1987) Selective immunocytochemical staining of mitral cells in rat olfactory bulb with high affinity purified antibodies against N-acetyl-aspartyl-glutamate. Brain Res 402:373-378

Cajal SR (1911) Histologie du systeme nerveux. Maloine, Paris

Dahlstrom A, Fuxe K, Olson L, Ungerstedt U (1965) On the distribution and possible function of monoamine nerve terminals in the olfactory bulb of the rabbit. Life Sci 4:2071-2074

Golgi C (1875) Sulla fina struttura dei bulbi olfattorii. In: Opera omnia, Vol I: Istologia normale. Ulrico Hoepli, Milano Greer CA (1987) Golgi analyses of dendritic organization among denervated olfactory bulb granule cells. J Comp Neurol 257:442-452

Halasz N (in press) The vertebrate olfactory system. Chemical neuroanatomy, function and development. Akad.Kiado, Budapest

Halasz N, Ljungdahl A, Hokfelt T (1978) Transmitter histochemistry of the rat olfactory bulb. II.Fluorescence histochemical, autoradiographic and electron microscopic localization of monoamines. Brain Res 154:253-271

Halasz N, Ljungdahl A, Hokfelt T (1979) Transmitter histochemistry of the rat olfactory bulb. III.Autoradiographic localization of (3H)GABA. Brain Res 167:221-240

Halasz N, Ljungdahl A, Hokfelt T, Johnsson O, Goldstein M, Park D, Biberfeld P (1977) Transmitter histochemistry of the rat olfactory bulb. I.Immunohistochemical localization of monoamine synthesizing emzymes. Support for intrabulbar, periglomerular dopamine neurons. Brain Res 126:455-474

Halasz N, Shepherd GM (1983) Neurochemistry of the vertebrate olfactory bulb. Neuroscience 10:579-619

Halasz N, Parry DM, Blackett NM, Ljungdahl A, Hokfelt T (1981) (3H)gamma-aminobutyrate autoradiography of the rat olfactory bulb: hypothetical grain analysis of the distribution of silver grains. Neuroscience 6:473-479

Kaplan MS, McNelly NA, Hinds JW (1985) Population dynamics of adult-formed granule neurons of the rat olfactory bulb. J Comp Neurol 239:117-125

Kishi K, Mori K, Ojima H (1984) Distribution of local axon collaterals of mitral, displaced mitral and tufted cells in the rabbit olfactory bulb. J Comp Neurol 225:511-526

Kosaka T, Hama K (1982-1983) Synaptic organization in the teleost olfactory bulb. J Physiol Paris 78:707-719

Kosaka T, Hataguchi Y, Hama K, Nagatsu I, Wu J-Y (1985) Coexistence of immunoreactivities for glutamate decarboxylase and tyrosine hydroxylase in some neurons in the periglomerular region of the rat main olfactory bulb: possible coexistence of gamma-aminobutyric acid (GABA) and dopamine. Brain Res 343:166-171

Liu C-j, Grandes P, Matute C, Cuenod M., Streit P (1989) Glutamate-like immunoreactivity revealed in rat olfactory bulb, hippocampus and cerebellum by monoclonal antibody and sensitive staining method. Histochemistry 90:427-445

Luskin MB, Price JL (1983) The topographic organization of associational fibers of the olfactory system in the rat, including centrifugal fibers to the olfactory bulb. J Comp Neurol 216:264-291

Macrides F, Davis BJ (1983) The olfactory bulb. In: Emson PC (ed) Chemical Neuroanatomy. Raven, New York, p391

Macrides F, Schneider SP (1982) Laminar organization of mitral and tufted cells in the main olfactory bulb of the adult hamster. J Comp Neurol 208:419-430.
McLean JH, Shipley MT (1987) Serotonergic afferents to the rat olfactory bulb. I.Origins and laminar specificity of serotonergic inputs in the adult rat. J Neurosci 7:3016-3028
McLean JH, Shipley MT (1988) Postmitotic, postmigrational expression of tyrosine hydroxylase in olfactory bulb dopaminergic neurons. J Neurosci 8:3658-3669
Mori K, Kishi K (1982) The morphology and physiology of the granule cells in the rabbit olfactory bulb revealed by intracellular recording and HRP reaction. Brain Res 247:129-133
Mori K, Kishi K, Ojima H (1983) Distribution of dendrites of mitral, displaced mitral, tufted, and granule cells in the rabbit olfactory bulb. J Comp Neurol 219:339-355
Mugnaini E, Wouterlood FG, Dahl A-L, Oertel WH (1984) Immunocytochemical identification of GABAergic neurons in the main olfactory bulb of the rat. Arch Ital Biol 122:83-113
Nickell WT, Shipley MT (1988) Two anatomically specific classes of candidate cholinoceptive neurons in the rat olfactory bulb. J Neurosci 8:4482-4491
Nieuwenhuys R (1967) Comparative anatomy of olfactory centers and tracts. Progr Brain Res 23:1-64
Orona E, Scott JW, Rainer EC (1983) Different granule cell populations innervate superficial and deep regions of the external plexiform layer in rat olfactory bulb. J Comp Neurol 217:227-237
Orona E, Rainer EC, Scott JW (1984) Dendritic and axonal organization of mitral and tufted cells in the rat olfactory bulb. J Comp Neurol 226:346-356
Pinching AJ, Powell TPS (1971a) The neuron types of the glomerular layer of the olfactory bulb. J Cell Sci 9:305-345
Pinching AJ, Powell TPS (1971b) The neuropil of the glomeruli of the olfactory bulb. J Cell Sci 9:347-377
Pinching AJ, Powell TPS (1971c) The neuropil of the periglomerular region of the olfactory bulb. J Cell Sci 9:379-409
Pinching AJ, Powell TPS (1972a) The termination of centrifugal fibers in the glomerular layer of the olfactory bulb. J Cell Sci 10:621-635
Pinching AJ, Powell TPS (1972b) Experimental studies on the axons intrinsic to the glomerular layer of the olfactory bulb. J Cell Sci 10: 637-655
Price JL, Powell TPS (1970a) The morphology of the granule cells of the olfactory bulb. J Cell Sci 7:91-123
Price JL, Powell TPS (1970b) The synaptology of the granule cells of the olfactory bulb. J Cell Sci 7: 125-155
Price JL, Powell TPS (1970c) An electron-microscopic study of the termination of the afferent fibers to the olfactory bulb from the cerebral hemisphere. J Cell Sci 7:157-187
Price JL, Powell TPS (1970d) The mitral and short axon cells of the olfactory bulb. J Cell Sci 7:631-651
Saito N, Kumoi K, Tanaka C (1986) Aspartate-like immunoreactivity in mitral cells of rat olfactory bulb. Neurosci Lett 65:89-93
Schneider SP, Macrides F (1978) Laminar distribution of interneurons in the main olfactory bulb of the adult hamster. Brain Res Bull 3:73-82
Schoenfeld TA, Macrides F (1984) Topographic organization of connections between the main olfactory bulb and pars externa of the anterior olfactory nucleus in the hamster. J Comp Neurol 227:121-135
Shepherd GM (1972) Synaptic organization of the mammalian olfactory bulb. Physiol Rev 52: 864-917
Shepherd GM (1979) The Synaptic Organization of the Brain. Oxford, New York.
Shute CCD, Lewis PR (1967) The ascending cholinergic reticular system: neocortical, olfactory and subcortical projections. Brain 90:497-520

Steinbusch HWM (1981) Distribution of serotonin-immunoreactivity in the central nervous system of the rat - cell bodies and terminals. Neuroscience 6:557-618

Watanabe K, Kawana E (1984) Selective retrograde transport of tritiated D-aspartate from the olfactory bulb to the anterior olfactory nucleus, pyriform cortex and nucleus of the lateral olfactory tract in the rat. Brain Res 296:148-151

White EL (1973) Synaptic organization of the mammalian olfactory glomerulus: new findings including an intraspecific variation. Brain Res 60:299-313

Zaborszky L, Carlsen J, Brashear HR, Heimer L (1986) Cholinergic and GABAergic afferents to the olfactory bulb in the rat with special emphasis on the projection neurons in the nucleus of the horizontal limb of the diagonal band. J Comp Neurol 243:488-509

Zheng LM, Ravel N, Jourdan F (1987) Topography of centrifugal acetylcholin-esterase-positive fibres in the olfactory bulb of the rat: evidence for original projections in atypical glomeruli. Neuroscience 23:1083-1093

OLFACTORY BULB PLASTICITY

Rémi Gervais, Anne-Marie Mouly,
Karim Elaagouby and Frédéric Lévy[1]

Laboratoire de Physiologie Neurosensorielle, UA CNRS 180
Université Claude Bernard, 69622 Villeurbanne, France

1) Laboratoire de Comportement Animal, INRA, 37380
Nouzilly, France

In neural network, potential for long-lasting modifications of functional characteristics is present. This capacity, known as neural plasticity, attracts attention of many neuroscientists. So there are numerous studies about the influence of pre- and post-natal experience on cellular differentiation and the formation of synaptic connexions as well as those about the neurobiology of learning and memory. Well-known examples of neural plasticity are the modifications in the visual cortex of kittens, appearing after a temporary unilateral eye closure (Wiesel and Hubel, 1963), and the long-term potentiation (LTP) in the hippocampus (Bliss and Lomo, 1973; Teyler and Di Scenna, 1987). In the latter case, early experiments showed that a brief high frequency stimulation of excitatory hippocampal afferences induces a long-lasting increase in synaptic transmission. Supported by findings of subsequent investigations, this form of cellular plasticity is now considered as a good candidate for transient storage of information. Surprisingly, even after extensive studies two major questions remain open. 1) Is the induction of hippocampal LTP required for acquisition?. 2) Is maintenance of the LTP necessary for retention of information? As it is not easy to find a learning paradigm in which hippocampal formation is specifically involved, these questions have been addressed in very few experiments. Recent advances suggest that hippocampus is involved in spatial memory (Barnes, 1988). In this context, some authors tried to correlate LTP and animal's performances (Mc Naugton et al., 1986; Morris et al., 1986). They found that normal improvement of animal's performances over days necessitates the development of LTP. Similar conclusions have been drawn from studies where electrical stimulation of the perforant path was used as a conditioning stimulus (Laroche et al., 1989). So, it took nearly 13 years before a direct link between

NATO ASI Series, Vol. H 39
Chemosensory Information Processing. Edited by D. Schild

hippocampal synaptic plasticity and acquisition of a behavioral response could be established.

Recent advances in the research of olfactory learning clearly show that neural plasticity is also present in olfactory structures. In newborn as well as in adult animals, long-lasting modifications of synaptic activity have been observed at the olfactory bulb level (Apfelbach and Wenler, 1985; Leon, 1987; Freeman and Schneider, 1982) and in the piriform cortex (Roman et al., 1987). In the olfactory bulb (OB), these experience-induced modifications of response to conditioning odor concern two layers : the glomerular layer, where neuroreceptor cells synapse with relay neurons, and the external plexiform layer (EPL) where relay neurons and deep interneurons establish reciprocal dendrodendritic synapses. Both layers receive central projections including those originating in the noradrenergic (NA) locus coeruleus and the cholinergic (Ach) basal forebrain. By this time, several experiments showed that centrifugal fibers are important in regulating OB responsiveness to odors according to their behavioral significance (Pager, 1986). Moreover, the central NA system is critical for the above mentioned olfactory plasticity (Gervais et al., 1988). In particular, long-lasting modifications of OB physiology were not observed when conditioning was performed after administration of a beta-adrenergic receptor antagonist, propranolol (Leon, 1987; Gray et al., 1986). This finding is in agreement with the current view that ascending NA fibres are not only involved in arousal and attentiveness (Aston-Jones, 1985) but also in cortical plasticity (Kasamatsu et al., 1984).

Olfaction is a very important sense in most animals used in experimental investigations. Thus studying olfactory learning offers the interesting possibility to test the importance of plasticity in a natural context. In the present paper we will focus attention on relation between olfactory learning and OB plasticity. We will summarize results from recents experiments concerning two questions:

First, does the central NA system, which seems to modulate OB plasticity, participate in olfactory learning? For that purpose, we examined the effect of OB beta-NA receptors blockade on olfactory learning. This had been done by using two different experimental models: the first one presents a form of imprinting in the context of sheep maternal behavior, while the second is based on a classical associative learning paradigm.

Second, does OB circuitry support LTP? In fact, if the OB makes part of the neural circuit underlying olfactory memory together with piriform cortex and hippocampus, it should similarly present the capability of storing information. In anesthetized rats, we have studied the effect of a short increase in extracellular calcium concentration on the excitatory synaptic transmission between olfactory relay neurons and granular cell interneurons.

OLFACTORY BULB PLASTICITY AND OLFACTORY LEARNING IN SHEEP

This experiment was carried out at the "Laboratoire de Comportement Animal (INRA, Nouzilly, France) in collaboration with Dr Frédéric Lévy.

After parturition, sheeps establishe a selective bound with their own lambs. Thus, from birth to weaning, the mother will provide maternal care exclusively to her own lamb and will reject any alien lamb. It is now well established that early recognition of lambs is exclusively based on olfactory cues. Moreover, olfactory learning is known to take place during the 2 to 4 hour period following parturition. A previous experiment investigated the importance of NA action on the OB for the establisment of maternal selectivity. Pissonnier et al (1985) were able to achieve selective NA depletion restricted to the OB by 6-hydroxydopamine injection into the medial olfactory stria. They found that when the level of NA depletion reached at least 70%, only 3 of 7 animals remained selective (43%). Since this very interesting observation is restricted to a small number of subjects, we decided further investigations. In the present experiment, NA transmission in the OB was altered by a transient blockade of beta-NA receptors using a local infusion of propranolol.

Eight weeks prior to parturition, the ewes (n=24) were cannulated bilaterally in the OB. While control animals received bulbar infusion of the drug vehicle (n=9), treated animals (n=15) received the same solution containing d-l propranolol-HCl (1 μM). Substances were infused via an osmotic minipump which delivered the solution continuously at a rate of 10 μl/h. Three days before parturition, the minipumps were implanted under the neck skin and connected to each bulbar cannula. At parturition the ewe and her offspring were left together for a 4-hour period to ensure that selectivity could take place. After separating the mother from her lamb for 2 hours, a test for maternal selectivity was performed. For that purpose, familiar and aliens lambs were given to the mother individually in a random order. A ewe was considered to provide non-selective care when one or more alien lambs were accepted at the udder for more than 10 seconds.

The results clearly showed that propranolol infusion was accompanied by a significant loss in selectivity. Indeed, only 7 of 15 propranolol-treated animals displayed selective maternal care (46%), while the majority of the control females remained selective (8 out 9; 89%, $p < 0.05$). One could argue that the effect of the propranolol infusion simply resulted from a sensorial deficit. Such interpretation is unlikely since another olfactory-guided behavior remained intact in treated animals. : As in normal animals, the odor of amniotic fluid still elicited a strong repulsion when presented 24 hours prior to parturition, and a strong attraction when presented within minutes following parturition (Lévy et al., 1985).

Thus, suppressing the action of the ascending NA system on the sheep OB either by depletion or blockade of beta-receptors induced a failure in establishing maternal selectivity. So, treated animals seem to be unable to learn the olfactory signature of their offspring. Similarly to what is known for visuocortical plasticity (Shirokawa and Kasamatsu, 1986) the activation of bulbar beta-receptors could be tightly associated with the induction of neural plasticity. Within the few hours following parturition, formation of an OB engram could represent one critical step for establishing maternal selectivity (Lévy et al., 1989).

OLFACTORY BULB PLASTICITY AND ASSOCIATIVE LEARNING IN RAT

Olfactory learning in the sheep occuring after parturition as well as olfactory learning in the mice following mating (Bruce, 1960; Keverne and De La Riva, 1982; Rosser and Keverne, 1985) is dependent on NA activation of the main and the accessory OB respectively. In both cases, learning occurs during a single 2 to 4 hours exposure to animal odors. This raises the question whether OB network is also involved in a more classical associative learning task which requires progressive improvement of performances over several days. In a serie of experiments, odors were replaced by patterned electrical stimulations of the OB and associated with reinforcement. Previous experiments showed that rats were able to solve the proposed task within a few days and remembered it for more than 2 weeks (Mouly et al., 1985; Mouly and Holley, 1986). In fact, rats learn and memorize information provided by electrical stimulations of the OB in a very similar fashion as they learn and memorize odors (Slotnick and Katz, 1974).

Eight stimulation electrodes were sticked regularly around a small cannula. The assembly was implanted unilaterally in the OB of adult male rats. The cannula was used for infusions of pharmacological agents near the electrode tips. Water-deprived animals were trained in the absence of infusion, for a first discrimination task (task 1; figure 1), where stimulation of a given pair of electrodes (example, electrodes 1+3) was associated with presentation of a palatable sucrose solution, whereas stimulation of another pair of electrodes (example, electrodes 2+4) signalled the presence of an unpalatable quinine solution. Daily sessions consisted of 12 trials. Under that condition, animals reached the learning criterion (at least 10 correct choices for 2 consecutive days) after 6 to 7 days of training. Then, animals were trained on a second discrimination task (task 2) under perfusion. Therefore, the intrabulbar cannula was connected to an osmotic minipump which continuously delivered a solution (0.5 μl/h) for a maximum of 14 days. Depending on the nature of the infused solution 3 experimental groups were defined:

group I: a propranolol solution (2×10^{-4} M)
group II: the vehicle (saline ascorbate 0.4%, pH 3)
group III: a standard saline solution (pH 7)

When animals reached the learning criterion for task 2, the training was interrupted for 5 days. Then, a retention test was performed.

The main results can be summarized as follows:

1) Perfusion of any of the three solutions during learning produced no detectable effects neither on animals perceptual abilities nor on learning velocity (figure 1). So, olfactory memory was not impaired during the daily training procedure.

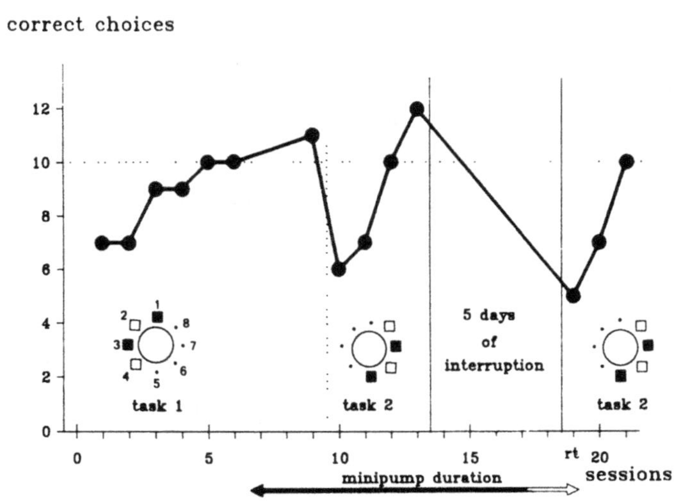

Figure 1 Effect of intrabulbar infusion of the saline-ascorbate solution on acquisition and retention of a discrimination task. The relative position of the stimulation electrodes is indicated in the lower part of the figure. The circle materializes the cannula. Full squares, stimulation associated with quinine; empty squares, stimulation associated with sucrose. In this exemple, the animal reached the learning criterion for task 1 within 6 days. The performance was not altered after the perfusion has been started for 2 days. Task 2 was learned normally while the OB was continously infused. However, after 5 days of training interruption, the animal's performance was severely impaired (rt, retention test). Finally, the animal was still able to relearn the task fairly rapidly.

2) Perfusion of propranolol or vehicle (groups I and II) during learning induced a deficit in the long-term retention of the task. This was shown by the fact that, when tested after 5 days of training interruption, animals made their choices randomly (figure 1). This deficit was not observed when a standard saline solution was perfused during training.
3) Perfusion of saline-ascorbate after completion of learning did not impair long-term retention.
4) The deficit in long-term retention cannot be ascribed to a specific blockade of beta-receptors. Most likely other effects of the ascorbic saline solution on bulbar synaptic activity should be taken into acccount. For instance, ascorbate could alter controls exerted on OB by central structures since it can modify the level of NA and acetylcholine release (Kuo et al., 1979) and induce a significant depletion of NA in rat OB (Doty et al., 1987).

In conclusion, these data bring further evidence for a an involvement of OB circuit in olfactory memory. Therefore, we suggest that the impairment of some form of neural plasticity <u>during</u> learning also impairs the long-term retention of the information (Mouly et al., 1989).

OLFACTORY BULB AND LONG-TERM POTENTIATION

The two behavioral studies we performed lead us to conclude that the OB is able to store information. So, one has to look for the neural basis supporting such a function. As stated in the introduction, LTP of excitatory neurotransmission is one of the best known property of neural tissue possibly involved in information storage. Physiological mechanisms underlying LTP have been extensively investigated (Gustafson and Wingström, 1988; Collingridge and Bliss, 1987). Induction of LTP involves the activation of N-methyl D-aspartate (NMDA) receptors, a subclass of excitatory amino acid receptors. Their activation leads to an important influx of Ca into the post-synaptic cell. The Ca entry triggers a sequence of events which culminates in enhanced synaptic efficacy. For that reason, artificial elevation of the extracellular and intracellular calcium concentration is sufficient to induce LTP in the hippocampus.

To our knowledge, LTP has not been demonstrated in the OB. In anesthetized rats we examined the effect of a transient increase in extracellular OB calcium concentration on the efficiency of the excitatory synaptic transmission between mitral and granular cell (Elaagouby and Gervais, 1989). This synaptic transmission was evaluated from evoked field potential obtained after antidromic stimulation of the lateral olfactory tract (Rall and Shepherd, 1968; Mori, 1987). The EPL region was perfused by using push-pull cannula which was coupled with recording electrodes. For 10 minutes the calcium concentration of the standard Krebs perfusion medium was elevated to 10 mM. In doing that, we observed an important increase in the amplitude of the evoked potential. However, this effect usually vanished within one hour. An addition of biccuculine (0.1 mM) to the medium with the high calcium concentration induced a long lasting potentiation of the mitral to granular cell transmission. The potentiation lasted for several hours (figure 2). Similarly to hippocampal and neocortical LTP, the bulbar LTP did not develop in presence of 2-amino-5-phosphonovaleric acid (APV), a specific NMDA receptor antagonist.

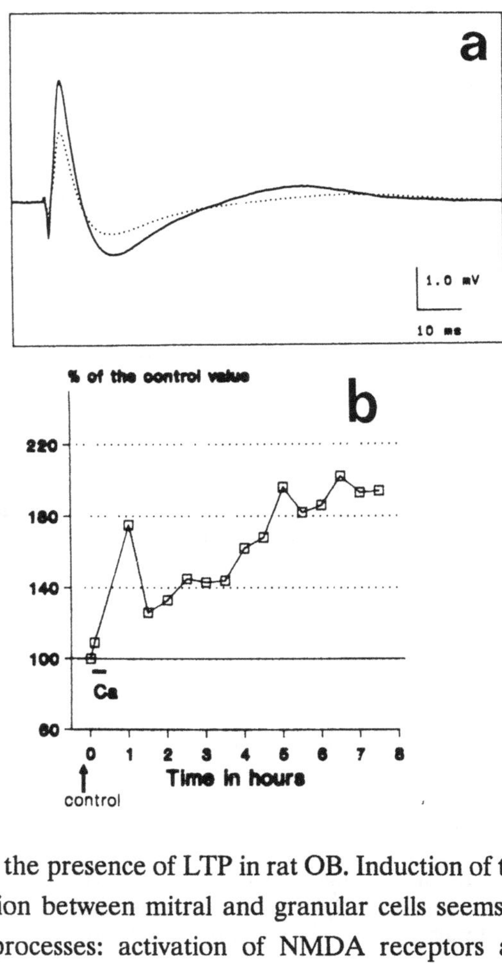

Figure 2 Evidence for long-term potentiation in rat OB. a, exemple of field potential recorded in the granular cell layer following stimulation of the lateral olfactory tract (0.1 ms, 4 times threshold). The reponse to 10 single stimulations were averaged by using microcomputer. Dotted line, control response; Full line, response obtained 1 hour after the introduction for 10 minutes of the high calcium solution containing bicuculline. Both responses were recorded while OB was perfused by the standard Krebs solution. Positivity upward. b, amplitude of the positive wave recorded throughout the experiment. In this case, the treatment induced a large increase in response amplitude which lasted for several hours.

So, these data provide first evidence for the presence of LTP in rat OB. Induction of the enhanced excitatory synaptic transmission between mitral and granular cells seems to be dependent on two simultaneous processes: activation of NMDA receptors and reduction of GABAergic inhibition. These characteristics are very similar to those governing LTP in rat visual cortex (Artola and Singer, 1987).

CONCLUDING REMARKS

The data which we have summarized support the view that the mammalian OB has to be integrated in forebrain circuits underlying olfactory memory. This strenghen the present interest in olfaction as a model for studying the neurobiology of learning and memory (Lynch, 1986) It also fits well in the emerging view that information storage is tied to the specific processing areas that are engaged during learning (Squire, 1986). One of our working hypothesis is that the action of central structures including the NA and the cholinergic systems authorizes plasticity in those OB networks coding biological relevant odors. Centrifugal controls could contribute to a reduction of powerfull GABAergic inhibition (Jahr and Nicoll, 1982), thus promoting induction of LTP in the OB, as well as in piriform cortex and hippocampus. This is in accordance with recent observations showing that intraventricular administration of APV (which blocks induction of LTP in OB and hippocampus) impairs acquisition but not retention of olfactory memory (Staubli et al., 1989). Several experiments remain to be done in order to understand how the OB is integrated in the processes leading to storage of olfactory information.

ACKNOWLEDGEMENT - The authors thank Dr. U. Kindermann for valuable discussion and improvement of the final version of the manuscript.

REFERENCES

Apfelbach R, Wenler E (1985) Olfactory deprivation enhances normal spine loss in the olfactory bulb of developping ferrets. Neurosci Letters 62 : 169-173

Artola A, Singer W, (1987) Long-term potentiation and NMDA receptors in rat visual cortex. Nature 330 : 649-652

Aston-Jones G, (1985) Behavioral functions of locus coeruleus derived from cellular attributes. Physiol Psychol 13 : 118-126

Barnes CA (1988) Spatial learning and memory processes : the search for their neurobiological mechanisms in the rat. TINS 11 : 163-169

Bliss TVP, Lomo T (1973) Long-lasting potentiation of synaptic transmission in the dentate gyrus of the anesthetized rabbit following stimulation of the perforant path. J Physiol (London) 232 : 331-356

Bruce HM, (1960) A block to pregnancy in the mouse caused by proximity to strange males. J Reprod Fertil 1 : 96-103
Collingridge GL, Bliss TVP (1987) NMDA-receptors - their role in long-term potentiation. TINS 10 : 228-293
Doty RL, Fergusson-Segall M, Lucki I, Kreider M (1987) Effects of intrabulbar injections of 6-hydroxydopamine on ethyl acetate odor detection in castrate and non-castrate male rats. Brain Res 444: 95-103
Elaagouby K, Gervais R (1989) Evidence for long-term potentiation in rat olfactory bulb (in preparation)
Freeman WJ, Schneider W (1982) Changes in spatial patterns of rabbit olfactory EEG with conditioning to odors. Psychophysiol 19 : 44-56
Gervais R, Holley A, Keverne, B (1988) The importance of central noradrenergic influences on the olfactory bulb in the processing of learned olfactory cues. Chem Senses 13 : 3-12.
Gray C, Freeman WJ, Skinner, JE (1986) Chemical dependencies of learning in the rabbit olfactory bulb : acquisition of the transient spatial-pattern change depends on norepinephrine. Behav Neurosci 100 : 585-596
Gustafson B, Wingström H (1988) Physiological mechanisms underlying long-term potentiation. TINS 11 : 156-162
Jahr CE, Nicoll RA (1982) Noradrenergic modulation of dendrodendritic inhibition of the olfactory bulb. Nature 297 : 227-228
Kasamatsu T, Itakura T, Jonnson G, Hegglund P, Pettigrew JD, Nakai K, Kuppermann BD, Ary M (1984) Neural plasticity in cat visual cortex : a proposed role for central noradrenaline system, In Descaries L, Reader TR, Jasper HH (eds) Monoamine innervation of cerabral cortex, AR Liss Press, New York, p 301
Keverne EB, De La Riva C, (1982) Pheromones in mice : reciprocal interaction between the nose and the brain. Nature 296 : 148-150
Kuo CH, Hata F, Yoshida H, Yamatodani A, Wada H (1979) Effects of ascorbic acid on release of acetylcholine from synaptic vesicles prepared from different species of animals and release of noradrenaline from synaptic vesicles of rat brain. Life Sci 24 : 911-916
Laroche S, Doyere V, Bloch V (1989) Linear relation between the magnitude of long-term potentiation in the dentate gyrus and associative learning in the rat. A demonstration using commissural inhibition and local infusion of n-methyl D-aspartate antagonist. Neurosci 28 : 375-386
Leon M (1987) Plasticity of olfactory output circuits related to early olfactory learning. TINS 10 : 434-438
Lévy F, Poindron P, Le Neindre P (1983) Attraction and repulsion by amniotic fluids and their olfactory control in the ewe around parturition. Physiol Behav 31 : 687-692
Lévy F, Gervais R, Kindermann U, Orgeur P, Pikety V (1989) Importance of beta-noradrenergic receptors in the olfactory bulb of sheep for recognition of lambs. (submitted for publication)
Lynch G (1986) Synapses, circuits and the beginnings of memory.,The MIT Press, Cambridge (Mass)
Mc Naugton BL, Barnes GA, Rao G, Baldwin J, Rasmussen M (1986) Long-term enhancement of hippocampal synaptic transmission and the acquisition of spatial memory. J Neurosci 6 : 563-571
Mori K (1987) Membrane and synaptic properties of identified neurons in the olfactory bulb. Progr Neurobiol 29 : 275-320
Morris RGM, Anderson F, Lynch G, Baudry M (1986) Selective impairment of learning and blockade of long-term potentiation by an N-methyl D-aspartate receptor antagonist, AP5. Nature 319 : 774-775
Mouly AM, Vigouroux M, Holley A (1985) On the ability of rats to discriminate between microstimulations of the olfactory bulb in different locations. Behav Brain Res 17 : 45-58

Mouly AM, Holley H (1986) Perceptive properties of the multi-site electrical microstimulation of the olfactory bulb in the rat. Behav Brain Res 21 : 1-12

Mouly AM, Gervais R, Holley A (1989) Neural plasticity and olfactory learning : a study with the model of multisite electrical stimulation of the olfactory bulb. (submitted for publication)

Pager J (1986) Neural correlates of odor-guided behaviors. Experientia 42 : 250-256

Pissonnier D, Thierry JC, Fabre-Nys C, Poindron P, Keverne EB (1985) The importance of olfactory bulb noradrenaline for maternal recognition in sheep. Physiol Behav 35 : 361-363

Rall W, Shepherd GM (1968) Theoretical reconstruction of field potentials and dendrodendritic synaptic interactions in the olfactory bulb. J Neurophysiol 31 : 884-915

Roman F, Staubli U, Lynch G (1987) Evidence for synaptic potentiation in a cortical network during learning. Brain Res 418 : 221-226

Rosser AE, Keverne EB (1985) The importance of central noradrenergic neurones in the formation of an olfactory memory in the prevention of pregnancy block. Neurosci 15 : 1141-1148

Shirokawa T, Kasamatsu T (1986) Concentration-dependent suppression of beta-adrenergic antagonists of the shift in ocular dominance following monocular deprivation in kitten visual cortex. Neurosci 18 : 1035-1046

Slotnick BM, Katz HM (1974) Olfactory learning set formation in rats. Science 185 : 796-798

Squire LR (1986) Mechanism of memory. Science 232 : 1612-1619

Staubli U, Thibault O, DiLorenzo M, Lynch G (1989) Antagonism of NMDA receptors impairs acquisition but not retention of olfactory memory. Behav Neurosci 103 : 54-60

Teyler TJ, Di Scenna P (1987) Long-term potentiation. Ann Rev Neurosci 10 : 131-161

Wiesel TN, Hubel DH (1963) Single-cell response in striate cortex of kittens deprived of vision in one eye. J Neurophysiol 26: 1003-1017

OLFACTORY BULB AND ANTENNAL LOBE

J. Boeckh, P. Distler, K.D. Ernst, M. Hösl, D. Malun

Institute for Zoology
University of Regensburg

INTRODUCTION

At first glance it may appear unrewarding to compare the functional morphology of two nerve centres so different in origin and construction as the vertebrate olfactory bulb and the antennal lobe of insects. Indeed, viewed as a venture in phylogenetically based comparative neuroanatomy, it does not even seem legitimate: one is a typical ganglion in the ropeladder nervous system of an arthropod, and the other is part of the cortex of the telencephalon of a vertebrate. However, each of these centres is the first central station in the olfactory pathway of animals on a par with one another in their ability to detect, discriminate and identify even complex scents, and with equivalent skills in locating and seeking out odour sources and in learning odours. It would be interesting to know whether these achievements are based on similar neural functions in the two animal groups. A number of questions arise. Do the two systems employ common priniciples of signal processing, based on similarities in function and connections of the neurons? Do such similarities extend to specific circuitry and functional types of neurons? Are the kinds of performance for which mammals require many million receptor cells and neurons actually the same as those that insects manage with a few hundred thousand receptor cells and a few thousand neurons?

With such questions in mind, characteristic features of the olfactory bulb and of the antennal lobe are here presented side by side. The presentation is by no means complete; more information is available in excellent reviews by experts (Shepherd, 1972; Holley and MacLeod, 1977; Christensen and Hildebrand, 1987; Kauer, 1987; Mori, 1987; Scott and Marrison, 1987; Rospars, 1988). The intention here is to compare the distinctive modes of connectivity and processing by reference to selected details. Furthermore, because other papers in this symposium provide many data on the

NATO ASI Series, Vol. H 39
Chemosensory Information Processing. Edited by D. Schild

olfactory system of vertebrates, it seems useful to concentrate here on results obtained in particular insects (including the American cockroach and the tobacco moth), referring to what is known about various vertebrates where it is relevant.

Terms that recur frequently in the text are abbreviated as follows: bulb = olfactory bulb; GC = granule cell; IN = local interneuron; lobe = antennal lobe of the deutocerebrum; MG = macroglomerulus; PG = periglomerular neuron; PN = projection neuron; RN = olfactory receptor neuron.

THE INPUT: RECEPTORS AND THEIR DISTRIBUTION IN THE SENSE ORGAN

In insects the olfactory receptor neurons (RNs) are enclosed in so-called sensilla-cuticular hairs, plates or cones. Most of these are located on the antennae; the receptor axons leaving them form the antennal nerve, which runs into the antennal lobe of the ipsilateral deutocerebrum (Fig. 1). Some of the RNs on other head appendages terminate in other ganglia. In mammals the RNs lie in the olfactory epithelium of the nasal mucosa or of the vomeronasal organ, and their axons pass to the medial or accessory ipsilateral bulb of the forebrain. The number of olfactory receptors has been found to be as large as 10^8 in vertebrates and up to 3×10^5 in insects. In all cases they are (primary) sensory nerve cells - which is an exception among vertebrate receptors. Whereas those of vertebrates are short-lived and are continuously being replaced by new cells, the insect RNs persist even through molts.

Within the olfactory organ, the small- and large-scale distribution of receptor cells with various odour specificities differs somewhat in the two animal groups. Local electro-olfactograms in a number of amphibian species have shown a tendency for receptor cells maximally sensitive to particular odour substances to be grouped together in, though not restricted to, certain regions of the mucosa (Mustaparta, 1971; MacKay-Sim et al., 1982; van Aas et al., 1985). Some insect antennae bear fields of homogeneous sensilla (Boeckh, 1962; Stocker et al., 1983), but usually sensilla of different morphological and hence physiological types are intermingled (Fig. 2, see, e.g., Schneider and Kaissling, 1957; Schaller, 1978; Sass, 1980).

The peripheral encoding of olfactory stimuli consists, in principle, in the formation of odour-specific across-fibre patterns by receptors with overlapping odour spectra (Fig. 2). In some insects distinct populations of receptors have been found, within each of which the cells respond identically, and each of these "receptor types" is situated in a particular morphological type of sensillum (cf. Kaissling, 1970; Sass, 1980; Altner and Prillinger, 1982). The cells give a simple tonic response to low- intensity stimuli and phasic to phasic-tonic responses to stronger stimuli. Inhibition is rarely observed.

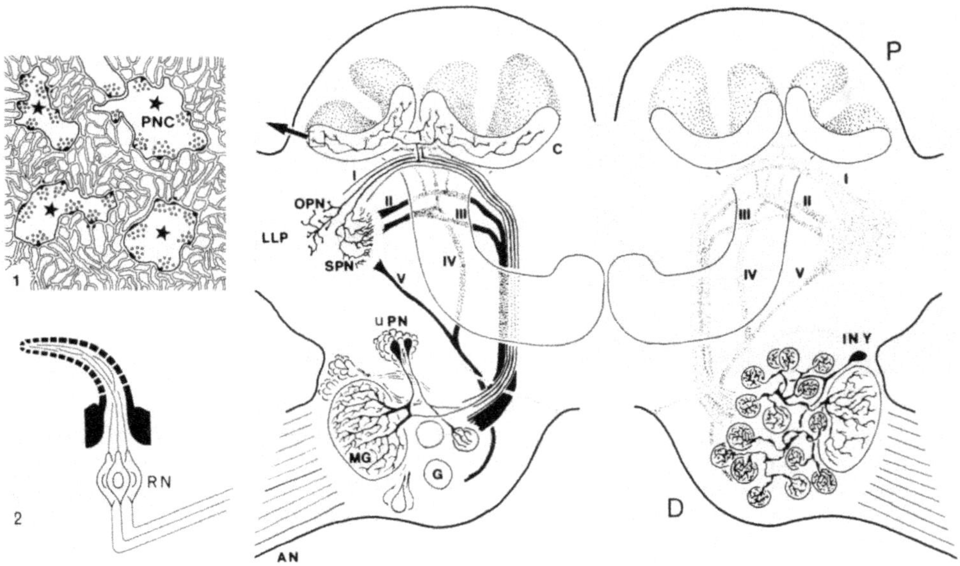

Fig. 1. Olfactory pathway of male Periplaneta americana. AN, antennal nerve; C, calyx of mushroom body with PN-projections; D, deutocerebrum; G, glomerulus; IN Y, multiglomerular local interneuron; LLP, lobus lateralis protocerebralis with PN-projections; MG, male specific macroglomerulus; OPN, olfactory PN not involved in female pheromone processing; P, protocerebrum; uPN, soma group of uniglomerular projection neurons with axons in I or II; PNC (inset 1), profiles of PN-terminals in C (note numerous output synapses); RN, antennal receptor neurons with axons in AN; SPN, terminals of pheromone-sensitive MG-PNs; I, II, tracts containing axons of uniglomerular PNs; III, IV other subtracts; V, tract containing axons of multiglomerular projection neurons. Inset 1, terminals of PN in C. Inset 2, antennal olfactory sensillum with RNs.

Although their spectra overlap broadly, the individual types are readily identifiable (see, e.g., Sass, 1978; Hartmann, 1987). In these animals, then, there are many receptors that respond to particular odours with predictable activity patterns. For some receptor types known to be associated with certain types of sensilla, histological methods can be used to determine their central projections and thus the input site of their activity pattern in the antennal lobe (see below). In vertebrates the number of differently tuned sense cells is enormous, the odour spectra of the cells overlap very broadly, and the time courses of the responses are extremely variable and complex. For these reasons it is practically impossible, for a given odour stimulus, to predict the particular pattern of excitation it will generate or to determine the site in the olfactory bulb that will receive this pattern (cf. Revial et al., 1982, 1983). Input/output functions are therefore difficult to formulate for the bulb.

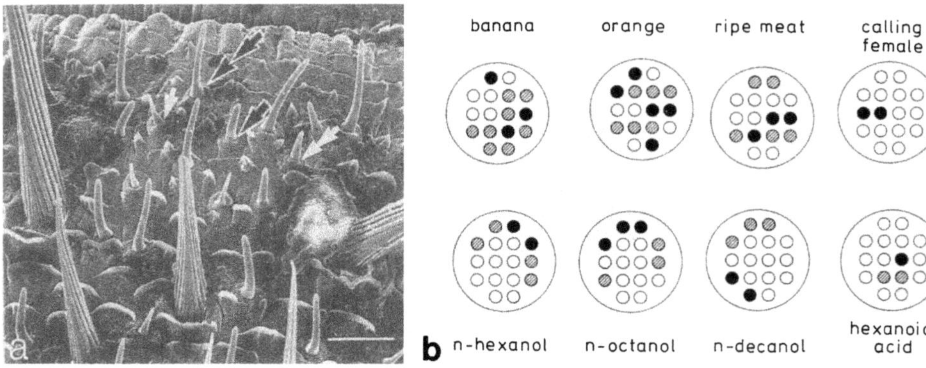

Fig. 2. Olfactory receptor organs of male Periplaneta americana ♂. a, part of surface of antennal segment with representatives of various sensillum- and receptor cell-types, identified in actual electrophysiological experiments and subsequent electron microscopy. Most effective stimuli indicated in brackets. Receptor cells in sensillum distinguished by different impulse amplitude and -shape. Black arrow, swA-type sensillum with 2 cells (certain alcohols, esters, fruit aromata); double black arrow, swB-type sensillum with 4 cells (pheromones); white arrow, dw-type sensillum with 3 cells (certain fatty acids, alcohols, "food"- aromata); double white arrow, dw-type sensillum with 4 cells (certain fatty acids, -aromata, 1 cell cold-sensitive), scale 20 μm (after Sass, 1976). b, responses of altogether 16 receptor types (each represented by circle) to aromata and compounds. Filled circle, more than 50% max. response; hatched circle, 25- 50% max. response; open circle, less than 25% max. response. (After data of Sass, 1976, 1978, and Selzer, 1984).

BULB AND LOBE

A. The glomeruli and the projections of the receptor cells

The bulb, with its layered arrangement, and the lobe, with somata enclosing a central neuropil, undoubtedly represent two quite different principles of nerve-centre construction (Fig. 3). A feature typical of both, however, is the presence of glomeruli, formed by the endings of the RN axons together with the processes of a certain set of interneurons. Indeed, glomeruli are characteristic structures in the first central station of the olfactory pathway of many crustaceans, molluscs and annelids as well as insects and vertebrates (cf. Cajal, 1911; Hanström, 1928). They are both morphological and functional subunits of the olfactory system (Leveteau and MacLeod, 1966; Shepherd,

1981; Ernst and Boeckh, 1983). Glomeruli are densely packed knots of very fine neuropil; in vertebrates they are separately ensheathed in glia and contain no neuron somata (Pinching and Powell, 1971). Each glomerulus receives the axons of a combination of RNs, which probably make the same pattern of connections with central neurons in all glomeruli of a given species. The central neurons, in turn, inhibit and excite one another by way of a connectivity pattern that is uniform within a species. In some insects and in mammals each glomerulus contains processes of one or more "private" projection neurons (PNs) and is thus represented at its own special site in the secondary projection regions of the olfactory pathway. The extent to which subdomains are produced within the glomeruli, by partial innervation of RNs, INs or PNs, is unclear (see below).

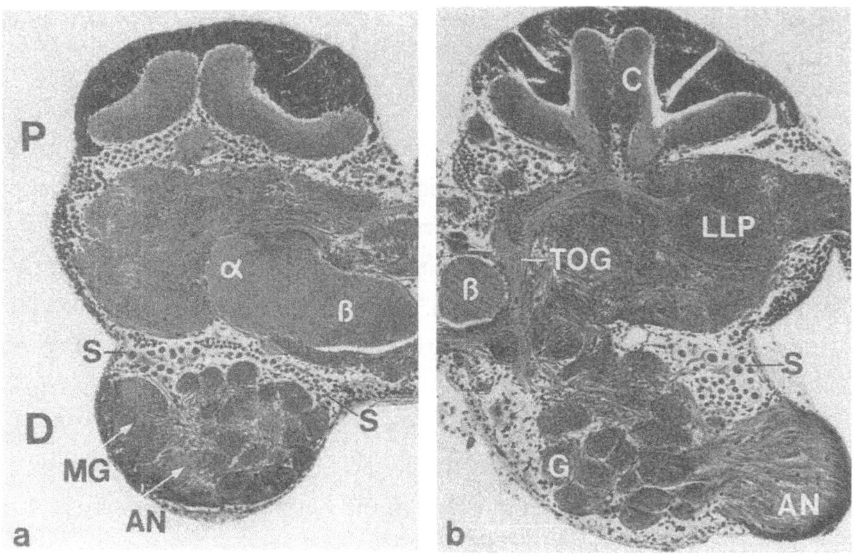

Fig. 3. Brain of Periplaneta americana ♂ in fronto-horizontal section, Bodian stain. a, superficial (ventral) area, b, deep (dorsal) area. AN, antennal nerve; C, calyx of corpus pedunculatum; D, deutocerebrum; G, glomerulus; α, β, α-β-lobes of corpus pedunculatum; LLP, lobus lateralis protocerebrals; MG, male specific macroglomerulus; P, protocerebrum; S, neuron somata; TOG, branch I of tractus olfactorio-globularis. Scale 100 μm.

Because there are only ten to hundreds of glomeruli in insects, it was easy to determine that their number and arrangement in the antennal lobe are species-specific (Rospars and Chambille, 1981; Rospars, 1983). They can be individually identified by their morphology, localized, and functionally characterized by electrophysiological studies of their PNs (and INs) (see below). The only glomerulus of known function so far identified in vertebrates is the MGC (modified glomerular complex) of young rats, which is clearly and selectively labelled with 2-deoxyglucose (2-DG) following presentation of the "maternal suckling pheromone" released by the teats of lactating females (Teicher et al., 1980). The size of the glomeruli is about the same in vertebrates as in insects (except for the several thousand very small glomeruli found in grasshoppers and wasps). In the vertebrates in which they have been counted there are about 2×10^3 glomeruli, whereas the number in insects ranges from 10 to 10^2.

In insects and mammals, the axon of a given RN branches in only one glomerulus (Fig. 4). Each axon makes an unknown number of synaptic contacts, but probably synapses with the processes of several neurons. In some cases the axon passes through the entire glomerulus; in others, it reaches only small regions. Electron micrographs show that the typical connection is dyadic, with one presynaptic RN profile and two postsynaptic profiles. Efferent innervation of receptor axons has been demonstrated in insects but not in vertebrates.

The central projections of the receptors exhibit both diveregence and convergence. Tracer and degeneration studies in vertebrates have shown that defined areas in the sensory epithelium project to large regions of the bulb, partially overlapping the projections of other areas. Some glomeruli within such a region are very densely innervated, while immediately adjacent glomeruli receive a comparatively sparse projection from the same area of mucosa. Conversely, a given glomerulus can receive inputs from large mucosal areas, and in some cases relatively small parts of such an area are massively represented in the glomerulus (Land, 1973; Costanzo and O'Connell, 1980; Greer et al., 1981; Kauer, 1981; Clancy et al., 1985; Mori et al., 1985). In bee-workers, each of the 11 antennal segments projects to every glomerulus in the antennal lobe, the distal segments being represented at the periphery of the glomerulus and the proximal segments at deeper levels (Pareto, 1972).

On average, in both insects and mammals, a glomerulus receives about 10^3 to 10^4 olfactory-cell axons. There is an exception, a particularly large glomerulus found in the deutocerebrum of the males of many insect species (Bretschneider, 1924; Jawlowski, 1948; Koontz and Schneider, 1987). The volume of this "macroglomerulus" (or macroglomerular complex) equals that of about 5-10 ordinary glomeruli. In American cockroach it contains the terminals of at least 70,000 axons of receptors specifically tuned to female pheromone (Schaller, 1978; Sass, 1980; Boeckh et al., 1984), and numbers of the same order of magnitude apply to other insects. No corresponding data are available for the MGC complex in the rat bulb.

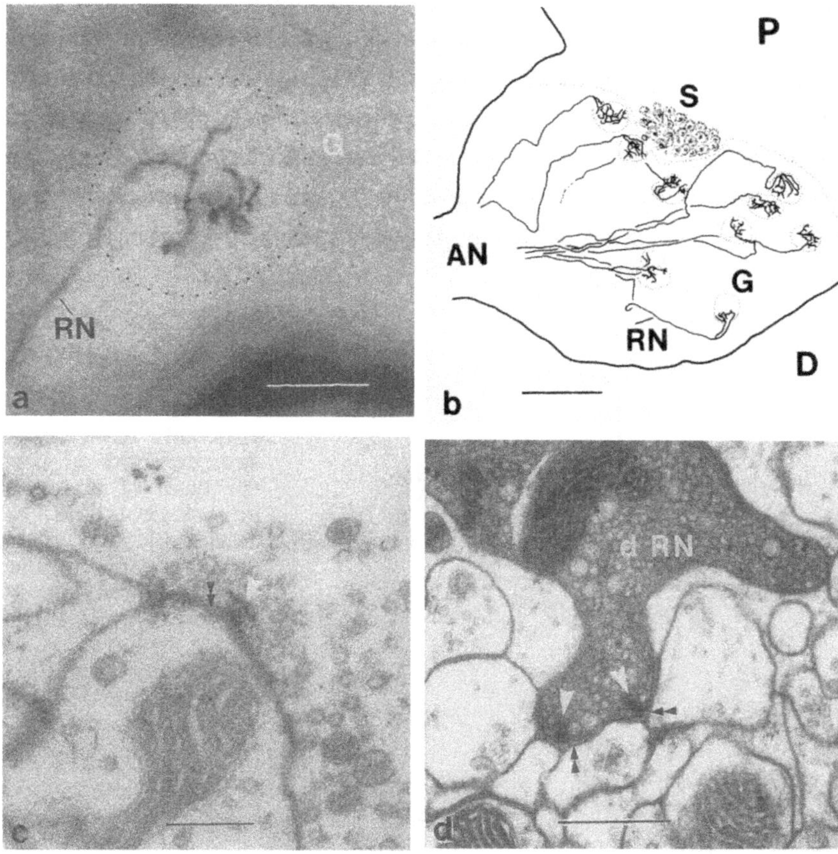

Fig. 4. Periplaneta americana ♂, central projection of antennal receptor axons in deutocerebral glomeruli. a, ending of cobalt- filled axon (RN) from cell underneath swB-type sensillum in glomerulus (dotted outline, G). b, 10 out of altogether 19 glomeruli innervated by RNs underneath swB-type sensilla. AN, antennal nerve; D, deutocerebrum; G, glomerulus; P, protocerecbrum; RN, receptor axon; S, neuron somata. Combined from individual preparations. c, synapse in glomerular neuropil between non identified profiles. d, dRN, RN, labelled by induced anterograde degeneration with 2 dyadic contacts to altogether 3 non identified profiles. White arrowheads, presynaptic densities; dark double arrowheads, synaptic clefts. Scales a, 50 μm; b, 100 μm; c, 0.25 μm; d, 0,4 μm.

Receptors of other types are distributed over several glomeruli (see also Stocker et al., 1983), so that different glomeruli receive input from different combinations of receptor types (Fig. 5, see also Selsam, 1987). In mammals, by means of the 2-DG method, it is possible clearly to distinguish small groups of glomeruli that are specially active when certain odours are presented (Steward et al., 1979; Jourdan et al., 1980; Hudson and Distel, 1988), which implies that RNs responsive to the odour concerned terminate in these glomeruli. But it should be kept in mind that stimulation with a given odour results in labelling at several such sites, sometimes widely separated from one another - which is in turn consistent with a divergent projection of particular regions of mucosa into the bulb (see above). Except in the case of the rat MGC, no one has demonstrated a clear association, constant in different individuals, between topographic activity distributions and specific odours. Experiments with 2-DG autoradiography in Diptera have given results that are basically similar (Rodrigues and Buchner, 1984). The local representation of specific odours can be tested in insects by recording from projection neurons of selected glomeruli.

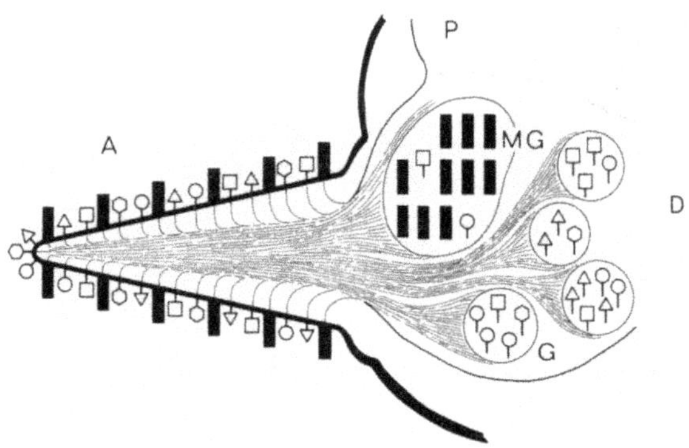

Fig. 5. Periplaneta americana ♂. Proposed central representation of different antennal RN-types in the deutocerebral glomeruli on the basis of anatomical and physiological data. Note the focal projection of receptors for female pheromone (filled columns) in MG, and "mixed" projections of other types (other symbols). A, antenna; D, deutocerebrum; G, glomerulus; MG, macroglomerulus; P, protocerebrum.

B. Glomerular circuitry and local interneurons

What elements receive the inputs that converge in a glomerulus, and how are particular stimulus parameters extracted? Is it possible to formulate input/output relationships, so as to draw conclusions about processing modes?

There are major differences in shape and arrangement of neurons in the bulb and in the lobe. As part of the telencephalic cortex, the bulb of tetrapods consists of discrete layers, within which the PNs (mitral cells and axon bearing tufted cells) synapse with RNs and interact with certain types of local interneurons, the periglomerular neurons (PGs) and granule cells (GCs). This characteristic manner of interaction is understood in principle; it involves longer or shorter inhibitory feedback loops. The interneurons, often GABAergic and in some cases also dopaminergic, mediate horizontal communication between neighboring glomeruli (or their PNs). It is also within these layers that centrifugal, probably modulatory efferents act (see the reviews listed above).

The antennal lobe, by contrast, is arranged in the way typical of an insect hemiganglion, with a peripheral soma region and an interior neuropil. The ventral neuropil region (the only region to be considered here) is partially organized into glomeruli, a specialization found only in the deutocerebrum. In holometabolous insects the glomeruli form a shell enclosing the central, coarse-fibred neuropil. The somata are assembled into characteristic clumps in several regions of the lobe, depending on the insect group (Fig. 3). Some of their processes pass through the neuropil in well-defined strands, with few if any synaptic contacts, and are distributed among the glomeruli. Insects are quite unlike vertebrates in that practically all the chemical synapses in the lobe are concentrated in the glomeruli (Schürmann and Wechsler, 1969; Boeckh et al., 1970; Tolbert and Hildebrand, 1981).

The connections of the RNs in the lobe also differ distinctly from those in the bulb (Figs. 6,7,8,9). The PN dendrites in the bulb receive massive monosynaptic excitatory inputs from RNs (Yamamoto et al., 1963; Andres, 1965; Mori and Takagi, 1975). There are also receptor inputs to PGs in the same glomerulus (Pinching and Powell, 1971; White 1972), which contribute to the inhibitory part of the spectrum and of the "receptive field" of the PN (cf. Kauer, 1987). In insects, all the available evidence indicates that monosynaptic connections of RNs to interneurons (INs) predominate (Figs. 6,8; Boeckh et al., 1989; Distler, 1989a), whereas monosynaptic RN-to-PN contacts are the exception (Fig. 7; Malun, 1989). The PNs have longer latencies than the INs and exhibit polysynaptic PSPs following electrical stimulation of the antennal nerve or stimulation of the antennae with odours, both of which argue against a monosynaptic RN-PN connection (Waldow, 1975; Christensen and Hildebrand, 1987). As far as the RN-IN-PN route is concerned, so far there are only suggestive indications (Fig. 8). Evidently, though, impulse activity of the receptors both excites PNs and - perhaps by way of other INs - inhibits them (see below). The net result is that the modes of activation of the PNs by the receptor inputs are similar in the lobe and the bulb.

Like the GCs of the bulb, the local interneurons of the lobe lack axons. Again like the GCs, they generate impulses, but, different to them, each usually innervates several to many glomeruli (Figs. 6,9). Many different branching and innervation patterns of INs have been described, and in Manduca, Periplaneta and Apis their responses to odourous or mechanical stimulation of the antenna have been studied (Matsumoto and Hildebrand, 1981; Ernst and Boeckh, 1983; Boeckh et al., 1983; Flanagan and Mercer, 1989; see also Christensen and Hildebrand, 1987). Both the INs and the PNs of the lobe in general respond only to ipsilateral inputs, an exception has been described by Kanzaki et al. (1989). Communication between the olfactory pathways on the two sides would be expected - for instance, to mediate left-right comparisons in osmotropotaxis - but it appears that such connections are made at a higher level.

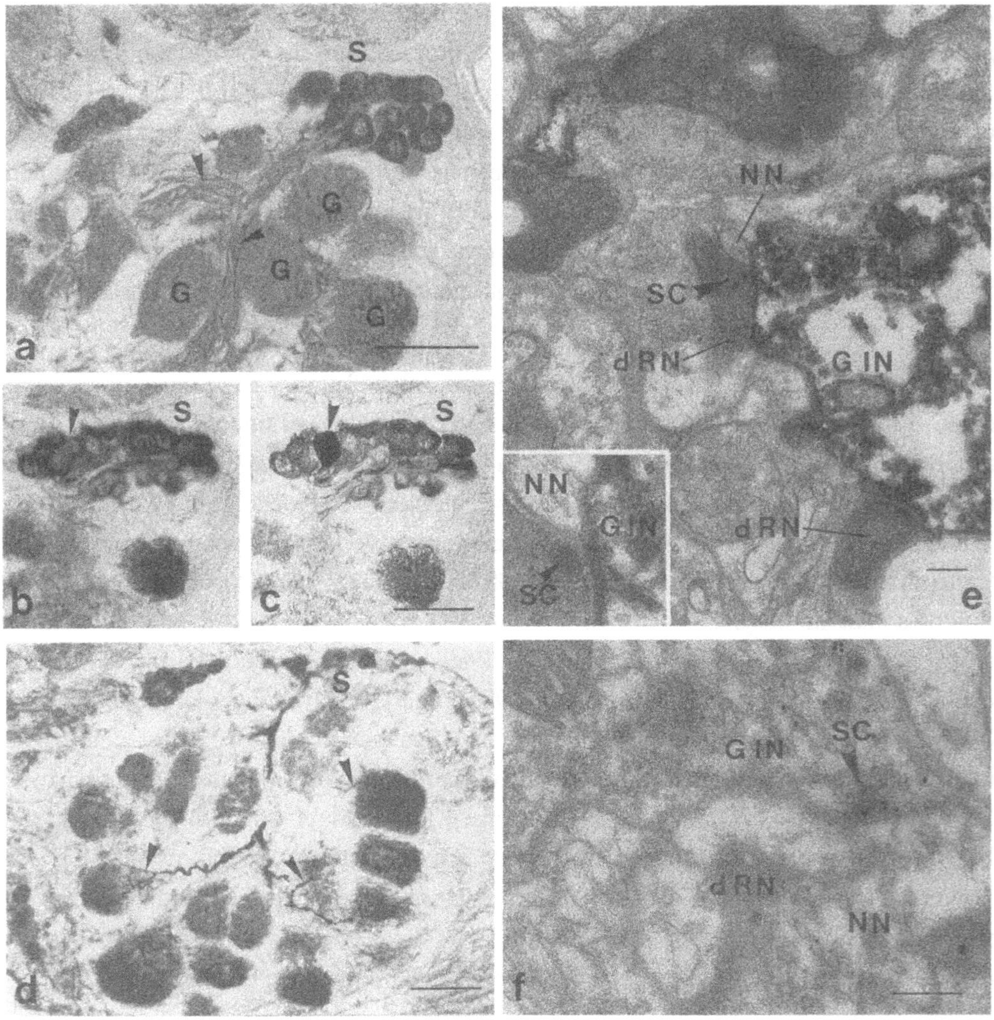

Fig. 6. Periplaneta americana ♂, deutocerebrum. a - d, demonstration of GABA-like immunoreactivity (GLIR) in INs; e, RN-IN- contact; f, IN-NN-contact. a, immunostained group of INs with somata (S) in antero-lateral soma group, processes in Y-shaped tract (arrows). G, glomerulus. b, soma group (S) with GLIR, immunostained after cobalt injection into individual IN (arrow) which, in c, shows-up after subsequent cobalt-silver intensification (arrows). d, branches of neuron after intensification (arrow). e, synaptic contact (SC) of GABA-immunolabelled IN (GIN) with receptor axon (dRN) labeled by anterograde degeneration. Preembedding immunostaining. Inset, enlarged RN-IN contact. f, synaptic contact (SC) of GABA-immunolabelled IN (GIN) with non- labelled neuron (NN). Postembedding immunogold staining. Scale a- d: 100 μm. e, f: 0,2 μm. (a-d: from Distler, 1989).

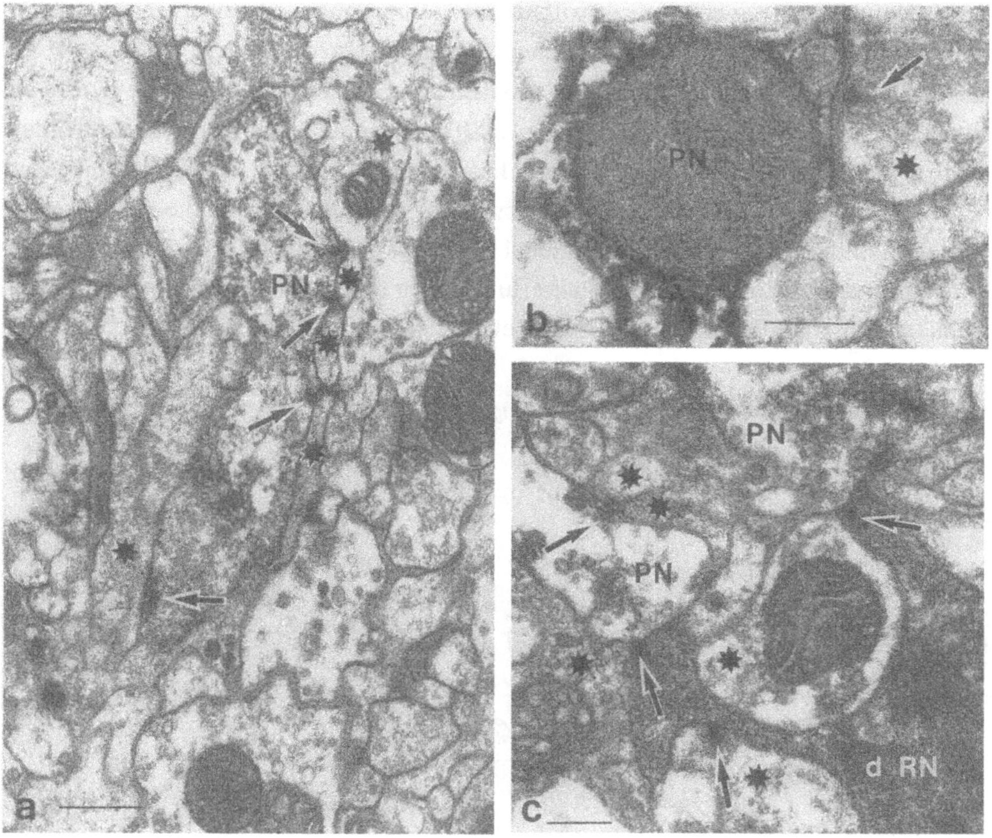

Fig. 7. Periplaneta americana ♂, deutocerebrum. Profiles of physiologically identified uniglomerular PNs injected with lucifer yellow and labelled with lucifer yellow antiserum. a, output contacts from pheromone-sensitive MG PN (PN) to non-identified profiles (asteriscs). b, input contact from non-identified profile (asterisc) to MG PN. c, input contacts from receptor axon (dRN) labelled by anterograde degeneration to the (olfactory) PN of "normal" glomerulus. Input contact from non identified profile to receptor axon, output contact from PN to non identified profiles (asteriscs). Arrows point to presynaptic densities. Scales, a, 0.5 μm; b, 0.2 μm; c, 0.23 μm.

A group of about 80 so-called Y-INs has been examined in some detail in Periplaneta (Fig. 6). Their somata are in the anterolateral group, from which the processes pass into the centre of the lobe as a Y-shaped bundle. Each neuron innervates between 20 and 30 glomeruli, with an approximately uniform array of thin branches; there is overlap with the innervation areas of other neurons. Each glomerulus receives processes of up to 15 such neurons. Histochemical tests reveal GABA-like immunoreactivity (GLIR), and in ca. 10% of the cells it is co-localized with dopamine-like immunoreactivity (Salecker and Distler, 1989). GLIR has also been demonstrated in INs of Manduca (Hoskins et al., 1986). The Y-INs make a number of characteristic connections, revealed by double labelling or histochemical labelling (Fig. 7). Their spontaneous activity, as recorded from the (unanaesthetized) animal with head capsule opened, is usually somewhat higher than that of the PNs, and their responses to stimulation of the antennae are smaller than the PN responses. Multimodal responses are common. The latencies of the local interneurons to antennal stimulation are shorter than those of any other neurons in the lobe (Ernst and Boeckh, 1983; Boeckh et al., 1983; cf. Waldow, 1975).

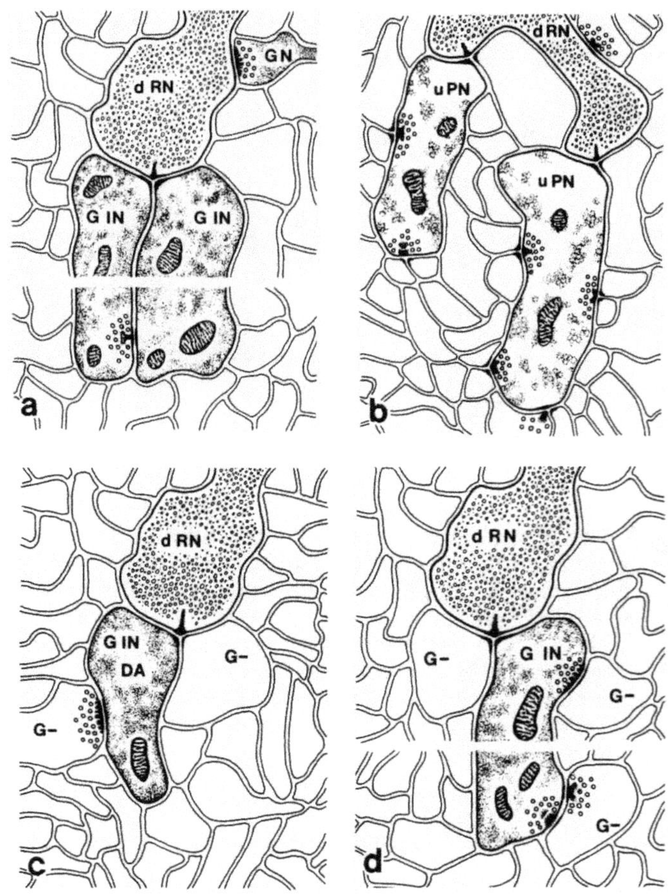

Figs. 8. Diagrammatic summary of modes of contacts within deutocerebral glomeruli of Periplaneta ♂, as revealed by electron microscopy of identified elements (cf. Figs. 6,7; Boeckh et al., 1989; Distler, 1989a; Malun, 1989). dRN, receptor neuron, labelled by induced anterograde degeneration; GN, non identified immunostained neuron with GABA-like immunoreactivity (GLIR); GIN, immunostained multiglomerular local interneuron with GLIR; G-, neuron without GLIR; uPN, uniglomerular projection neuron, labelled by lucifer yellow and subsequent lucifer yellow antiserum; DA, dopamine-like immunoreactivity. a, RN makes dyadic contact to two INs with GLIR, and receives contact from neuron with GLIR. Lower part: GIN-GIN-contact. b, single RN makes contacts to two processes of uPN. c, RN makes dyadic contact to IN with GLIR and colocalized DA; IN receives contact from non-identified GLIR- negative neuron. d, RN makes dyadic contact to non-identified GLIR-negative neuron and GIN which in turn contacts non-identified GLIR-negative neuron ("serial" synapses). Lower part, "reciprocal" contacts between other GIN and GLIR-negative non-identified neuron.

The extent to which such multiglomerular INs are involved in "horizontal" communication within the lobe - like such bulbar elements as the PGs or the widespread basal dendrites of the PNs, linked by GCs - is not clear. As impulse-generating neurons, they should be able to transmit signals reliably over fairly large distances. INs certainly play a fundamental role in mediating between receptor input and PNs, but it is uncertain how much they affect the odour spectra by filtering and modifying actions (see below), or whether a given IN provides both an RN-PN link and a connection to other glomeruli. Nor is it known whether there are functional subgroups of INs and/or whether various domains of such an IN are responsible for different functions, partially independent of one another - for instance, within a glomerulus on one hand and between glomeruli on the other.

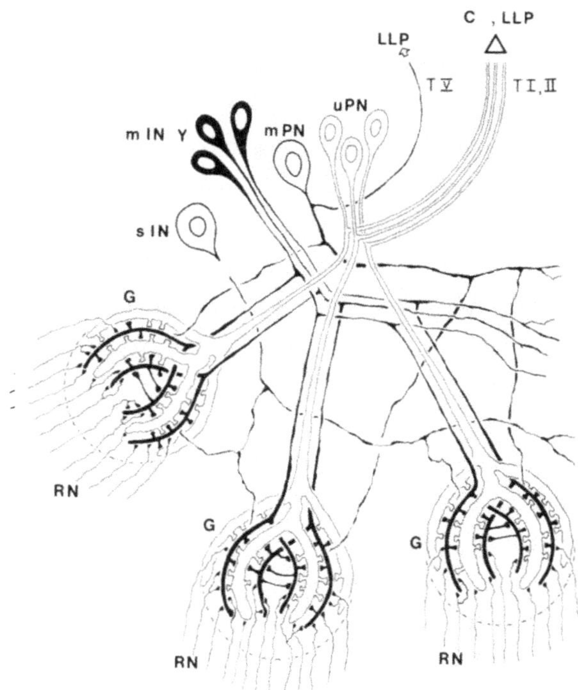

Fig. 9. Diagrammatic summary of circuitry in deutocerebrum of Periplaneta (cf. Fig. 8). C, calices of ipsilateral corpus pedunculatum; G, glomerulus; mIN Y, multiglomerular (GABAergic) INs with processes in Y-shaped tract; LLP, ipsilateral lobus lateralis protocerebralis; mPN, multiglomerular PN with axon in T V; uPN, uniglomerular (probably cholinergic; Salecker and Distler, 1989) PNs with axons in T I or T II; RN, antennal receptor cell axons; sIN, multiglomerular serotonergic neuron with processes in deuto- and protocerebrum (Salecker, pers. comm.); T I, II, V, different tracts resp. subtracts between deuto- and protocerebrum (cf. Fig. 1).

The large and crucial gap in our information is ignorance of the IN activity in the glomerulus and of the connections among INs and between INs and PNs. Here research on the lobe falls far short of what would be desirable for comparison with the bulb. But certain formal parallels, at least, come to mind:

1. The above-mentioned projections of the RNs onto GABAergic INs in Periplaneta and onto GABAergic PGs in the bulb.

2. The role of GABAergic INs in inhibitory modulation of the odour-specific time course of the PN responses in lobe and in bulb (see below and Kauer, 1987).
3. The existence of reciprocal "dendro-dendritic" synapses between GABAergic INs and other (unidentified) profiles in the glomeruli of the cockroach (Fig. 8); such synaptic configurations have also been observed between (unidentified) profiles in Manduca (Tolbert and Hildebrand, 1981). Connections of this sort are frequently found in the bulb, between PGs and PNs as well as between GCs and PNs.
4. "Horizontal" connections between glomeruli or their PNs by way of GABAergic INs, both in the lobe and in the bulb.
5. The colocalization of GABA-like and dopamine-like immunoreactivity in both INs of the lobe (see above) and those of the bulb (Gall et al., 1987).

C. Projection neurons and the coding of odours

The PNs of the bulb have been most thoroughly studied with respect to their electrical properties and their synaptic influences. The various mitral and tufted cells have been classified by morphological, and to some extent physiological, criteria (cf. Mori, 1987), but the individual cell types have not been systematically characterized in terms of their olfactory functions. The response is determined in part by monosynaptic excitatory inputs from the RNs, at the distal dendrites, and by ultimately inhibitory inputs mediated by the PGs; the time course of PN activity can be more or less complex, depending on the odour stimulus and the area of mucosa active in each case (Kauer, 1974; Kauer and Moulton, 1974; Kauer and Shepherd, 1977; Duchamp, 1982; Hamilton and Kauer, 1985; Schild, 1987). The degree to which this complexity reflects the variable time courses of the RN responses remains to be determined. Other inputs are probably mainly inhibitory. These come from neighboring glomeruli (or their PNs), either directly or mediated by the INs, and from the feedback loops.

The odour spectra of the PNs are said to differ from those of individual RNs, though it remains unclear how the combination of receptor spectra at the input can, in some circumstances, be modulated and narrowed by "lateral" or feedback interactions (cf. Kauer, 1987). The odour spectra of the PNs in a given glomerulus resemble one another fairly closely, with relatively little similarity to those in neighboring glomeruli (Chaput, this volume).

Considering together the results of the projection studies, the 2-DG experiments, the most recent measurements of activity with voltage-sensitive dyes (cf. Kauer, this vol.) and the electrophysiological recordings from single PNs and whole glomeruli, one can arrive at the following inferences. An odour stimulus elicits a response of a large number of PNs, depending on the type of molecule or mixture of which an odour is composed, and on the intensity and time course of the stimulus; these responses have a stimulus-specific time course. Because of the broad overlap of the working ranges of the individual PNs, the result is a very finely graded across-fibre pattern over a large population of cells, which provides differentiated encoding of the stimulus parameters. The overall pattern can change during the stimulus. It, in turn, spreads over certain regions of the bulb in a manner characteristic of the stimulus, producing a particular spatial distribution of the across-fibre pattern. Because many "cooperative" neurons and bulb regions are involved, the odour situation is multiply represented, both spatially and in the temporal pattern of the responses; that is, there is considerable redundancy (cf. Kauer, 1987; Hudson and Distel, 1988). A comprehensive hypothesis for the encoding and discrimination of odours at the levels of the glomeruli and of the bulb as a whole has recently been presented by Schild (1988).

Although generalities like the above can be derived from the available data, the specifics are still out of reach; we cannot say, for any particular odour, what happens in particular neurons, glomeruli or regions of the bulb.

In insects, because of the clear-cut responses of particular types of olfactory cell and because reproducible morphological and physiological identification of single neurons is possible, response patterns of the PNs can be interpreted as the result of processing specific inputs. Thus one has at least a foundation on which to construct hypotheses about the interactions involved, even though details of the circuitry remain obscure.

An unusually good opportunity is provided by the sex- specific MGs and their pheromone-sensitive PNs, the responses of which can also readily be related to behavioural responses. Accordingly, much attention has been devoted to such neurons, especially in various moths and in the American cockroach (see the reviews by Boeckh et al., 1984; Boeckh and Ernst, 1987; Christensen and Hildebrand, 1987; Light, 1986; and, as the most recent comprehensive contribution, Kanzaki et al., 1989).

In the operation of these PNs one has a typical example of the processing of pheromone stimuli, but that is not all. At least in Periplaneta, and to a lesser extent in Manduca, the working range of the MG-PNs includes odour stimuli present in food aroma but entirely absent from the pheromone and from the spectra of the pheromone receptors (see below). Hence the responses of MG-PNs described in the following paragraphs can be taken as examples of glomerular signal processing in general.

1. Receptor convergence (Fig. 10). Periplaneta PNs, the processes of which extend through the entire MG, respond to pheromone at concentrations so low (1 pg/stimulus source; that is, probably no more than 10^3 molecules per cm^3 of air) that only a fraction of the receptors are likely to receive, statistically, one molecule per cell - a situation demonstrated for Bombyx some time ago (Kaissling and Priesner, 1970). Not until the concentration is raised a hundredfold are enough molecules present to constitute a suprathreshold stimulus, on average, for all the receptors. Projection studies and neurophysiological tests show that most if not all of the receptors for female pheromone terminate in this MG. In Periplaneta this amounts to ca. 70,000 receptor axons, for a total of ca. 15 MG-PNs. Because of this enormous convergence, just-suprathreshold activity in any random combination of even very few receptor cells will produce a signal sufficiently above the noise at the PN. This efficient utilization of small signal magnitudes is typical of insect female attractants. The characteristic curves of such PNs cover wider ranges of concentration than those of single RNs, which is sometimes a factor in orientation within an odour gradient (Boeckh et al., 1984).

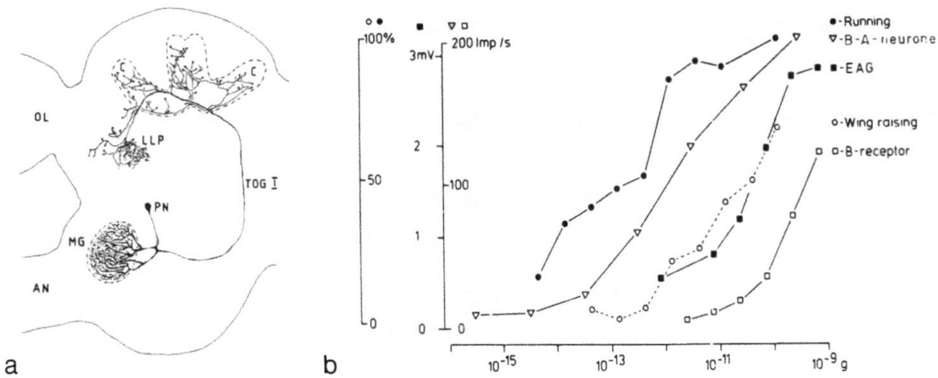

Fig. 10. Periplaneta ♂, deutocerebrum. a, camera lucida reconstruction of uniglomerular MG-PN, responding to components B and A of female pheromone, identified by cobalt injection and recording. AN, antennal nerve; C, calyx of corpus pedunculatum; LLP, lobus lateralis protocerebralis; MG, macroglomerulus; OL, optic lobe; PN, projection neuron; TOG I, subtract I of tractus olfactorio-globularis. b, reactions to component B of female pheromone: percentage of reacting males in behavior test (left ordinate, dots, circles), mean curve for 8 MG-PNs (right ordinate, triangles), mean curve for 6 B-type antennal receptor cells (right ordinate, open rectangles), electroantennogram (EAG, right ordinate, EAG-amplitudes in mV). Stimulus intensity given in amount of odourant at source (after Boeckh et al., 1984).

2. Receptive fields (Fig. 11). Whereas some neurons respond in the same way to odour stimuli applied anywhere on the antenna, others respond only to

stimulation in a region of the antenna specific to the neuron, with some overlap of the effective regions of different neurons (Hösl, 1989). Small clouds or filaments of odours, such as those an insect encounters when walking or flying in the lee of an odour source (cf. Murlis, 1986), will excite only a certain constellation of PNs, depending on the antennal region affected. Given that the two antennae span more than 10 cm, the insect is thus able to sample quite a large odour space. The processes of each PN arborize most densely in a different longitudinal region of the MG, which suggests that here the PN receives (probably via INs) its major synaptic input from receptors in a particular region of the antenna.

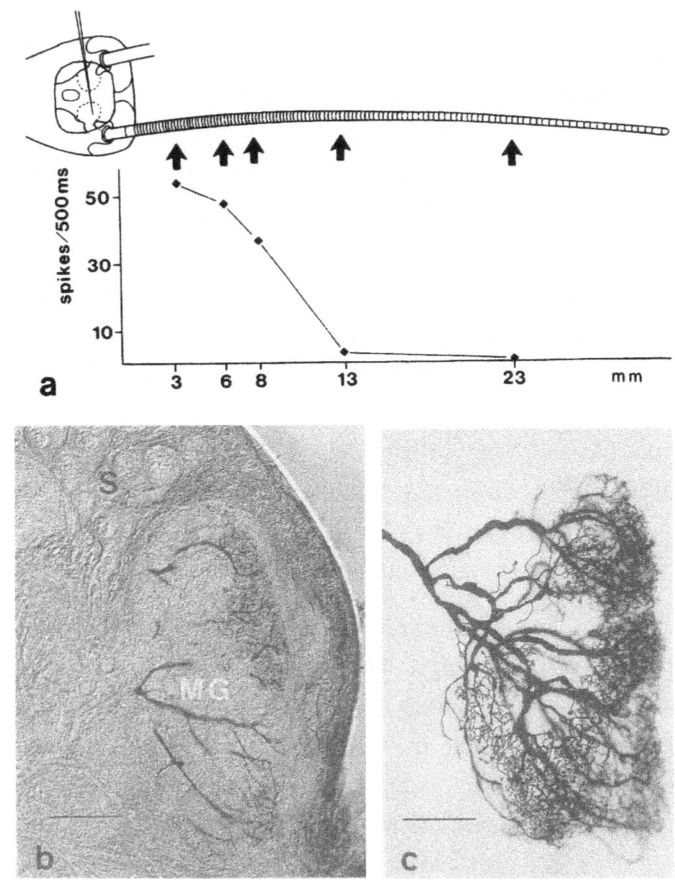

Fig. 11. Periplaneta ♂, deutocerebrum. a, responses of macro glomerular projection neuron to local stimulation of antenna with 5 x 10^{-13} g B-component of female pheromone at source, odour current of ca. 2 mm diameter at antenna. Abscissa, distance of stimulus site from antennal base. b, dendritic tree of neuron in macroglomerulus (MG) with most dense arborizations in a band tangential to glomerulus outline. Cobalt stained processes of PN in MG. Microphotograph of 20 μm section, Nomarski-optics. S, neuron somata. c, reconstruction of processes in MG from series of 20 μm sections. Scale 50 μm.

3. Temporal resolution. In Manduca MG-PNs follow trains of pheromone pulses applied to the antenna up to a repetition rate of 10 Hz, firing bursts in synchrony with the pulses. This response enables temporal resolution of odour clouds or edges of odour filaments encountered in rapid succession, which - according to the behavioural

physiologists - greatly facilitates combined chemo-anemotaxis (see Kramer, 1986). IN- and GABA-mediated inhibition, probably in the form of a delayed feedback loop, is responsible for the sharp temporal separation of these bursts from one another (Christensen and Hildebrand, 1988).

4. Odour mixtures, components and convergence of different receptor types. Behavioural data show clearly that a female pheromone is more than a combination of substances that together, mixed in species-specific proportions, constitute a triggering stimulus. Often individual components are significant with respect to individual subroutines in the behaviour sequence (long-distance attraction, stopping when the source is reached, and so on) or to interspecific interaction (cf. Seelinger, 1987). The diverse responses of the various PNs in an MG are to be understood in this light. Some respond only to one pheromone component, others to several (Boeckh and Selsam, 1984; Christensen and Hildebrand, 1987), and still others behave antagonistically, being excited by some components and inhibited by others, and giving mixed responses to the whole pheromone (Kanzaki et al., 1989). It is especially interesting that some neurons respond maximally only when the species-specific mixture is presented, and not to individual components (Christensen et al., 1989; Mustaparta, 1989). The implication for the glomerular circuitry is that receptors of a given type make different kinds of connection with PNs: excitatory, inhibitory, excitatory in combination, and so on up to a "synergistic" arrangement such that combined input from two receptor types has a stronger effect than even the most massive input from either alone. The INs involved here are unknown. In Manduca, however, GABAergic inhibitory INs have been found to be responsible for a considerable number of PN responses (Waldrop et al., 1987; Christensen and Hildebrand, 1987; cf. also Distler, 1989).

5. Odour discrimination. Although the MG probably has the special function of processing female attractant odour, it is not an isolated subcentre of the lobe, for it is linked to other glomeruli by way of multiglomerular INs and PNs. Furthermore, at least in Periplaneta the MG-PNs are affected by odour substances that have nothing to do with the spectra of the receptors for female pheromone; these include fruit aromas and their components, such as octanol, citral, fenchone and so on, as well as an extract of dorsal glands of male Periplaneta, which acts as an "aphrodisiac", causing the female to accept copulation. But the responses to such stimuli are not tonic or phasic-tonic, like those to the female pheromone or its components, but rather inhibitory, excitatory-on, followed by inhibition or on-off (Fig. 12; Waldow, 1975; Burrows et al., 1982; Boeckh et al., 1984). A simple interpretation of this phenomenon would be that the MG contains the endings of receptors of one or more types that are also represented in other glomeruli (or INs) (see above). In this case an excitatory

influence could have brief access to the PN before the same stimulus (by way of other INs) causes delayed inhibition. Disappearance of the inhibition after the end of the stimulus could then cause rebound activation of the PN. In this way female-pheromone and non-female-pheromone stimuli could produce PN responses with different time courses. Similarly, in the bulb different odours give rise to different time courses of the PN response.

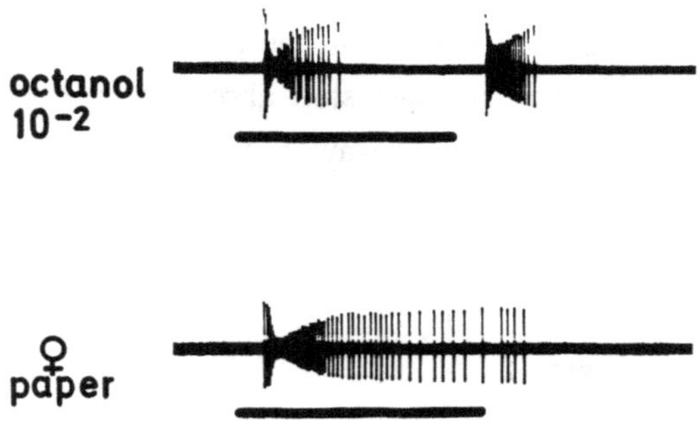

Fig. 12. Periplaneta americana ♂, deutocerebrum. Reactions of pheromone-sensitive neuron to female pheromone and to substance from spectrum of other than pheromone sensitive receptor type. Note the steady phasic-tonic response to pheromone but on-off to octanol (from Waldow, 1977).

In addition to the MG, a few other glomeruli and their PNs have been characterized sufficiently to permit fairly general inferences. Analysis is made much easier by the fact that the "ordinary" glomeruli, at least in Periplaneta, are each innervated by only one uniglomerular PN (Ernst and Boeckh, 1984, 1987).

The PN of the so-called C-glomerulus (CG) of Periplaneta, antero-vetrally of the MG, responds to a number of odour substances and aromata of potential food sources, but also to such compounds as terpenes present in wood and the extract of male dorsal glands (see above). As might be expected, differences in response time course similar to those found for MG-PNs are observed here: the dorsal-gland extract produces phasic-tonic excitation, whereas the responses to other stimuli are phasic (Waldow, 1987). When this neuron was examined repeatedly in 20 animals, it was found to have an interindividually invariant "core spectrum" of substances such as fenchone or cineol, plus a "side spectrum" varying from one animal to another, which in two cases even included female pheromone at very high concentration. Substances from both the "core" and the "side" spectra of this EG-PN act on PNs of other glomeruli, but in different combinations and orders of effectiveness.

The same applies to quite a large number of PNs of other glomeruli in Periplaneta, which respond to a broad palette of aromas of potential foods and other natural odour sources, and to many of their aromatic components (Fig. 13; Boeckh, 1978; Boeckh et al., 1984; Selsam, 1987). These responses are often complex sequences of excitation and inhibition. Responses to aromas are more often excitatory, whereas individual compounds often cause inhibition. As yet no PNs with identical spectra have been found; each has an individual spectrum with a distinct order of effectiveness of the stimuli, though there is considerable overlap with the spectra of other PNs. No PN spectrum is identical to one of the known RN spectra or to the sum of several such RN spectra, although certain substances from the spectra of various types of RN are effective on PNs (Boeckh, 1974; Selsam, 1987; Waldow, 1987). It follows that parts of the spectra represented in the combination of receptor inputs to a glomerulus are "cut out", perhaps (as suggested above for the MG) by interaction between glomeruli or within a glomerulus by way of INs.

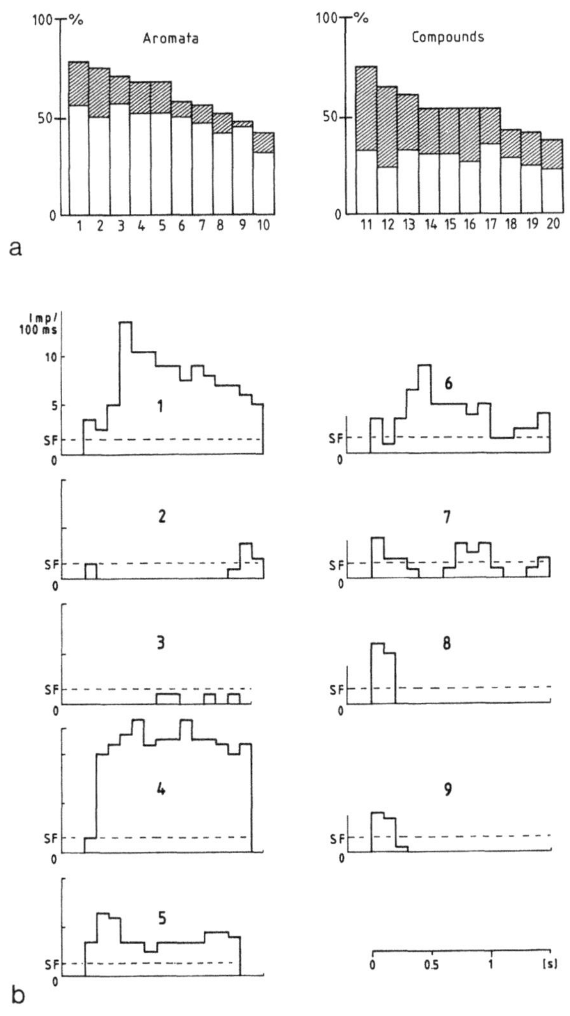

Fig. 13. Periplaneta americana ♂, deutocerebrum. Responses of uniglomerular PNs innervating other glomeruli than MG. a, percentage of altogether 68 identified neurons responding to different aromata and/or single compounds (inhibition indicated by hatched areas): 1, apple; 2, banana; 3, wheat; 4, orange; 5, lemon; 6, meat; 7, potato; 8, cheese; 9, bread; 10, artificial diet; 11, butyric acid; 12, series of homologous aliphatic aldehydes, (C2-C8); 13, series of aliphatic esters; 14, octanol; 15, terpineol; 16, hexanoic acid; 17, cinnamon aldehyde; 18, fenchone; 19, hexylamine; 20, eucalyptole. Aroma sources were 5 x 5 x 5 mm pieces of natural sources, compounds were given in 10^{-2} rel. dil. at source. b, reactions of individual uniglomerular PN to various odor stimuli. 1, banana; 2, lemon; 3, apple; 4, cheese; 5, artificial diet; 6, fenchone; 7, octanol; 8, cinnamon aldehyde; 9, hexanoic acid. Note large responses to aromata in comparison to compounds. Ordinate, nr. of impulses per 100 ms (= counting intervals), time scale in s, stimulus period from 0 s to 1.5 s. SF, background frequency of neuron (after Selsam, 1987).

PNs often give distinctly stronger responses to aromas than would be expected from their responses to the components. In fact, some of them respond considerably more strongly to a mixture of two components, each at very low concentration, than to high concentration of one component alone (Boeckh, 1976). The marked overlap of the spectra of different PNs clearly reveals the multiple representation of an odour in the lobe. For every odour apart from the female pheromone, there seem to be at least a dozen or more glomeruli in Periplaneta (out of a total of 130) in which the odour is fairly high in order of effectiveness.

Like the "side spectra" of the PN in the E-glomerulus, those in other glomeruli exhibit interindividual variation. Correlation studies of certain reliably identified PNs that have been tested on several occasions show that there are just as many differences or similarities between a given neuron and the "morphologically identical" neuron in the next animal as between the PNs of any two glomeruli in the same region of the lobe of a given animal (Fig. 14; Selsam, 1987; Boeckh et al., 1987). It is not clear whether the variation among individuals arises from differences in the combination of RN axons that grow into a glomerulus, in the interneuronal connectivity, or in the modulatory influences present in the (always acute) experiments.

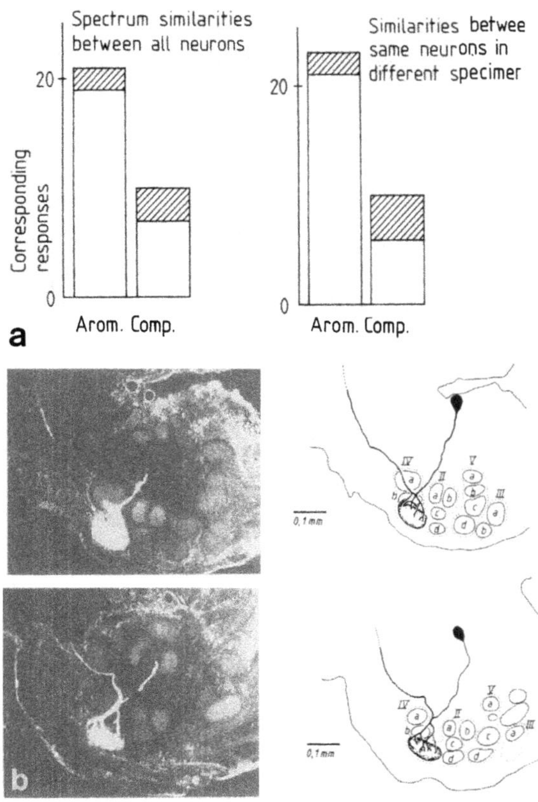

Fig. 14. Periplaneta americana ♂, deutocerebrum. a, left: similarities of odour spectra of 68 identified uniglomerular PNs (neurons and odorants same as in Fig. 13) as revealed from correlation- and probability tests for all possible pairs of stimuli and neurons. Right: Same calculation for PN of same glomeruli in 7 different experimental animals. Arom., odours from natural sources; Comp., single compounds (cf. Fig. 13). Open parts of columns, excitatory response; hatched areas, inhibition. b, morphological identification of glomerulus, innervated by lucifer yellow filled and physiologically identified PN in deutocerebra of two different males. Left, microphotographs from 20 μm resin sections of the right hand deutocerebra of two different males; right, camera lucida reconstructions of PNs with columns (roman numbers) of corresponding glomeruli around the labelled one (after Selsam, 1987).

SUMMARY

On the whole, it appears that the findings from the bulb and the lobe could be assembled to give a unified picture of the processing of olfactory inputs, as follows. Due to convergence and divergence of receptors, whether with the same or similar specificities or with different specificities, the individual glomeruli receive different spectral combinations, each odour substance and each aroma being represented in many glomeruli and in their PNs (in the case of the MG, various MG-PNs). The PN spectra therefore overlap broadly, although each is individually identifiable by a distinct rank ordering of effectiveness of the odour substances it comprises. By way of IN-mediated interactions, often inhibitory, between glomeruli (or PNs), the PN spectrum is "tailored" from the inputs to the glomerulus, and odour-specific temporal patterning of the PN response is produced as an additional means of odour coding.

The MG gives a clear example of the way the various items of information in a complex odour stimulus with a certain time course can be extracted centrally on the above principles: multiple representation of female pheromones and their components in the various MG-PNs, and different combinations of receptor input that give rise to a diversity of PN responses. Furthermore, the MG is a system in which the PN responses can be interpreted in the context of behaviour. Other "simple" cases of central odour coding - for instance, coding of chemical social signals or aromas of the food of oligophagous insects (cf. de Jong, 1988) - could show whether the present working hypothesis applies more widely.

For the reasons given above, the vertebrate olfactory bulb is probably less amenable to the single-identified-neuron approach. However, it offers good opportunities for system analysis-backed up, ideally, by specific tests of associated behavioural performance. In mammals analysis has already been developed so as to take account of modulation and plasticity, and even the possibility of manipulation by pharmacological means. And the successful application of voltage-sensitive staining promises undreamed - of progress in the study of rapid global and individual responses in the bulb. It is entirely conceivable that studies of insects and, of course, other invertebrates such as crustaceans (cf. Ache and Derby, 1987; Girardot and Derby, 1988) on one hand, and those of vertebrates, on the other, will complement one another more usefully than has previously been supposed.

REFERENCES

Ache BW, Derby CD (1985) Functional organization of olfaction in crustaceans. Trans Neurosci 8:356-360

Altner H, Prillinger L (1980) Ultrastructure of invertebrate chemo-, thermo-, and hygroreceptors and its functional significance. Int Rev Cytol 67:69-139

Andres KH (1975) Neue morphologische Grundlagen zur Physiologie des Riechens und Schmeckens. Arch Oto-Rhino-Laryng 210:1-41

Arbas A, Humphreys CJ, Ache BW (1988) Morphology and physiological properties of interneurons in the olfactory midbrain of the crayfish, Procambarus clarkii. J Comp Physiol A 164:231- 241

Boeckh J (1962) Elektrophysiologische Untersuchungen an einzelnen Geruchsrezeptoren auf den Antennen des Totengräbers (Necrophorus, Coleoptera). Z vergl Physiol 46:212-248

Boeckh J (1974) Die Reaktionen olfaktorischer Neurone im Deutocerebrum von Insekten im Vergleich zu den Antwortmustern der Geruchssinneszellen. J Comp Physiol 90:183-205

Boeckh J, Sandri C, Akert K (1970) Sensorische Eingänge und synaptische Verbindungen im Zentralnervensystem von Insekten. Z Zellforsch 103:429-446

Boeckh J, Ernst KD, Sass H, Waldow U (1976) Zur nervösen Organisation antennaler Sinneseingänge bei Insekten unter besonderer Berücksichtigung der Riechbahn. Verh dt Zool Ges 69:123-139

Boeckh J, Ernst KD, Polz H (1983) Convergence of inputs from different sense organs upon neurones of the central antennal pathway in insects. In: Horn E (ed) Multimodal convergences in sensory systems. G Fischer, Stuttgart New York, pp 143-148

Boeckh J, Ernst KD, Sass H, Waldow U (1984) Anatomical and physiological characteristics of individual neurones in the central antennal pathway of insects. J Insect Physiol 30:15- 26

Boeckh J, Ernst KD (1987) Contribution of single unit analysis in insect to an understanding of olfactory function. J Comp Physiol A 161:549-565

Boeckh J, Ernst KD, Selsam P (1987) Neurophysiology and neuroanatomy of the olfactory pathway in the cockroach. Ann NY Acad Sci 510:39-43

Boeckh J, Ernst KD, Selsam P (1989) Double labeling reveals monosynaptic connection between antennal receptor cells and identified local interneurons of the deutocerebrum in the American cockroach. Zool Jb Anat, in press

Bretschneider F (1924) Über die Gehirne des Eichenspinners und des Seidenspinners (Lasiocampa quercus L. und Bombyx mori). Jena Z Naturw 60:563-78

Buchner E, Rodrigues S (1983) Autoradiographic localization of (^{3}H) choline uptake in the brain of Drosophila melanogaster. Neurosci Lett 42:25-31

Burrows M, Boeckh J, Esslen J (1982) Physiological and morphological properties of interneurones in the deutocerebrum of male cockroaches which respond to female pheromone. J Comp Physiol 145:447-457

Cajal SR (1911) Histologie du Systeme Nerveux de L'Homme et des Vertebres, Vol. II, Maloine, Paris

Christensen TA, Hildebrand JG (1987) Functions, organisation, and physiology of the olfactory pathways in the Lepidopteran brain. In: Gupta AP (ed) Arthropod brain: its evolution, development, structure and functions. Wiley & Sons, New York

Christensen TA, Hildebrand JG (1987) Male-specific, sex pheromone-selective projection neurons in the antennal lobes of the moth Manduca sexta. J Comp Physiol A 160:553-569

Christensen TA, Mustaparta H, Hildebrand JG (1989) Discrimination of sex pheromone blends in the olfactory system of the moth. Chemical Senses 14:463-477

Clancy AN, Schoenfeld TA, Macrides F (1985) Topographic organization of peripheral input to the hamster main olfactory bulb. Chem Senses 10:399-400

Costanzo RM, O'Connell RJ (1980) Receptive fields of second order neurons in the olfactory bulb of the hamster. J Gen Physiol 76:53-68

De Jong R (1988) Host odour recognition by the colorado potato beetle. Ph.D-thesis, Agricultural University Wageningen, Netherlands

Distel H, Martin H, Hudson R (1987) The modified glomerular region and pattern of 2-DG uptake in the olfactory bulb of rabbit pups exposed to nipple-search pheromone. Chem Senses 12:181-182

Distler P (1989a) Histochemical demonstration of GABA-like immunoreactivity in cobalt labeled neuron individuals in the insect olfactory pathway. Histochemistry 91:245-249 Distler P (1989b) Synaptic contacts of GABA-immunoreactive neurons in the antennal lobes of the cockroach Periplaneta americana. In: Elsner N, Singer W (eds) Dynamics and Plasticity in Neuronal Systems. Thieme Stuttgart New York, p 70

Duchamp A (1982) Electrophysiological responses of olfactory bulb neurons to odour stimuli in the frog. A comparison with receptor cells. Chem Senses 7:191-210

Ernst KD, Boeckh J (1983) A neuroanatomical study on the organization of the central antennal pathways in insects. Cell Tissue Res 229:1-22

Flanagan D, Mercer AR (1989) Morphology and response characteristics of neurones in the deutocerebrum of the brain in the honeybee Apis mellifera. J Comp Physiol A 164:483-494

Gall GM, Stewart HC, Seroogy KB, Jones EG, Haycock JW (1987) Evidence for coexistence of GABA and dopamine in neurons of the rat olfactory bulb. J Comp Neurol 266-307-318

Girardot M-N, Derby CD (1988) Neural coding of quality of complex olfactory stimuli in lobsters. J Neurophysiol 60:303-324

Greer CA, Stewart WB, Kauer JS, Shepherd GM (1981) Topographical and laminar localization of 2-deoxyglucose uptake in the rat olfactory bulb induced by electrical stimulation of the olfactory nerves. Brain Res 217:279-293

Hamilton KA, Kauer JS (1985) Intracellular potentials of salamander mitral/tufted neurons in response to odor stimulation. Brain Res 338:181-185

Hanström B (1928) Vergleichende Anatomie des Nervensystems der wirbellosen Tiere. Springer, Berlin

Hartmann N (1987) Function and development of sensory cells in 'attractant sensilla': physiological differences between larvae and adults of Periplaneta americana L. Chem Senses 12:210

Hösl M (1989) Receptive fields on the antenna for deutocerebral pheromone-sensitive projection neurons in the American cockroach. In: Elsner N, Singer W (eds) Dynamics and Plasticity in Neuronal Systems. Thieme Stuttgart New York, p 67

Holley A, MacLeod P (1977) Transduction et codage des informations olfactives chez les vertebres. J Physiol (Paris) 73:725-828

Hoskins SG, Homberg U, Kingan TG, Christensen TA, Hildebrand JG (1986) Immunocytochemistry of GABA in the antennal lobes of the sphinx moth Manduca sexta. Cell Tissue Res 244:243-252

Jawlowski H (1948) Studies on the insect brain. Ann Univ M Curie Sklodowska C 3:1-30

Jourdan F, Duveau A, Astic L, Holley A (1980) Spatial patterns of 2-deoxyglucose uptake in the olfactory bulb of rats stimulated with two different odors. Br Res 188:139-154

Kaissling KE, Priesner E (1970) Die Riechschwelle des Seidenspinners. Naturwissenschaften 57:23-28

Kaissling KE (1971) Insect olfaction. In: Beidler LM (ed) Handbook of sensory Physiology. Springer, Berlin, Vol 4, pp 351- 431

Kanzaki R, Arbas EA, Strausfeld NJ, Hildebrand JG (1989) Physiology and morphology of projection neurons in the antennal lobe of the male moth Manduca sexta. J Comp Physiol A 165:427-453

Kauer JS (1974) Response patterns of amphibian olfactory bulb neurones to odour stimulation. J Physiol (Lond) 243:695-715

Kauer JS (1981) Olfactory receptor cell staining using horse- radish peroxidase. Anat Rec 200:331-336

Kauer JS (1987) Coding in the olfactory system. In: Finger TE, Sievers WL (eds) Neurobiology of taste and smell. Wiley & Sons, New York, pp 205-231

Kauer JS, Moulton DG (1974) Responses of olfactory bulb neurones to odour stimulation of small nasalareas in the salamander. J Physiol (Lond) 243:717-737

Kauer JS, Shepherd GM (1977) Analysis of the onset phase of olfactory bulb unit response to odour pulses in the salaman der. J Physiol (Lond) 272:495-516

Koontz MA, Schneider D (1987) Sexual dimorphism in neuronal projections from the antennae of silk moths (Bombyx mori, Antheraea polyphemus) and the gypsy moth (Lymantria dispar). Cell Tissue Res 249:39-50

Kramer E (1986) Turbulent diffusion and pheromone-triggered anemotaxis. In: Payne TL, Birch MC, Kennedy CEJ (eds) Mechanisms in insect olfaction. Clarendon Press Oxford, pp 59-68

Land LJ (1973) Localized projections of olfactory neurons to the rabbit olfactory bulb. Brain Res 63:153-166

Levetau J, MacLeod P (1966) Olfactory discrimination in the rabbit olfactory glomerulus. Science 153:175-176

Light DM (1986) Central integration of sensory signals: An exploration of processing of pheromonal and multimodal information in lepidopteran brains. In: Payne TL, Birch MC, Kennedy CEJ (eds) Mechanisms in insect olfaction. Oxford University Press, London, pp 287-301

MacKay-Sim A, Shaman P, Moulton DG (1982) Topographic coding of olfactory quality: Odorant specific patterns of epithelial responsivity in the salamander. J Neurophysiol 48:584-596

Malun D (1989) Fine structure of morphologically and physiologically identified projection neurons in the antennal lobe of Periplaneta americana. In: Elsner N, Singer W (eds) Dynamics and Plasticity in Neuronal Systems. Thieme Stuttgart New York, p 68

Matsumoto SG, Hildebrand JG (1981) Olfactory mechanisms in the moth Manduca sexta: response characteristics and morphology of central neurons in the antennal lobes. Proc R Soc Lond B 213:249-277

Mori K (1987) Membrane and synaptic properties of identified neurons in the olfactory bulb. Progr Neurobiol 29:275-320

Mori K, Takagi SF (1975) Spike generation in the mitral cell dendrite of the rabbit olfactory bulb. Brain Res 100:685-689

Mori K, Fujita SC, Imamura K, Obata K (1985) Immunohistochemical study of subclasses of olfactory nerve fibers and their projections to the olfactory bulb in the rabbit. J Comp Neurol 242:214-229

Murlis J (1986) The structure of odor plumes. In: Payne TL, Birch MC, Kennedy CEJ (eds) Mechanisms in insect olfaction. Clarendon Press, Oxford, pp 59-79

Mustaparta H (1971) Spatial distribution of receptor responses to stimulation with different odours. Acta Physiol Scand 82:154-166

Pareto A (1972) Die zentrale Verteilung der Fühlerafferenz bei Arbeiterinnen der Honigbiene, Apis melifera L. Z Zellforsch 131:109-10

Pinching AJ, Powell TPS (1971) The neuropil of the glomeruli of the olfactory bulb. J Cell Sci 9:347-377

Revial MF, Sicard G, Duchamp A, Holley A (1982) New studies on odour discrimination in the frog's olfactory receptor cells. I. Experimental results. Chem Senses 7:175-190

Revial MF, Sicard G, Duchamp A, Holley A (1983) New studies on odour discrimination in the frog's olfactory receptor cells. II. Mathematical analysis of electrophysiological responses. Chem Senses 8:179-190 Rospars JP (1983) Invariance and sex-specific variations of the glomerular organization in the antennal lobes of a moths, Mamestra brassicae and a butterfly, Pieris brassicae. J Comp Neurol 220:80-96

Rospars JP (1988) Structure and development of the insect antennodeutocerebral system. Int J Insect Morphol & Embryol 17:243-294

Rospars JP, Chambille I (1981) The deutocerebrum of the cockroach Blaberus craniifer Burm. Quantitative study and automated identification of the glomeruli. J Neurobiol 12:221-47

Sass H (1978) Olfactory receptors on the antenna of Periplaneta americana: REsponse constellations that encode food odours. J Comp Physiol 128:227-233

Sass H (1980) Physiological and morphological identification of olfactory receptors on the antenna of male Periplaneta americana. In: van der Starre H (ed) Olfaction and Taste VII. IRL, London, pp 194

Salecker I, Distler P (1989) Immunohistochemical localisation of dopamine-, GABA- and choline acetyltransferase (Chat)-reactivity in neurons of the antennal lobes in the American cockroach, Periplaneta americana. In: Elsner N, Singer W (eds) Dynamics and Plasticity in Neuronal Systems. Thieme Stuttgart New York, p 69

Schaller D (1978) Antennal sensory system of Periplaneta americana L. Cell Tissue Res 19:121-139

Schild D (1988) Principles of odor coding and a neural network for odor discrimination. Biophys J 54:1001-1011

Schneider D, Kaissling KE (1957) Der Bau der Antenne des Seidenspinners Bombyx mori L. II. Sensillen, cuticulare Bildungen und innerer Bau. Zool Jb Anat.76:223-250

Schürmann FW, Wechsler W (1969) Elektronenmikroskopische Untersuchung am Antennallobus des Deutocerebrum der Wanderheuschrecke Locusta migratoria. Z Zellforsch 95:223-48

Scott JW (1986) The olfactory bulb and central pathways. Experientia 42:223-231

Scott JW, Harrison TA (1987) The olfactory bulb: Anatomy and Physiology. In: Finger TE, Sievers WL (eds) Neurobiology of taste and smell. Wiley & Sons, New York, pp 151-178

Selsam P (1987) Duftcodierung, Morphologie und synaptische Verschaltung identifizierter Neurone im Deutocerebrum der Schabe. PhD-thesis, University of Regensburg, FRG Seelinger G, Gagel S (1985) On the function of sex pheromone components in Periplaneta americana: improved odour source localization with periplanone-A. Physiol Ent 10:221-234

Shepherd GM (1972) Synaptic organization of the mammalian olfactory bulb. Physiol Rev 52:864-917

Shepherd GM (1981) The olfactory glomerulus; its significance for sensory processing. In: Katsuki Y, Norgren R, Sto M (eds) Brain Mechanisms of Sensation. Wiley, New York, pp 209-223

Stewart, WB, Kauer JS, Shepherd GM (1979) Functional organization of rat olfactory bulb, analyzed by the 2-deoxyglucose method. J Comp Neurol 185:715-734

Stocker RF, Singh RN, Schorderet M, Siddiqi O (1983) Projection patterns of different types of antennal sensilla in the antennal glomeruli of Drosophila melanogaster. Cell Tissue Res 232:237-248

Teicher MH, Stewart WB, Kauer JS, Shepherd GM (1980) Suckling pheromone stimulation of a modified glomerular region in the developing rat olfactory bulb revealed by the 2-deoxyglucose method. Brain Res 194:530-535

Tolbert LP, Hildebrand JG (1981) Organization and synaptic ultrastructure of glomeruli in the antennal lobes of the moth Manduca sexta: A study using thin sections and freeze- fracture. Proc R Soc London B 213:279-301

Van As W, Kauer JS, Menco BPhM, Koster EP (1985) Quantitative aspects of the EOG in the tiger salamander. Chem Senses 10:1-21

Waldow U (1975) Multimodale Neurone im Deutocerebrum von Periplaneta americana. J Comp Physiol 101:329-341

Waldow U (1987) Untersuchungen am Ausgangsneuron eines Glomerulus im Deutocerebrum von Periplaneta americana. In: Elsner N, Creutzfeldt O (eds) New Frontiers in Brain Research (Proc 15 Göttingen Neurobiology Conference), Thieme Stuttgart, p 76

Waldrop B, Christensen TA, Hildebrand JG (1987) GABA-mediated synaptic inhibition of projection neurons in the antennal lobes of the sphinx moth, Manduca sexta. J Comp Physiol A 161:23-32

White EL (1972) Synaptic organization in the olfactory glomerulus of the mouse. Brain Res 37:69-80

Yamamoto C, Yamamoto T, Iwama K (1963) The inhibitory system in the olfactory bulb studied by intracellular recording. J Neurophysiol 26:403-415

PROCESSING OF PHEROMONE INFORMATION FROM RECEPTOR CELLS TO ANTENNAL LOBE NEURONS IN HELIOTHIS MOTHS

Hanna Mustaparta and Tor Jørgen Almaas

Department of Zoology
University of Trondheim
Norway

The pheromone blends of insects consist of certain ratios of compounds where the geometrical and optical configurations play an important role. Pheromone compounds, primarily produced as intraspecific signals, may also influence neighbouring species. This effect across species is well developed in bark beetles, where interspecific interruption and synergism, as well as attraction of predators to prey odours have been demonstrated in field studies (Birch 1984). The chemoreception of pheromones and interspecific signals involved in these interactions have been studied by recordings from single receptor cells. The results have revealed that all investigated beetle-produced compounds are received by specialist types of receptor cells, i.e. cells tuned to one particular compound (Mustaparta 1984). This also applies to optical isomers (Mustaparta et al 1980). Furthermore, the receptors exhibit largely identical responses as regards stimulus threshold, maximum response levels, dynamic range and reaction to analogues; characteristics which are independent of taxonomical belonging, geographical distribution, and kind of message mediated by the compound (Mustaparta 1989). These observations have lead to the tentative conclusion that pheromone receptors to a great extent are conserved through evolution. Thus, the changes in the speciation process may preferably be concerned with the manner in which the CNS processes the information. Against this background, we have undertaken to compare in moth species of <u>Heliothis</u> the specificity of the olfactory receptor cells with that of the neurons projecting from the antennal lobe (Christensen et al 1989; Almaas and Mustaparta 1989 a,b).

NATO ASI Series, Vol. H 39
Chemosensory Information Processing. Edited by D. Schild

THE SPECIFICITY OF THE RECEPTOR CELLS

The pheromones in H.virescens and H.zea consist of 7 and 4 components, respectively. In each species, a major and a secondary component, in a certain ratio, are sufficient for triggering the whole sequence of behavioural response (Roelofs et al 1974; Teal et al 1986; Vetter and Baker 1984). The major compound (Z-11-16:Al) is shared by the two species, whereas the secondary compound in H.virescens (Z-9-14:Al) acts as an interspecific interruptant in H.zea (Klun et al 1980). The minor components seem to have subtle effects on the pre-copulatory behaviour of the males (Vetter and Baker 1983, 1984).

We have recorded responses to these pheromone components from single receptor cells on the male antennae, primarily in H.virescens. Here, the receptor cells are located within two different types of sensilla trichodea (types 1 and 2) (Fig 1). The male specific type 1 sensilla are located on the 50 most proximal segments of the antennae and possess long hairs with a characteristic circumficial distribution (Almaas and Mustaparta 1989 a). The type 2 sensilla, having shorter hairs, are found on all 80 antennal segments in both sexes.

In H.virescens the majority of the receptors of the type 1 sensilla were found to be tuned either to the major (Z-11-16:Al) or to the secondary (Z-9-14:Al) pheromone component (Almaas and Mustaparta 1989 a). The only minor compound found to be represented by a few specialist cells was the Z-11-16:OH. These three specialist types of cells also responded weakly to some other pheromone components. However, these reponses did apparently not interfere with the much stronger reaction to the key compound.

Since no specialists for the other minor components were found in the type 1 sensilla, it was of interest to investigate also the character of the receptors present in the type 2 sensilla. Also here, pheromone specialists were found, all of them tuned to the major, or to the secondary pheromone component, but exhibiting lower sensitivitiy than the receptor cells of the type 1 sensilla, especially as regards cells located toward the antennal tip (Almaas and Mustaparta 1989b).

Since a receptor cell exhibiting relatively low sensitivity to a particular compound, in reality might be specialized for some other compounds, it was of interest to investigate whether cells of type 2 sensilla were indeed mediating pheromone information. The antennae of different males were cut at segment no 35 and 75, respectively. The cut end of the antennae were placed in cobalt solution, allowing cobalt to diffuse along the primary axons for 2 days. Both preparations showed cobalt staining in the macroglomerular complex (MGC), demonstrating that also axons from the short hair sensilla on the antennal tip project into the MGC (Fig.1).

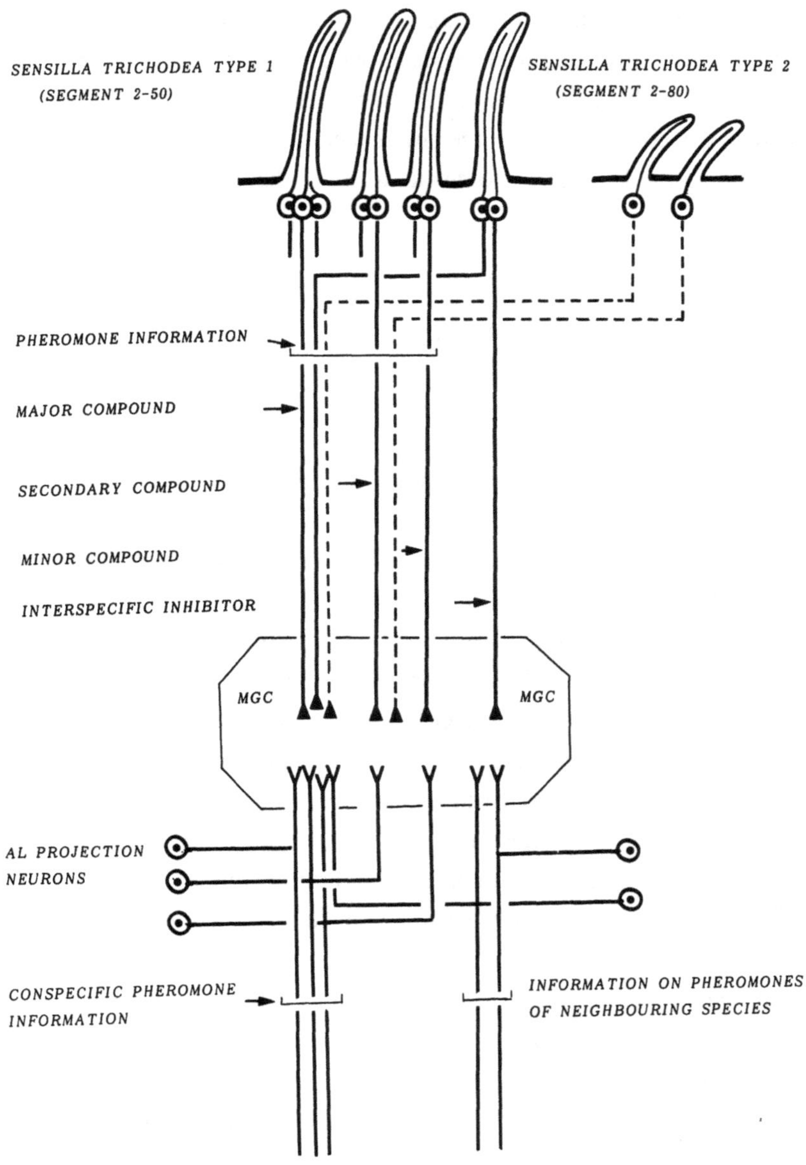

Fig. 1. Tentative scheme, indicating the processing of pheromone and interspecific signal information in Heliothis species. Local interneurons of the antennal lobe, conveying information from the receptor cells to the projection neurons are not indicated.
MGC=Macroglomerulus complex.

Only the receptor cells responding to host odours showed some reactions to the minor pheromone components. Thus, the integrated results of the receptor cell recordings in H.virescens are in general agreement with previous observations in other species of moths as well as in bark beetles (Priesner 1979; Kaissling 1979; Mustaparta 1984), implying that the pheromone information is conveyed to the antennal lobe via labelled-lines. That this also holds true for H.zea, is implied by most recordings so far carried out in this species (unpublished). However, a few recordings showing responses to both the major pheromone compound and the interspecific inhibitor, need to be analyzed by more advanced methods, in order to decide whether they originate from a single, or two adjacent receptor cell(s).

CHARACTERISTICS OF ANTENNAL LOBE PROJECTION NEURONS

Intracellular recordings from antennal lobe (AL) neurons, combined with staining for morphological characterizations, have been made in the two species of Heliothis, H.virescens and H.zea (Christensen et al 1989). In this study, some characteristics of the output neurons have been identified. Like concerns receptor neurons, the response type was mainly excitatory. Inhibition was rearly seen. Many AL neurons followed pulsed stimulation (up to 8HZ) with intermittant inhibition. In Manduca sexta this kind of inhibition is ascribed to input from the GABA-ergic local interneurons (Christensen and Hildebrand 1988; Waldrop et al 1989). Of particular interest is the specificity of the AL neurons, conveying information from the antennal lobe to protocerebrum. In both species of Heliothis many of these neurons did not discriminate between pheromone volatiles of the female glands from conspecifics and neighbouring species (Christensen et al 1989). In many of the recordings, the single pheromone components were not tested. However, in other experiments, where the single pheromone compounds were used as stimulants, it was found that the majority of AL projection neurons were exclusively or mainly driven by Z-11-16:Al, which is the major pheromone component in both Heliothis species. It suggests that these neurons may be involved in the first excitation of the males when detecting a pheromone plume. Furthermore, it was found that some of the output neurons in H.zea, responding mainly to the major pheromone compound, also exhibited weak responses to the secondary component. Such neurons may either receive input from two types of receptor cells (tuned to Z-11-16:Al and Z-9-16:Al, respectively in H.zea) or from one type of cells tuned to Z-11-16:Al but also responding weakly to the secondary pheromone component. However, more detailed quantitative studies are needed in order to determine which types of receptor cells converge onto these AL neurons.

Another group of AL neurons in H.zea, identified in the same study, were mainly driven by the secondary pheromone component of H.virescens, Z-9-14:Al which blocks the pheromone attraction in H.zea. In addition, weaker responses of these neurons to the major pheromone component were also seen. Furthermore, the neurons were more strongly activated by the pheromone blend of the neighbouring species than by that of the conspecific females. Summated responses to the mixture of the interspecific inhibitor and the major pheromone compound was observed in some AL neurons as well as occasional synergistic responses (Christensen et al 1989). As regards the summated responses, the AL neurons may be driven either by one type of receptor cells, tuned to the interspecific inhibitor (Z-9-14:Al) and weakly responding to the major pheromone component, or by two types, tuned to the inhibitor and the major pheromone, respectively. In the case of the synergistic responses, it seems likely that such neurons receive input from both types of receptor cells. The AL neurons, discussed here, can be regarded as two major groups of projection neurons, one being predominately activated by the conspecific pheromones, with the major pheromone component (Z-11-16:Al) playing the most important role, and the other being activated by the pheromones of the neighbouring species, with the inhibitor (Z-9-14:Al in H.zea) playing the main role. Apparently, these two groups of projection neurons convey (along separate pathways) information about presence of pheromones from conspecifics and neighbouring females, respectively. We might therefore expect that interactions between these two information pathways takes place at higher orders of neurons in the protocerebrum. The major difference between the two investigated species, concerning processing of information on the various pheromone components, might therefore be present at the protocerebrum level.

The convergence of numerous pheromone receptor cells on the antennae onto much fewer AL neurons might reflect integration of information from the same types of receptor cells and result in higher sensitivity of the antennal lobe neurons. Obviously, however, also some integration of information on the major and the secondary pheromone compounds occurs in the AL neurons as well as of information on the major pheromone and the interspecific interruptant. What remains to be elucidated is the reception and processing of information on the minor pheromone components. For these, no specific receptor cells or responding AL neurons have as yet been identified.

ACKNOWLEDGEMENTS - The studies reviewed were supported by grants from the Norwegian Research Council for Sciences and Humanities (project no D.65.46.112 and 452.89/006) and from NATO (RG. 0149/88).

REFERENCES

ALMAAS, T.J. and MUSTAPARTA, H. 1989a. J.Chem.Ecol. (in press)

ALMAAS, T.J. and MUSTAPARTA, H. 1989b. Proceedings from ISOT 1989

BIRCH, M.C. 1984 In: Chemical Ecology in Insects (W.J. Bell and R.T. Cardé eds) Chapman and Hall. London New York. pp 331-353

CHRISTENSEN, T.A. and HILDEBRAND, J.G. 1988. Chem.Senses 13, 123-130

CHRISTENSEN, T.A., MUSTAPARTA, H. and HILDEBRAND, J.G. 1989. Chem. Senses 14, 202-217

KAISSLING, K.-E. 1979 In: Chemical Ecology: Odour Communication in Animals (J.F. Ritter ed) Elsevier/N.Holland, Amsterdam, pp 43-56

KLUN, J.A., BIERL-LEONARDT, B.A., PLIMMER, J.R., SPARKS, A.N., PRIMIANI, M., CHAPMAN, O.L., LEPONE, G. and LEE, G.H. 1980. J.Chem. Ecol. 6,177-183

MUSTAPARTA, H. 1984. In: Chemical Ecology in Insects (W.J.Bell and R.T. Cardé eds) Chapman and Hall, London New York, pp 37-70

MUSTAPARTA, H., ANGST, M.E. and LANIER, G.N. 1980. J.Chem.Ecol. 6,689-701

MUSTAPARTA, H. 1989. Proceedings from ISOT, Oslo 1989

PRIESNER, E. 1979. In: Chemical Ecology: Odour Communication in Animals (J.F. Ritter) Elsevier/N.Holland, Amsterdam, pp 57-71

ROELOFS, W.L., HILL, A.S., CARDÉ, R.T. and BAKER, T.C. 1974. Life Sci. 14, 1555-1562

TEAL, P.E.A., TUMLINSON, J.H. and HEATH, R.R. 1986. J.Chem.Ecol. 12,107-126

VETTER, R.S. and BAKER, T.C. 1983. J.Chem. Ecol. 9,747-759

VETTER, R.S. and BAKER, T.C. 1984. J.Chem. Ecol. 10,193-202

WALDROP, B., CHRISTENSEN, T.A. and HILDEBRAND, J.G. 1987. J.Comp.Physiol.A. 161,23-32

SINGLE CELL ACTIVITIES AND THE OLFACTORY CODE

André HOLLEY and Patricia DUCHAMP-VIRET

Laboratoire de Physiologie Neurosensorielle, UA CNRS 180
Université Claude Bernard, 69622 Villeurbanne, France

This presentation is aimed at discussing what is known about odour representation in neural unitary activities of olfactory structures and especially in the olfactory bulb. Odour molecules flowing in the nasal cavity cause an alteration of the olfactory bulb neuron activity which is communicated to other neuronal populations which react according to their own properties and their particular connection patterns, resulting in perception. Usually, the altered activity is called a response and this response is thought to specify relevant characteristics of the stimulus, that is information. This information is coded in the sense that its multiple representations in different structures are connected to each other through transformation rules which are what we term codes. This way of describing the function of a sensory system presents several advantages, even though other expressions can be preferred (see for instance Skarda and Freeman, 1987 for a description in terms of self-organizing processes).

We are interested in disclosing the nature of the neural code utilized by olfactory bulb neurons, that is the relationship between the odour molecules and the patterns of altered unitary activity in the bulb. At this stage several remarks must be made.

The first remark is that analysing the code must not be confounded with the description of the stimulus-evoked activity. It can be a priori predicted that neural activity contains both relevant and not relevant components and that separating them is not an easy task. The problem is all the more acute as most of the available data were obtained in experimental conditions which were far from the natural use of the olfactory system. For example, in acute preparations, when olfactory stimulation and breathing are uncoupled, retronasal suction of odour molecules triggers complex and long-lasting discharge patterns which would undoubtedly not be expressed if a normal breathing activity had imposed a periodical resetting through the modulated input from receptor cells.

NATO ASI Series, Vol. H 39
Chemosensory Information Processing. Edited by D. Schild

Along this line, we are indebted to Michel Chaput for developing single unit recordings in the olfactory bulb of normally-breathing, awake animals. I would like to emphasize two significative observations from his studies. The first one is concerned with the time-patterning of the mitral cell activity during odour presentation (Fig. 1) (Chaput and Holley, 1985). In normally-breathing animals, the main feature of a mitral cell excitatory response is not necessarily an increase in the total number of spikes emitted during odour exposure. It is rather a restructuration of the spontaneous discharge in such a way that the firing rate is increased at the end of each inhalation phase and decreased to virtually zero during the exhalation phase.

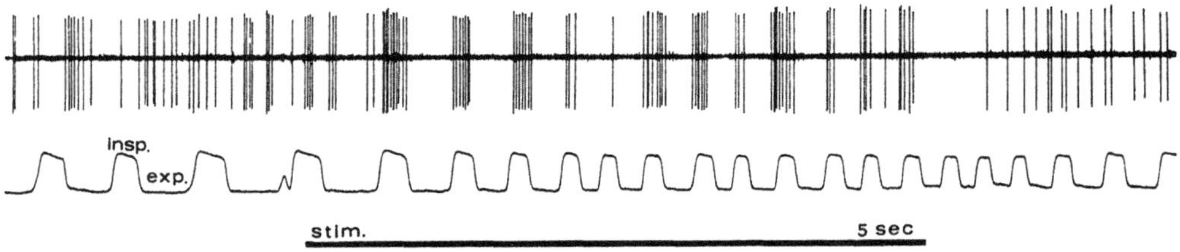

Fig. 1 Mitral cell excitatory responses to a 5-sec odour pulse recorded in an awake freely-breathing rabbit.
Upper trace: spike discharge
lower trace: respiratory activity. INSP: inspiratory phase.
EXP: expiratory phase.

Thus, in presence of an odour stimulus an excited neuron sees its activity periodically dropping below its resting level, and the bulbar overall output integrated over several seconds undergoes little change. We are inclined to assume that the most relevant part of the discharge pattern lies in the abrupt surge of activity occurring during the inhalation phase of each breathing cycle. Nevertheless, we cannot reject the view that some information is contained in the subsequent decreased activity occurring during the exhalation phase. Whether or not such a pattern is relevant to odour coding cannot be decided until one demonstrates that the projection area, the piriform cortex, is or not able to make the most of it. Nevertheless, Chaput (1986) demonstrated that the exhalation phase-related activity taken alone contained discriminative information about stimuli (Fig. 2). A principal component analysis of exhalation-related response profiles collected from 31 mitral cells stimulated by 5 odorants, 3 of which were repeatedly delivered 5 times, showed odorants well-discriminated from each other.

A second kind of interesting result obtained by Chaput and Lankheet (1987) is the observation that in awake rabbits mitral cell inhibitory responses were almost exceptional, in contrast with several earlier reports of a large proportion of such responses in other preparations. These inhibitory responses were hardly expressed in a

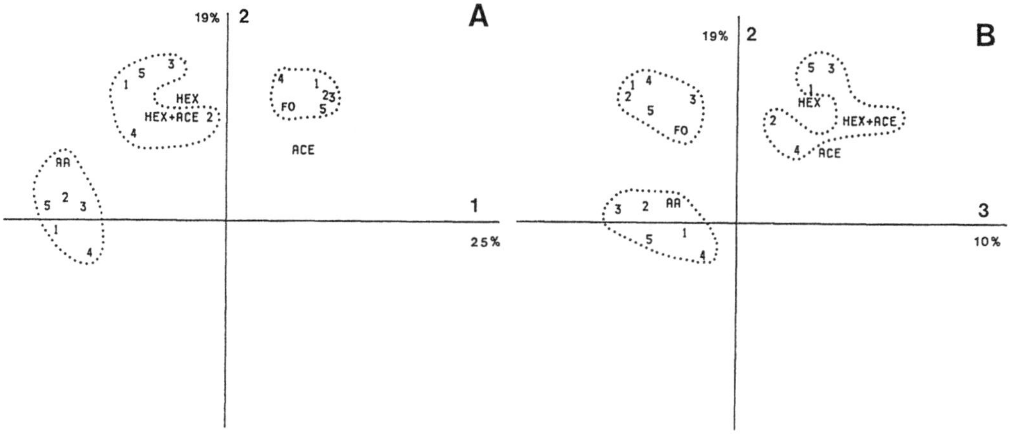

Fig. 2 Principal component analysis of olfactory bulb responses. Two dimensional space defined by factors 1 and 2 (A) and by factors 2 and 3 (B).
Odour code: HEX: hexenol; ACE: acetophenone; AA: amyl acetate; FO: food odour; HEX + ACE: mixture of hexenol and acetophenone.
Bulb responses elicited by the first presentation of a given stimulus (symbolized by its code) and by 5 further presentations of the same stimulus (symbolized by numbers 1 to 5) are found to be clustered together (dotted lines).

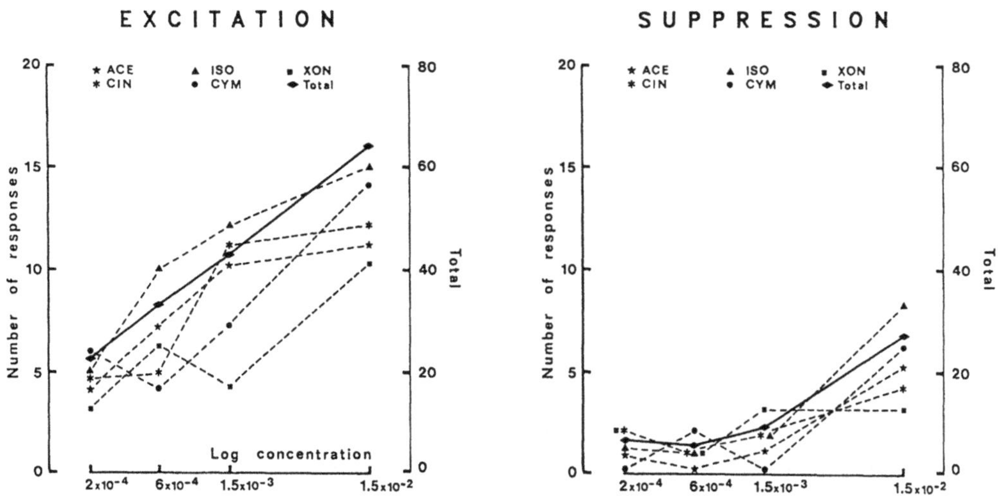

Fig. 3 Number of excitatory and suppressive bulb responses as a function of concentration (as a fraction of saturated vapor). In each graph, the left axis of ordinate corresponds to dashed-line curves obtained with the different stimuli: acetophenone (ACE); cineole (CIN); isoamyl acetate (ISO); cymene (CYM), cyclohexanone (XON). The right axis of ordinate corresponds to the full-line curve which represents the cumulative number of responses elicited by all stimuli considered together.

concentration range which is that of the normal use of the olfactory system. They suddenly increased in number when the stimulus concentration reached about 10^{-2} of the saturated vapour (Fig. 3). A quite different concentration-response relationship was exhibited by excitatory responses which progressively increased as a function of the concentration log. This does not mean that inhibitory circuits are not involved at low stimulus intensity. More likely, these circuits act to stabilize neural activity without reaching an actual inhibition state, which is artificially obtained at highest concentrations and in anaesthetized preparations.

These considerations must be kept in mind when discussing the involvement of inhibition in odour coding. I shall conclude this first remark, saying that when looking at the cell activity evoked in different experimental conditions, it is useful to distinguish between the information one collects on intrinsic properties of neuron networks from that bearing on the code.

The second remark is that the code cannot be understood unless one disposes of informations on the parameters which are concerned in the translation process. Discovering these parameters leads first to investigate the relevant properties of the stimulus. Now it is obvious that molecules which interact with the olfactory epithelium are indowed with a lot of properties, most of which are not at all involved in chemoreception. Unfortunately it is not possible to make an a priori distinction between what is relevant and what is not. A systematic analysis of odorant physicochemical properties does not permit to predict what dimensions are actually taken into account by the sensory organ and the nervous system. One is therefore led to look for some guide line in the final stage of the information processing, that is in the performance of the organism. Unfortunately, this strategy seems to have a limited outcome. From introspection, behavioural experiments and psychophysics it is possible to learn that human beings and animals are apt to recognize in the stimulus an intensity dimension. However, it is also well-known that this dimension is not completely and universally independent from other attributes of olfactory perception. The most conspicuous proprty of the olfactory performance lies in the ability that organisms have to discriminate a very large number of molecules species, an ability that reaches a theoretical maximum as each molecule is potentially discriminable from any other, provided it fulfils some conditions to be perceived. Finally it must be also mentioned that odour perceivers establish similarity-dissimilarity relations between odours and that these relations take place in a multidimensional system acccording to a non euclidian metrics. It must be added that the olfactory qualities are not organized according to some hierarchical principle. One is confronted with a type of sensory experience which is not reducible to a combinaison of basic features. This situation contrasts with that of the visual experience which offers the experimentalist a large variety of permutable and combining characteristics such as spatial coordinates, angles, borders, orientation, movement, color ..... Analysing the task that visual neurons have to achieve delivers a large number of components that are valuable guide lines for the neurophysiologist in search of codes.

Another way of looking at the coding properties of the olfactory bulb neurons is to compare their output with their input. Very few data are available on this subject. The most relevant work is that by Duchamp-Viret (1988) who recorded the changes in activity induced by identical olfactory stimulations in receptor cells and olfactory bulb neurons. Four odorants available at 20 discrete concentration values over a range going from 10^{-6} to $10^{-1.25}$ of saturated vapour were applied to the nasal mucosa of frogs while single unit activities were recorded from equal-sized populations of receptor cells and olfactory bulb relay neurons.

One of the results of this comparative study brings interesting data on temporal aspects of information processing in the two populations. Cell discharges were analysed on the basis of 200-ms time bins and the responses of all cells in each population to each stimulus delivered for 2 sec, were pooled (Fig. 4).

This summated activity provides a picture of the spike outflow in each structure. It appeared that despite variations in discharge patterning shown by individual receptors, the total population exhibited a peak of maximum activity comprised within 5 sec following stimulation onset, with a delay depending on the nature of the stimulus and its

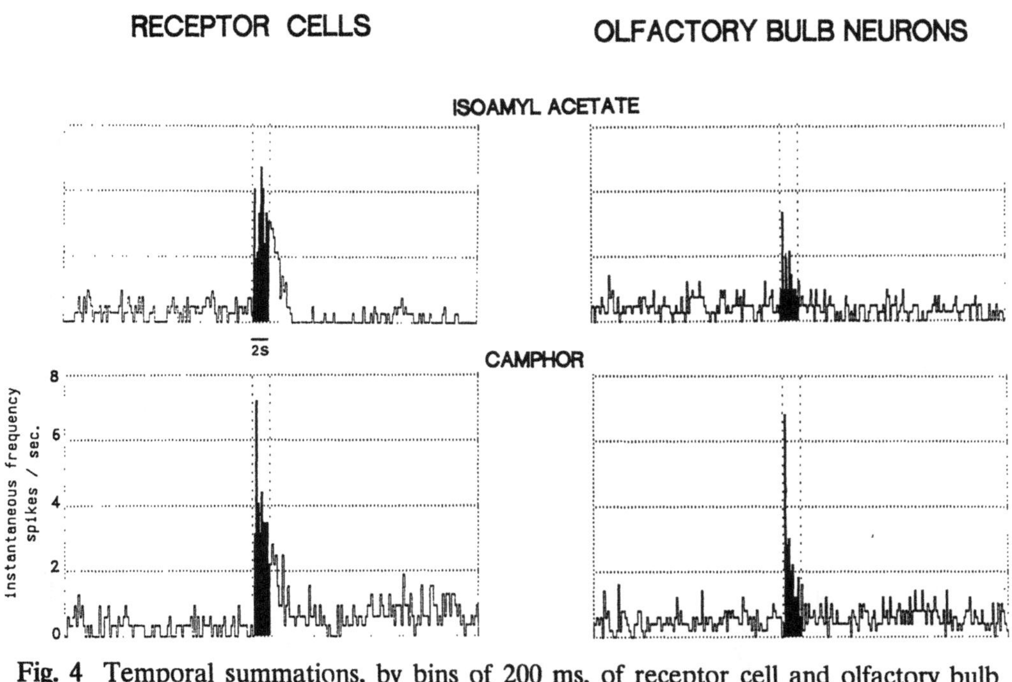

Fig. 4 Temporal summations, by bins of 200 ms, of receptor cell and olfactory bulb neuron activities. Comparison of response discharge patterns elicited in the two populations. Stimulations consisted of 2-sec pulses of isoamyl acetate and camphor at $10^{-1.25}$ of saturation.

concentration. The same analysis achieved on second order neurons showed that the stimulus-evoked discharge alteration was shorter and culminated sooner.

These results are in agreement with those reported by Kauer and Shepherd (1977) who investigated the timing of excitatory activity in the olfactory bulb. They provide the additional information that the olfactory bulb modifies the temporal structure of the afferent message since the late and sustained increase in activity as recorded from receptor cells was not coincident with the earlier and briefer surge of activity delivered by the bulb. It is therefore possible that receptor cell spikes have different degrees of incidence on the structure of the bulbar output and different degrees of participation in odour-coding, depending on their time of occurrence. The olfactory bulb seems to combine an amplifying role for earlier components of the afferent discharge with a filtering action on later events.

The amplifying function of the bulb was clearly put into evidence by Duchamp-Viret et al. (1989) in the same experimental conditions. Comparing receptor cell and bulb neuron sensitivities showed two different distributions (Fig. 5). Response thresholds were statistically lower for bulbar neurons than for receptor cells.

This was reflected in the cumulated curves of cell recruitment for both kind of cells in response to increasing concentration steps (Fig. 6). At the lowest concentration used in the study, which was around 10^{-6} of the satured

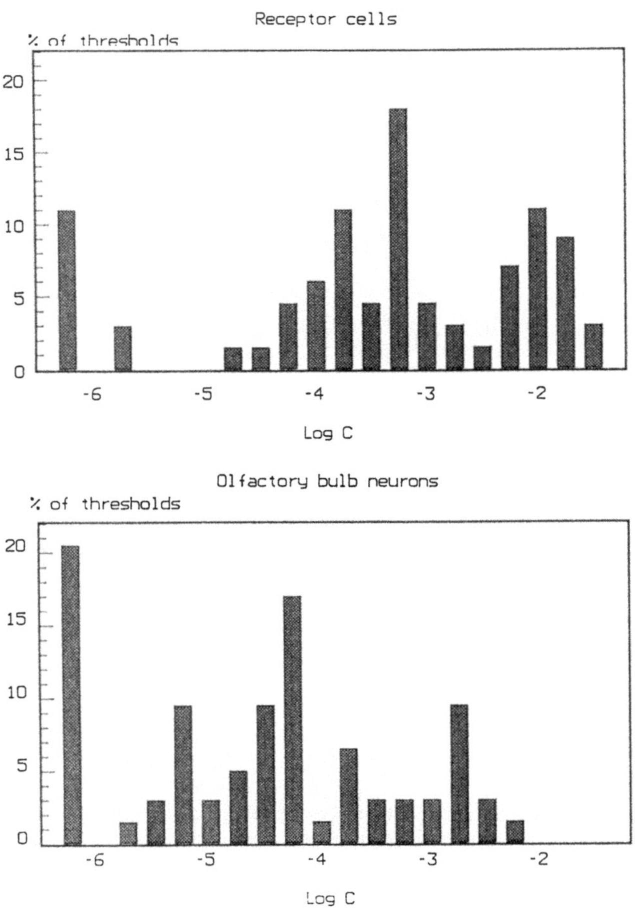

Fig. 5 Excitatory response threshold distributions of receptor cells and olfactory bulb neurons as a function of concentration $\log_{10}$ (as a fraction of saturated vapor).

vapour, 20% of the potentially excitable neurons had reached their response threshold while the corresponding percentage was only 10 for receptor cells. With increasing concentrations the recruitment curves remained well separated from each other. Seventy per cent of the recruitable bulbar neurons responded to 10^{-4} of saturated vapour, while the same percentage of receptor cells was not yet recruited at a 10-times higher concentration level. These findings can be taken as an experimental demonstration of the amplifying role of the glomerular convergence of receptor afferents, as proposed by Van Drongelen (1978) and Van Drongelen et al. (1978).

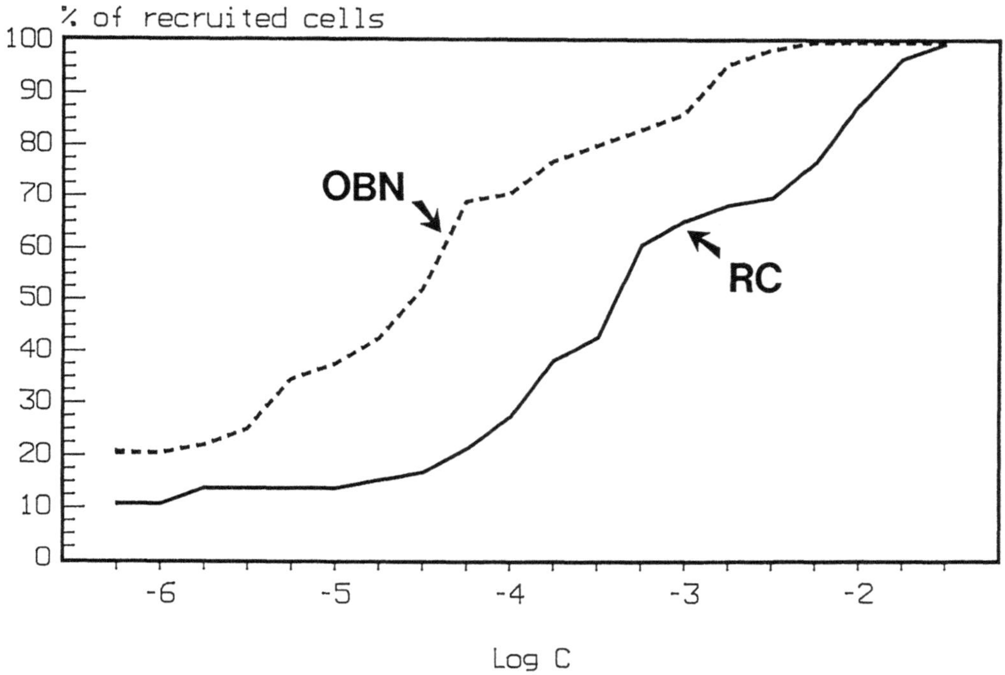

Fig. 6 Dynamics of receptor cell (RC) and olfactory bulb neuron (OBN) recruitments as a function of concentration $\log_{10}$.

Receptor cells and relay-neurons could also be compared with respect to the discriminability of the across-neuron patterns of activity evoked by odour stimuli. Even if there still is some uncertainty on the question of what stimulus properties are to be encoded in the olfactory bulb and elsewhere, there is at least one subject of general agreement: Individual discriminability of odours is a basic property of olfactory perception and finds its correlate and support in the discriminability of odour-evoked activity patterns.

In Duchamp-Viret's experiments (Duchamp-Viret et al., submitted), odour-evoked patterns were studied at both epithelial and bulb levels, using 4 odorants at different concentration levels. When applied to response discharge frequencies profiles evoked by each stimulus in the two cell populations, correlation coefficients showed marked differences between receptor cells and olfactory bulb neurons. At the receptor cell level, significant correlation coefficients were found between stimuli of different nature while low and high concentrations of a same odorant were not systematically correlated (Fig. 7).

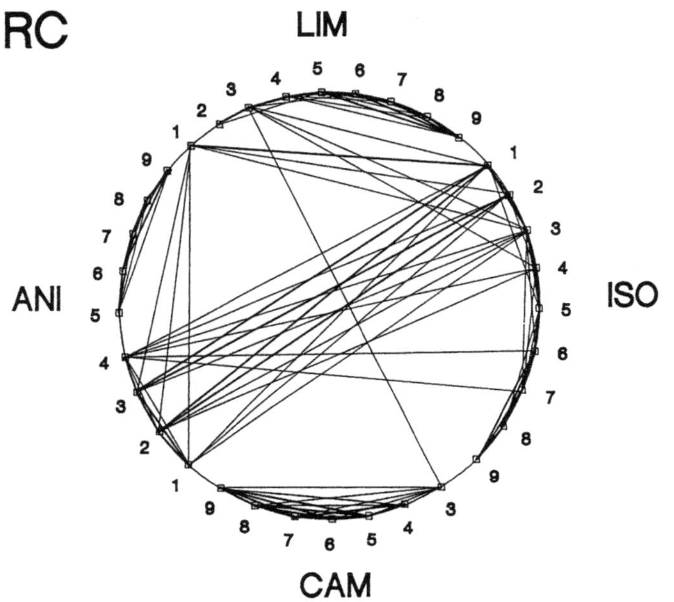

Fig. 7 Correlations between the responses of receptor cells (RC) evoked by limonene (LIM); isoamyl acetate (ISO); camphor (CAM) and anisole (ANI) at different concentrations coded from 1 to 9.

Correspondence between codes and concentrations (as a fraction of saturated vapor): 1 = 3.16×10^{-6}; 2 = 5.62×10^{-5} ; 3 = 10^{-4}; 4 = 1.7×10^{-4}; 5 = 5.62×10^{-4}; 6 = 1.7×10^{-3}; 7 = 5.62×10^{-3}; 8 = 1.7×10^{-2}; 9 = 5.62×10^{-2}.

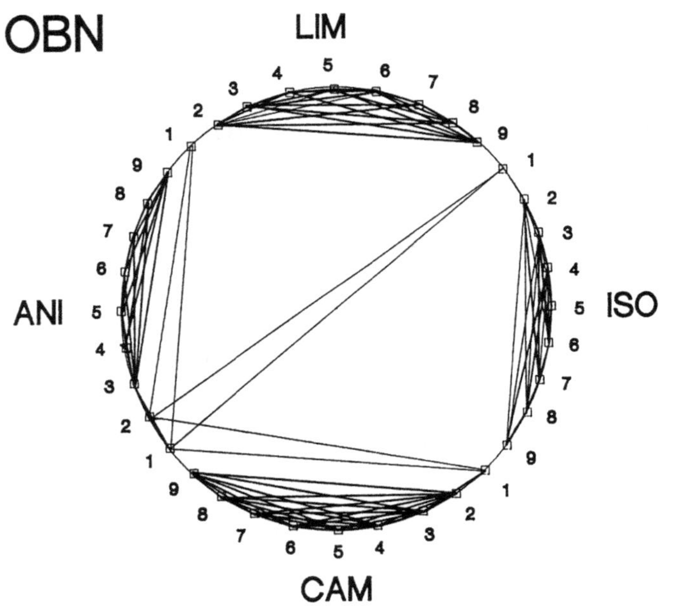

Fig. 8 Correlations between the responses of olfactory bulb neurons (OBN) evoked by limonene (LIM); isoamyl acetate (ISO); camphor (CAM) and anisole (ANI) at different concentrations coded from 1 to 9 (see correspondence between codes and concentrations in legend of Fig. 7).

In a relay-neuron population of approximately equal size, the picture appeared to be different (Fig. 8). All concentrations of the same odorant evoked highly correlated response profiles and no significant values of the correlation coefficient were found between chemically different odorants, except at the lowest concentrations of the compounds.

Interesting differences appeared also in a principal component analysis performed on the same data. For receptor cells, the first factor, which represented a substantial part of the variance (32%) was essentially associated with stimulus intensity and did not participate in discriminating stimuli. This discrimination was performed by other factors which had a smaller contribution to the variance. The same analysis made on olfactory bulb data showed a different pattern (Fig. 9). This time, the first component representing 37% of the variance was active in terms of quality discrimination since the different concentrations of camphor were separated from those of isoamyl acetate along this axis. It must be noted, in addition, that in both receptor cell and olfactory bulb neuron factorial spaces, the four sets of points corresponding to the 4 odorants were disposed along 4 directions. From the centre to the periphery the points were ranked according to increasing concentrations.

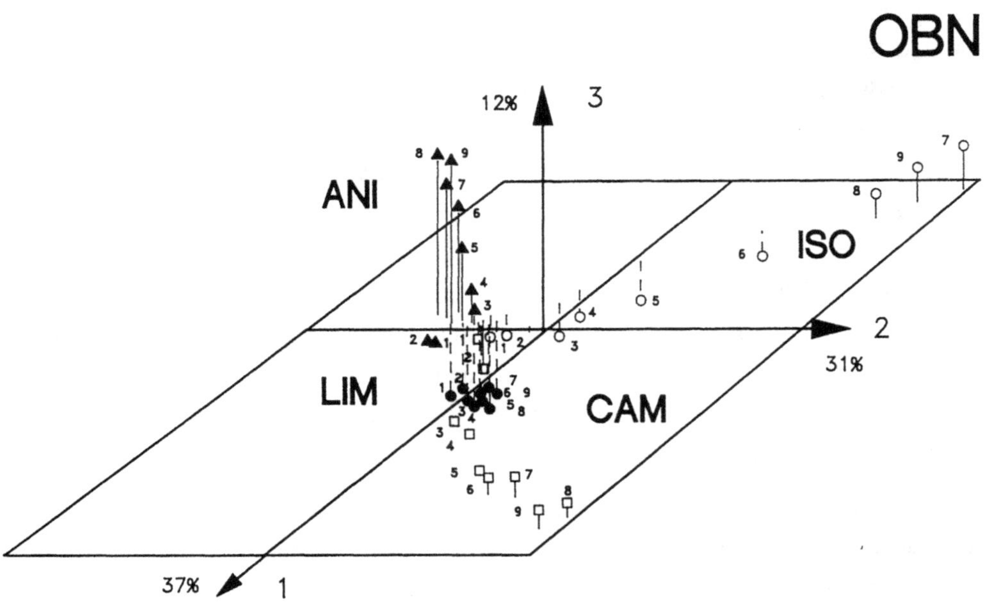

Fig. 9 Principal component analysis of olfactory bulb neuron responses (OBN): three dimensional space defined by factors 1, 2 and 3. Odorant and concentration codes are the same as in Fig. 7 and Fig. 8.

From this study it may be concluded that two samples of receptor cells and relay neurons made of equal numbers of elements, have different abilities to represent odour quality dependently on odour concentration. At the olfactory bulb level the information on intensity is clearly present but a greater proportion of the total information seems to be devoted to specify the nature of stimuli, that is odour quality.

There is another potentially important dimension of neural coding which has not yet been discussed, I mean its spatial dimension. When one claims that an odour is encoded in the olfactory bulb as a spatial pattern of neural activity, one means that the location of excited neurons - and not only their identity and their address - plays some role in the information processing. There are at least two ways for topography to interfere with sensory processing. First, interactions are more likely to occur among neurons located close to each other than among neurons separated by large distances. For this reason, an activation pattern formed by neurons concentrated in a few areas is not equivalent with a pattern of neurons dispersed in the whole structure. The importance of a topographical representation lies also in the opportunity it offers to prepare a selected transfer of the relevant information to the appropriate projection structures.

Regarding the olfactory bulb, the basic finding is that the odour-evoked afferent activity reaching glomeruli is not evenly distributed over the layer. The main support for this view is provided by 2-deoxyglucose (2-DG) mapping of metabolic activity (Stewart et al., 1979; Jourdan et al, 1980). The afferent activity appears to be concentrated in several foci according to a spatial pattern which is reproducible and odour-specific (Royet et al., 1987). However it must be acknowledged that we lack imformation on the actual relationship between the local density of 2-DG labelling in glomeruli and the output activity of the corresponding relay-neurons. The only study of some relevance to this question (Wilson and Leon, 1988) surprisingly indicated that an increased proportion of inhibitory responses was found in a region of the olfactory bulb where 2-DG uptake was enhanced following olfactory learning, in young rats. I am not aware of single unit recording studies reporting that the probability of collecting responses from a given region of the bulb is odour-specific.

Should a topographical organization of relay-neuron sensitivity be demonstrated, the coding significance of this organization could not be directly inferred. In fact, this problem cannot find a solution without considering the fate of the olfactory bulb message in the projection areas. In other words the coding value of the postulated spatial distribution in the bulb will not be fully understood until one reaches a good understanding of odour processing in the olfactory cortex. There are anatomical and electrophysiological arguments (Haberly, 1985) suggesting that the cortical projections from the olfactory bulb do not follow a topographical principle. In addition, well-developed horizontal association fibers contribute to spread the afferent message, still reducing traces of spatial organization remaining in direct projections. Nevertheless, it is not yet definitively demonstrated that the across fiber patterns of activity which encode odours in the tertiary projections have lost all vestiges of spatial dimension. Even diffuse, a persisting topography such as reported by Scott et al. (1980) would be appropriate to diversify the activity patterns expressed in different sub-areas of the cortex and transmitted to other brain structures. It is likely that the multiple parallel outputs of the cortex transmit highly redundant informations received from the whole olfactory bulb. However we cannot rule out the assumption that each set of cortical output has an additionnal specialization for optimally representing some types of odours, because the corresponding input originates preferentially (obviously not exclusively) from a part of the olfactory bulb where these odours activate mitral cells. This would confer a full significance to the concept of spatial coding of odours in the olfactory bulb.

REFERENCES

Chaput MA (1986) Respiratory-phase-related coding of olfactory information in the olfactory bulb of awake freely-breathing rabbits. Physiol Behav 36 : 319-324

Chaput MA, Holley A (1985) Responses of olfactory bulb neurons to repeated odour stimulations in awake freely-breathing rabbits. Physiol Behav 34 : 249-258

Chaput MA, Lankheet MJ (1987) Influence of stimulus intensity on the categories of single-unit responses recorded from olfactory bulb neurons in awake freely-breathing rabbits. Physiol Behav 40 : 453-462

Duchamp-Viret P (1988) Le codage intensitif du stimulus olfactif : Rôle des neurorécepteurs et des deutoneurones. Doctoral thesis, University Claude Bernard, Lyon, France

Duchamp-Viret P, Duchamp A, Vigouroux M (1989) The amplifying role of convergence in olfactory system. A comparison of receptor cell and second order neuron sensitivities, J Neurophysiol 61 : 1061-1094

Duchamp-Viret P, Duchamp A, Sicard G (submitted 1989) Olfactory discrimination over a wide concentration range. Comparison of receptor cell and bulb neuron abilities.

Haberly LB (1985) Neuronal circuitry in olfactory cortex : anatomy and functional implication. Chem Senses 10 : 219-238

Jourdan F, Duveau A, Astic L, Holley A (1980) Spatial distribution of (14C)2-deoxyglucose uptake in the olfactory bulbs of rats stimulated with twó different odours. Brain Res 188 : 139-154

Kauer JS, Shepherd GM (1977) Analysis of the onset phase of olfactory bulb unit responses to odour pulses in the salamander. J Physiol 272 : 495-516

Royet JP, Sicard G, Souchier C, Jourdan F (1987) Specificity of spatial patterns of glomerular activation in the mouse olfactory bulb : computer-assisted image analysis of 2-DG autoradiograms. Brain Res 417 : 1-11

Scott JW, Mc Bride RL, Schneider SP (1980) The organization of projections from the olfactory bulb to the piriform cortex and olfactory tubercle in the rat. J Comp Neurol 194 : 519-534

Skarda CA, Freeman WJ (1987) How brains make chaos in order to make sense of the world. Behav Brain Sci 10 : 161-173

Stewart WB, Kauer JS, Shepherd GM (1979) Fonctional organization of rat olfactory bulb analysed by the 2-deoxyglucose method. J Comp Neurol 185 : 715-734.

Van Drongelen W (1978) Unitary recordings of near threshold responses of receptor cells in the olfactory mucosa of the frog. J Physiol 277 : 423-435

Van Drongelen W, Holley A, Doving KB (1978) Convergence in the olfactory system : quantitative aspects of odour sensitivity. J theor Biol 71 : 39-48

Wilson DA, Leon M (1988) Spatial patterns of olfactory bulb single-unit responses to learned olfactory cues in young rats. J. Neurophysiol. 59 : 1770-1782

IS THE OLFACTORY BULB FUNCTIONALLY ORGANIZED IN PARALLEL COLUMNS ?

MA CHAPUT

Physiologie Neurosensorielle
CNRS - Univ. Claude Bernard LYON I
69622 VILLEURBANNE cedex
FRANCE.

INTRODUCTION

The olfactory bulb is the first relay of the olfactory pathway. It receives inputs from the olfactory epithelium and sends outputs to the cortex. Contacts between the axons of olfactory receptor cells and the dendrites of the bulbar output cells, mitral and internal tufted cells, occur in characteristic neuropils called glomeruli. Since output cells are known to send their single primary dendrite to the glomerulus which is located in a different laminae, but in register with their somata, the groups of output cells connected to different glomeruli appear as anatomical units and may form functional columns.

My purpose here is to analyze the possibility that the olfactory bulb will be organized in functional columns. I will first briefly review anatomical arguments in support of such an organization. Then, I will present in details recent results that may be interpreted as a manifestation of a unitary functioning of glomerular efferences. Lastly I will discuss the consequences of this functional columnar organization for sensory processing.

OLFACTORY BULB ANATOMICAL ORGANIZATION

The olfactory bulb presents the same relatively-simple cytoarchitecture throughout virtually all species of Vertebrates. Its organization has been reviewed in recent years by

NATO ASI Series, Vol. H 39
Chemosensory Information Processing. Edited by D. Schild

Shepherd (1972, 1977, 1979), Pinching (1972), Holley and Mac Leod (1977), Scott and Harrison (1987) and Mori (1987) while special emphasis was laid on the glomerular region by Pinching in 1970, Pinching and Powell in 1971 (a,b) and Shepherd in 1981.

In the mammalian olfactory bulb, most of the cellular and synaptic elements are segregated in distinct layers. Receptor cell axons form the outermost layer of the bulb. Then, from the surface inward, the major layers are: 1) the glomerular layer composed of spherical-shaped regions of neuropil arranged in a nearly single layer and surrounded by regions of neuron somata; 2) the external plexiform layer which contains in its deep one-third the cell bodies of the internal tufted cells and where the primary dendrites of the bulbar output cells extend radially towards the glomeruli while the secondary dendrites of these cells extend for several hundred micrometers in the plane of the layer; 3) the mitral cell layer formed by the bodies of the mitral cells arranged in a thin sheet; and 4) the granule cell layer which contains mainly the bodies of the granule cells, deep bulbar interneurons.

An increasing body of evidence suggests that besides its particularly patent organization in parallel layers, the mammalian olfactory bulb may also be organized in vertical columns, each composed of a glomerulus and the group of output cells connected to this glomerulus.

A major argument is the discrete and well-circumscribed nature of the glomeruli. Indeed, glomeruli appear as roughly spherical structures well-separated from one another and from the surrounding groups of neuronal somata. As can be seen in electron microscopy studies, their core is made up largely of receptor cell axons, periglomerular extensions, and bulbar output cell primary dendrites. Furthermore they are enclosed by a glial wrapping of several thin lamellae (Pinching and Powell, 1971b), so that they appear as anatomical units.

Another argument is given by the organization of the dendritic connections of the bulbar output cells. Macrides and Schneider (1982) observed that mitral and internal tufted cells give rise to a single primary dendrite which extends radially through the external plexiform layer and forms a tuft within a single glomerulus. Since output cells receive their peripheral afferent input from the glomerulus generally situated in register with their somata, the group of output cells connected to a single glomerulus may form a anatomical column.

OLFACTORY BULB FUNCTIONAL ORGANIZATION

On the basis of the findings obtained to date, there are three possibilities for the olfactory bulb to be functionally organized in parallel columns.

First, the functional homogeneity of glomerular efferences may result from a functional homogeneity of glomerular afferences. Indeed, results from 2DG studies show that during odor stimulation, metabolic activity do not widespread throughout the glomerular layer, but remains localized in a small number of glomeruli (Greer and Stewart, 1981; Jourdan, 1982, 1984; Jourdan et al., 1980; Lancet et al., 1982 a,b; Sharp et al., 1975, 1977; Skeen, 1977; Stewart et al., 1979). This suggests that stimulation with a given odor activates a set of receptors that has input to a spatially defined set of glomeruli within the bulb. On the other hand, immuno-histo-chemical studies have revealed the existence of subclasses of receptor cell axons which project preferentially to certain glomeruli (Fujita et al., 1985; Mori 1987a; Mori et al., 1985; Schwob and Gottlieb, 1986). If one assumes that the common expression of certain antigenic properties reflects to a certain extent a similarity of response properties, inputs to individual glomeruli may be supposed to share certain common response properties.

Second, the columnar organization of the bulb may result from the morphology of the glomerulus itself. The glomerulus appears as an almost completely closed structure (Le Gros Clark, 1957; Pinching and Powell, 1971a,b; Willey, 1973) within which K+ and/or neurotransmitters released by the excited afferents may be expected to diffuse (Khayari et al., 1988). One may therefore envisage that this diffusion within the whole glomerular volume will generate a generalized depolarization of all the efferences leaving the glomerulus. It will result in the activation of all output cells connected to the same glomerulus (Freeman, 1974; Jahr and Nicoll, 1981).

Last, the synaptic organization within the glomerulus may be the source of the functional columns. Several authors have reported that the primary nerve terminals and the dendritic arbors of bulbar output cells tend to fill the whole glomerulus and that they contact each other extensively (Kishi et al., 1982; Le Gros Clark, 1957; Macrides and Schneider, 1982; Orona et al., 1984; Pinching and Powell, 1971a,b; Price and Powell, 1970) through "en passant" synapses (Willey, 1973). On the other hand, when broadly examined, labelled glomeruli generally show a relatively uniform level of uptake throughout their neuropil (Jourdan, 1984; Jourdan et al., 1980; Lancet et al., 1982b). This implies that axons of receptor cells with a given odor specificity converge onto some individual glomeruli and terminate throughout their volumes. One may therefore hypothesize that through the extensive contacts of the receptor cells axons onto the output cell dendrites, the peripheral excitation reaching a glomerulus will be transmitted to all the output cells connected to this glomerulus, even if this excitation was conveyed by some afferents only.

FUNCTIONAL ARGUMENTS FOR A COLUMNAR ORGANIZATION OF THE BULB

Although much has been learned about the response properties of olfactory bulb output cells to odorants, far less is known about the role of individual glomeruli in the organization of information processing. For example, indirect arguments are only available on the possibility of a functional organization of the olfactory bulb in parallel columns centered on the glomeruli (for an exhaustive review, see Kauer, 1987; Mori, 1987; Shepherd, 1981; Scott and Harrison, 1987;). Indeed, the differential odorant-evoked 2DG uptake observed by several authors in the deep bulbar layers facing the most heavily-labelled glomeruli (Jourdan, 1984; Lancet et al., 1982a,b) did not concern the mitral cell bodies, but the granule cells. On the other hand, the inter-cell distances explored by Meredith (1986) and by Wilson and Leon (1987) in their electrophysiological studies of the mitral cell reactivity to odorants during electrode penetrations were short enough to analyze intra-bulbar inhibition, but too large to investigate the reactivity of output cells receiving their input through the same glomerulus.

By comparing the reactivity of pairs of mitral cells recorded close together or further apart to a variety of odorants, Buonviso and Chaput (submitted) obtained results which may be interpreted as a manifestation of a columnar functioning of the bulb.

Briefly, simultaneous recordings were made in the olfactory bulbs of anaesthetized rats with paired microelectrodes either close together in the mitral cell layer (less than 40 μm) or at wider distances apart (150 to 200 μm) across this laminae. Stimulations were performed with 5 odorants: Acetophenone (ACE), Cineole (CIN), Isoamyle acetate (ISO), Cymene (CYM) and Cyclohexanone (CYC). The odorants were delivered at 2×10^{-2} of the saturated vapor. They were presented during 10-sec at 1-min intervals at least.

Determination of the reactivity of each cell to each stimulation was performed using a computer. A significant decrease (one sided t- test, level 0.05) in the mean interspike interval with respect to the 30-sec pre-stimulation period was classified as an activation (+). A significant increase was classified as a suppression (-) and a lack of change was classified as a null response (0). This allowed us to construct the response profile of each cell by collecting the succession of its response types to the 5 odorants.

Examples of the response associations the most frequent in close (A and B) and in distant (C) cell pairs are shown in Fig. 1. As can be seen in the pairs of traces presented in A and B, the close cells responded both by excitation to the 10-sec odor stimulations.

One of the distant cells (upper trace in C) also responded by activation while the other was suppressed.

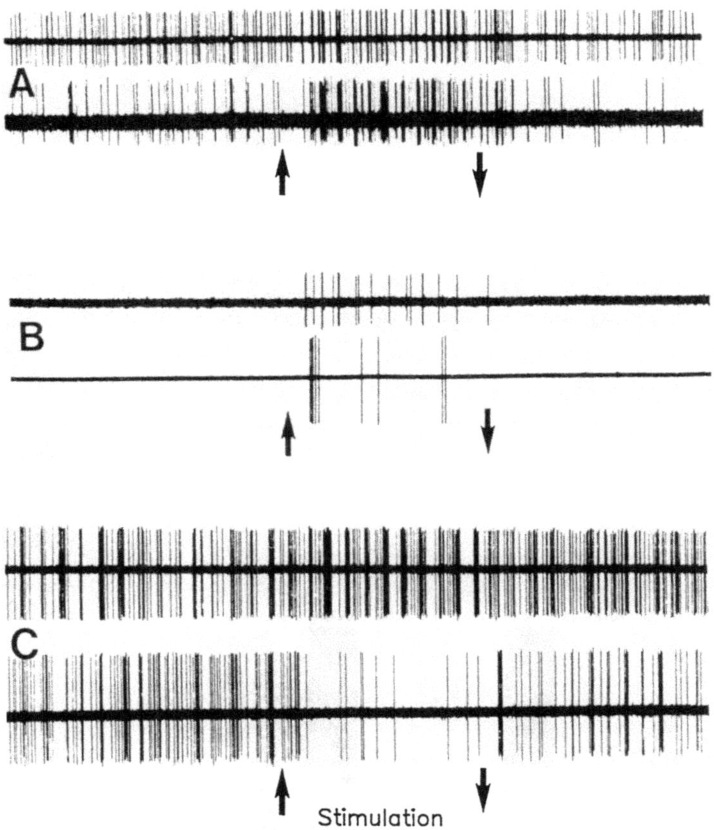

Fig.1: Examples of response associations the most frequently observed in close (A and B) and distant cell pairs (C).

The reactivity of close and distant cells was first analyzed by calculating the probability of observing at least 1, then 2, ... up to 5 pairs of similar-type responses in the response profiles of the paired cells. The analysis was performed with reference to the reactivity of control cell pairs, whose twin profiles were obtained by pairing the response profile of each close cell with the response profile of each distant cell. These twin profiles were considered to be representative of the response profiles we would have obtained from simultaneous recordings performed in far distant regions of the mitral cell layer. Fig. 2 displays results of this analysis. As can be seen, the probability of observing an increa-

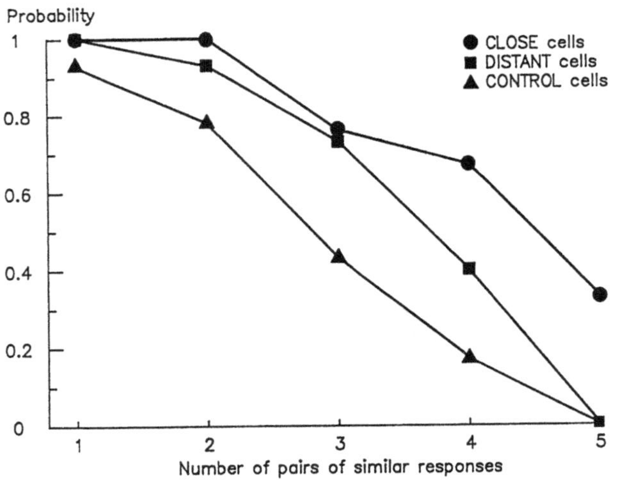

Fig.2: Probability of observing an increasing number of pairs of similar responses (++, 00 or−−) in close−, distant− and control cells.

sing number of pairs of responses of the same type decreases in the three groups, but it decreases less rapidly in the close cell group than in the other two. For 5 odorants, this probability reaches a zero value in the distant and control cell groups while it remains significantly higher in the close cell group. Thus, close cells only displayed the same response profiles to the 5 odorants.

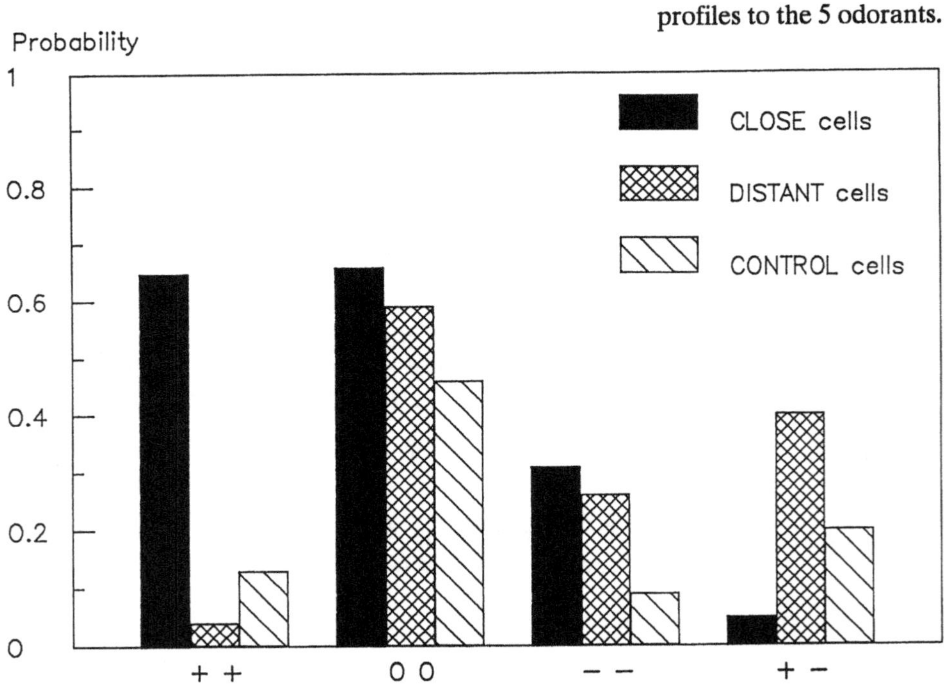

Fig.3: Probability of observing a double excitation (++), a double null response (00), a double suppression (−−) and an excitation paired with a suppression (+−), in each cell group.

Further analyses showed that the similitude of reactivity of close cells resulted not exclusively from a high proportion of double null responses (00), but also from a high proportion of double excitatory (++) or supressive (--) responses (Fig. 3). Indeed, given a cell that was excited by an odorant, the probability of finding another cell that was also excited by this odorant was 5 time greater in the close cell group than in the control one. Finding two cells simultaneously suppressed was also much more frequent in the former group than in the latter.

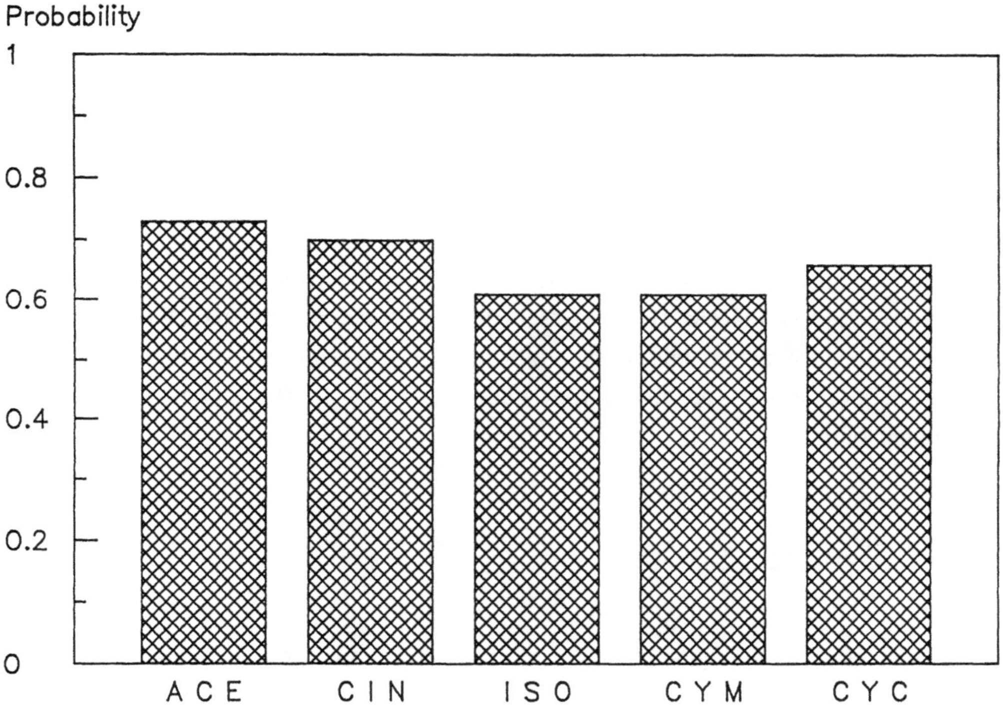

Fig.4: Probability of observing a double excitation (++) in close cells as a function of the odorant.

This similitude of reactivity did not depend on the odorant. Indeed, Fig. 4 shows that, given a cell that responded by excitation to a stimulation, the probability that the other cell will also respond by excitation to this stimulation did not differ significantly among the odorants.

Lastly, the close cells rarely displayed opposite response properties. As shown in Fig. 3, given a cell that was excited by an odorant, the probability that the second cell of the pair will be be suppressed was lower in the close cell group than in the control group and much lower than in the distant cell group. Pairs of distant cells also showed often opposite response properties. Indeed, it was also most unlikely to observe in this group an excitatory response paired with another excitatory response, as shown by the left-histograms in Fig. 3, while it was frequent to observe an excitation paired with a suppression, as shown by the right-hand histograms of the same figure.

Regarding the numerous anatomical arguments presented above, all these results were considered to reflect a step-function of response similarity, with the steps located at glomerular boundaries, rather than a continuum of response similarity, independent of the glomerular boundaries. Indeed, the glomeruli have a average diameter of 130 μm (Royet at al., In press) and there are arranged in an almost single layer. Furthermore, it is known that all olfactory bulb output cells receive their excitatory input through the glomerulus which is situated in register with their somata. Thus it may be argued that close cells, recorded through electrode tips located considerably closer than the average glomerular diameter, were likely connected to the same glomerulus, while distant cells, recorded through electrode tips separated by a distance wider than one average glomerular diameter and lower than two diameters, were likely connected to adjacent glomeruli. Since close cells showed essentially the same reactivity to the variety of odorants and since distant cells tended to have opposite responses, it may be concluded that each glomerulus and the output cells connected to this glomerulus form a functional column.

FUNCTIONAL IMPLICATIONS

Turning to the possible consequences of a similar reactivity of the bulbar output cells connected to the same glomerulus, it may be hypothesized that a columnar functioning of the bulb will insure a further reduction of the dimensions of the olfactory message and permit the simultaneous transfer of the same information to the different zones of bulbar projection.

It is well known that a key aspect of the relationships between the olfactory mucosa and the bulb is the convergence of thousands of receptor cell axons onto a much smaller number of glomeruli. In the rat for example, olfactory fibers converge onto about 3000 glomeruli (Meisami and Safari, 1981) themselves connected to about 70000 mitral cells. Since neither olfactory fibers nor mitral cell primary dendrites ramify before entering the glomeruli, it is about 25000 primary fibers which converge on a single glomerulus and 24 mitral cells which are connected to this glomerulus. Therefore, what happens to the dimension of the olfactory message depends first on the degree of convergence of olfactory fibers onto the glomeruli, but also on what is assumed about the mitral cell functioning. If one assumes that mitral cells function as independent units, then the olfactory message will be encoded on a maximum of 70000 points, corresponding each to a mitral cell. But if one assumes, in view of the results presented above, that the mitral cells connected to the same glomerulus function as a unit, then the bulbar information will be encoded among 3000 foci only, corresponding each to the output of a glomerulus. Instead of being reduced by a factor of about 1000, due to the glomerular convergence alone, the dimension of the olfactory message will be reduced by a factor of about 25000. In our opinion, this reduction of the number of independent bulbar channels encoding information to the number of glomeruli contained in each bulb would not impair the ability of the olfactory system to encode a large number of odorants.

At this state of discussion, it should be stressed that a functional organization of the bulb in discrete, mutually exclusive channels would permit a simultaneous transfer of the same information to the different areas of bulbar projections. One may conceive that the bulbar output cells connected to individual glomeruli may form functionally-independant channels, each of them gathering together a number of elements encoding the same information, but reaching different projection areas. Such an hypothesis is in agreement with the existence of different cell categories within the olfactory bulbar output cell population. Indeed, a growing body of evidence supports the view that the two kinds of bulbar output cells, mitral and internal tufted cells, on the one hand, and the mitral cells on the other hand, do not form an homogeneous population, but are comprised into several morphologically and physiologically distinct categories which differ mainly in the location of their soma, in the position and extension of their secondary dendrites and in the target areas of their axons (For review, see Macrides et al., 1985; Mori, 1987). However, further studies are needed to determine to what extent the various morphologically and/or physiologically distinguished populations of bulbar output cells are implicated in a simultaneous transfer of the olfactory information.

REFERENCES

Buonviso N, Chaput MA (Submitted) Similar responsivity to odors in olfactory bulb output cells presumed to be connected to the same glomerulus: Electrophysiological study using simultaneous single-unit recordings.

Freeman WJ (1974) Relation of glomerular neuronal activity to glomerular transmission attenuation. Brain Res 65: 91-107.

Fujita SC, Mori K, Imamura K, Obata K (1985) Subclasses of olfactory receptor cells and their segregated central projections demonstrated by a monoclonal antibody. Brain Res 326: 192-196.

Greer CA, Stewart WB, Kauer JS, Shepherd GM (1981) Topographical and laminar organization of 2-DG uptake in rat olfactory bulb induced by electrical stimulation of olfactory nerves. Brain Res 217: 279-293

Holley A, Mac Leod P (1977) Transduction et codage des informations olfactives chez les vertebres. J Physiol Paris 73: 725-828.

Jahr CE, Nicoll RA (1981) Primary afferent depolarization in the in vitro frog olfactory bulb. J Physiol London 318: 375-384.

Jourdan F (1982) Spatial dimension in olfactory coding: A representation of the 2-deoxyglucose patterns of glomerular labeling in the olfactory bulb. Brain Res 240: 341-344.

Jourdan F (1984) Le codage spatial de l'information olfactive. Etude anatomo-fonctionnelle. Thesis, Lyon.

Jourdan F, Duveau A, Astic L, Holley A (1980) Spatial distribution of 14C-2DG uptake in the olfactory bulb of rats stimulated with two different odors. Brain Res 188: 139-154.

Kauer J (1987). Coding in the olfactory system. In: Northcutt G (ed) Neurobiology of Taste and Smell. Wiley Series in Neurobiology, pp 205-231.

Khayari A, Math F, Davrainville JL (1988) Electrical stimulation of primary olfactory nerve induces two types of variations in extracellular potassium activity within the glomerulus of the rat olfactory bulb in vivo. Brain Res 457: 188-191.

Kishi K, Mori K, Tazawa Y (1982) Three-dimensional analysis of dendritic trees of mitral cells in the rabbit olfactory bulb. Neuroscience Letters 28: 127-132.

Lancet D, Greer CA, Kauer J.S, Shepherd G.M (1982a) Mapping of odor-related neuronal activity in the olfactory bulb by high-resolution 2-DG autoradiography. Proc Natl Acad Sci USA 79: 670-674.

Lancet D, Greer CA, Kauer JS, Shepherd GM (1982b) Autoradiographic mapping of neuronal activity in the olfactory system of behaving vertebrates. In: Steiner JE, Ganchrow JR (eds) Chemoreception and the determination of behavior. IRL, London.

Le Gros Clark WE (1957) Inquires into the anatomical basis of olfactory discrimination. Proc Roy Soc London Ser B 146: 299-319.

Macrides F, Schoenfield TA, Marchand JE, Clancy AN (1985) Evidence for morphologically, neurochemically and functionally heterogeneous classes of mitral and tufted cells in the olfactory bulb. Chemical Senses 10: 175-202.

Macrides F, Schneider SP (1982) Laminar organization of mitral and tufted cells in the main olfactory bulb of the adult hamster. J Comp Neurol 208: 419-430.

Meisami E, Safari L (1981) A quantitative study of the effects of early unilateral olfactory deprivation on the number and distribution of mitral and tufted cells and of glomeruli in the rat olfactory bulb. Brain Res 221: 81-107.

Meredith M (1986) Patterned response to odor in mammalian olfactory bulb: The influence of intensity. J Neurophysiol 56: 572-597.

Mori K (1987a) Monoclonal antibodies (2C5 and 4C9) against lactoseries carbohydrates identify subset of olfactory and vomeronasal receptor cells and their axons in the rabbit. Brain Res 408: 215-221.

Mori K (1987b) Membrane and synaptic properties of identified neurons in the olfactory bulb. Prog in Neurobiology. Vol. 29, pp 275-320.

Mori K, Fujita SC, Imamura K, Obata K (1985) Immunohistochemical study of subclasses of olfactory nerve fibers and their projections to the olfactory bulb in the rabbit. J Comp Neurol 242: 214-229.

Orona E, Rainer EC, Scott JW (1984) Dendritic and axonal organization of mitral and tufted cells in the rat olfactory bulb. J Comp Neurol 226: 346-356.

Pinching AJ (1970) Synaptic connections in the glomerular layer of the olfactory bulb. J Physiol London 210: 14-15.

Pinching AJ (1972) Spatial aspects of the neuronal connections in the rat olfactory bulb. In: Schneider D (ed) Olfaction and taste IV, Viessenschaftliche Verlagsgesellschaft, Stuttgart.

Pinching AJ, Powell TPS (1971a) The neuron types of the glomerular layer of the olfactory bulb. J Cell Sci 9: 305-345.
Pinching AJ, Powell TPS (1971b) The neuropil of the glomeruli of the olfactory bulb. J Cell Sci 9: 347-377.
Price JL, Powell TPS (1970) The mitral and short axon cells of the olfactory bulb. J Cell Sci 7: 631-651.
Royet JP, Jourdan F, Ploye H, Souchier C (In Press) Morphometric modifications associated with early sensory experience in the rat olfactory bulb. II. Stereological study of the population of olfactory glomeruli. J Comp Neurol.
Schwob JE, Gottlieb DI (1986) The primary olfactory projection has two chemically distinct zones. J Neurosci 6: 3393-3404.
Scott JW, Harrison T (1987) The olfactory bulb: anatomy and physiology. In: Northcutt G (ed) Neurobiology of Taste and Smell. Wiley Series in Neurobiology pp 151-178.
Sharp FR, Kauer JS, Shepherd GM (1975) Local sites of activity- related glucose metabolism in rat olfactory bulb during olfactory stimulation. Brain Res 98: 596-600.
Sharp FR, Kauer JS, Shepherd GM (1977) Laminar analysis of 2- deoxyglucose uptake in olfactory bulb and olfactory cortex of rabbit and rat. J Neurophysiol 40: 800-813.
Shepherd GM (1972) Synaptic organization of the mammalian olfactory bulb. Physiol Rev 52: 864-977.
Shepherd GM (1977) The olfactory bulb: a simple system in the mammalian brain. In: Handbook of Physiology: The Nervous System. Vol I/1, American Physiol Society, Bethesda, pp 945-968.
Shepherd GM (1979) The synaptic organization of the brain. Oxford Univ Press, Oxford.
Shepherd GM (1981) The olfactory glomerulus: its significance for sensory processing. In: Katsuki Y, Norgren R, Sato M (eds) Brain Mechanisms of Sensation. Wiley series: New York, pp 209-223.
Skeen LC (1977) Odor-induced patterns of deoxyglucose consumption in the olfactory bulb of the Treeshrew, Tupaia glis. Brain Res 124: 147-153.
Stewart WB, Kauer JS, Shepherd GM (1979) Functional organization of rat olfactory bulb analysed by the 2-deoxyglucose method. J Comp Neurol 185: 715-734.
Willey TJ (1973) The ultrastructure of the cat olfactory bulb. J Comp Neurol 152: 211-232.
Wilson DA, Leon M (1987) Evidence of lateral synaptic interactions in olfactory bulb output cell responses to odors. Brain Res 417: 175-180.

PERCEPTUAL PERFORMANCE IN PERIPHERALLY REDUCED OLFACTORY SYSTEMS

R. Hudson[1], H. Distel[1] and H. P. Zippel[2]

1) Institut für Medizinische Psychologie
Universität München, Goethestr. 31, D-8000 München 2, FRG

2) Physiologisches Institut der Universität
Humboldtallee 23, D-3400 Göttingen, FRG

The relationship between structure and function is one of the enduring themes of biology, and one which has greatly influenced efforts to understand the neural basis of odour perception. In attempting to answer such basic questions as how odour signals are transmitted and coded, neurobiologists have repeatedly turned to the anatomical organization of the peripheral olfactory system for clues. Cajal was among the first to be impressed by the clear structural organization of the olfactory bulb, an impression of order which was later enhanced by reports that the receptor cells within the nasal cavity project to the bulb in a broadly topological fashion, and that groups of receptor axons then appear to converge onto discrete glomerular units (Kauer, 1987). Not surprisingly, this attractive picture of anatomical orderliness has come to dominate the way we think about olfactory function, and most attempts to explain how odour stimuli are perceived now make reference to some form of primary spatial coding (Shepherd, 1985).

Traditionally it has been thought that molecules of different odourants maximally excite different regions of the receptor sheet, and via the topological projection to the bulb, give rise to distinctive spatiotemporal patterns of activity which represent the basic coding for different odours. More recently, renewed impetus has been given to these ideas by the finding of apparently odour-specific patterns of enhanced 2-deoxyglucose (2-DG) uptake in the glomerular layer of the bulb (Stewart et al., 1979). However, attractive though such concepts of spatial coding are, there have in fact been very few investigations of the behavioural significance of the electrophysiological and neuroanatomical findings on which they are primarily based. And yet this would not appear to be particularly difficult. Challenging the structural integrity of the peripheral

NATO ASI Series, Vol. H 39
Chemosensory Information Processing. Edited by D. Schild

olfactory system in controlled and definable ways and then testing the effects of this on the animal's behaviour should provide a direct means of investigating the relationship between broad features of anatomical organization on the one hand and function at the whole animal level on the other.

In this paper we would therefore like to review the effects of performing two kinds of lesions on odour-guided behaviour: (1) partial bulbectomy, a procedure involving the quantifiable reduction of neural space; and (2) olfactory nerve transsection followed by bulbar reinnervation, a procedure which presumably disrupts the integrity of previously established connections. In doing so, we will concentrate primarily on findings from two very different animal models, namely the newborn rabbit and the goldfish. However, before coming to the actual results, it may be useful to describe the behavioural paradigms used in testing the perceptual performance of these two species.

THE BEHAVIOURAL PARADIGMS

Newborn rabbits posses a number of behavioural specializations making them particularly suitable subjects for investigating mammalian olfactory function. Rabbit pups are only nursed for about 3 minutes once a day, and to locate nipples in this short time, they are totally dependent on a short-ranging pheromone on the mother's belly which releases and guides a highly stereotyped nipple-search behaviour. This is the case even for pups delivered by caesarian section and thus without postnatal suckling experience. The emission of the pheromone is stimulated by pregnancy, long photoperiod, or the administration of estradiol and thus appears to be under hormonal control (Hudson and Distel, 1986).

To assess responsiveness to the nipple-search pheromone in a standardized way a procedure has been developed in which pups are tested in an arena enclosing the belly of a doe rabbit (Fig. 1). Each pup is given five, 30-second trials per condition, and the number of successful trials, the time spent searching and the latency to make the first

Fig. 1 Arena used for testing nipple-search performance in newborn rabbits with a 5-day-old pup at a nipple of an anaesthetized test doe.

vigorous probing movement with the snout, that is to start searching, scored. To control for the possible effects of general rabbit odours, pups are also tested on an ovariectomized, non-pheromone producing doe. In addition, suckling performance under the normal, highly competitive conditions in the nest is assessed by weighing pups before and after nursing (Distel and Hudson, 1984).

Although rabbit pups do not need to learn postnatally the odour cues governing this nipple-search behaviour, they can nevertheless rapidly learn to associate novel odours with suckling. Pups nursed by a doe whose ventrum has been scented with an odourant such as the perfume Chanel No.5, citral or camphor, come to respond to the novel scent as to the pheromone itself. After only one such pairing, newborn pups show the full sequence of search behaviour, nipple attachment and vigorous sucking when tested in the arena on the belly of a non-pheromone producing doe, or even a female cat, scented with the odour. In contrast, control pups which have not been conditioned will not respond to these odours and quickly come to rest (Hudson, 1985).

Thus, in testing the effect of an intervention on odour perception in newborn rabbits one may ask the following three questions: 1) can they perceive the naturally occurring pheromone; 2) can they learn to associate a novel odour with suckling, and 3) will they continue to respond to an odour signal learned prior to the intervention?

The second behavioural paradigm, represented by the goldfish, also makes use of responsiveness to natural and artificial odourants in the context of feeding (Zippel et al., 1988). Goldfish placed in a large aquarium equipped with opaque feeding funnels at either end, will spontaneously orient to the funnel containing an extract of a preferred food such as tubifex worms. By pairing a reward of worms with otherwise neutral amino acids or synthetic odourants such as amyl acetate or cumarine, the fish can be trained to

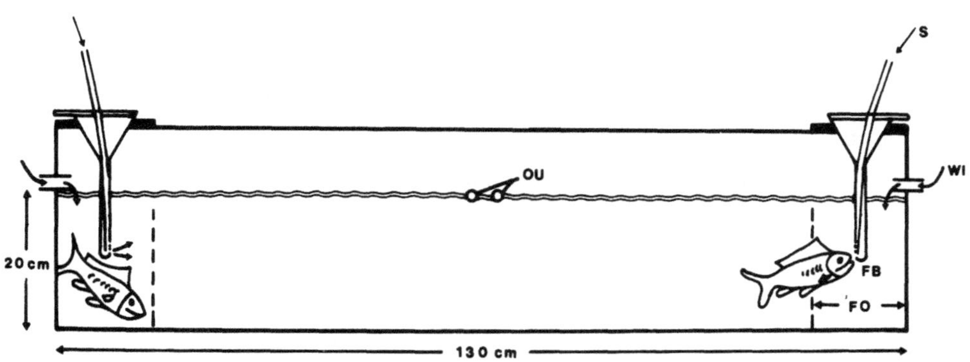

Fig. 2 Aquarium used for training and testing odour-guided food location in goldfish. FO funnel orientation; FB funnel biting; S stimulus application; WI water inlet; OU water outlet.

associate these with food and use them to correctly identify which of two funnels will be used for the presentation of food in an aquarium choice test (Fig. 2).

Training is generally preceded by a 4-5 week familiarization period during which the animals learn to associate the funnels with food. By the end of this time the fish have developed a complex behavioural repertoire, the two main quantifiable components of which are funnel orientation - that is the time spent by an animal in the immediate vicinity of a particular funnel, and funnel biting. Before each daily training or testing session, a 3 minute control test is conducted during which the animal's reaction to the funnels in the absence of odour stimuli is recorded. The training or test stimulus is then presented through the funnel to which least activity was directed during the control registration, and the animal's response recorded for a further 3 minutes. Whereas training sessions end with food being presented through the appropriate funnel, at no time during or immediately following a test session is the animal's behaviour reinforced. Using this procedure, the fish may be trained either to respond to a single stimulus or to discriminate between two different odourants presented simultaneously and, as in the newborn rabbit, their ability to perceive such odours following lesioning tested (Zippel et al., 1970; 1981; 1988).

DISRUPTION OF SPATIAL MAPS

When considering the functional interpretation of maps such as those generated by electrophysiologists interested in the distribution of potential receptor cell classes, or by anatomists using the 2-DG method to plot the distribution of metabolic activity in the glomerular layer of the bulb, three broad possibilities come to mind: 1) that such maps or patterns represent a gestalt, the whole of which is necessary for recognition of a particular odour, 2) that the maps contain considerable 'noise' in which the essential information is locally embedded, or 3) that they reflect a certain redundancy with adequate information being represented in any part.

In a first investigation of these possibilities, the ability of rabbit pups to respond to pheromonal and artificial odour cues associated with suckling has been tested following extensive medial or lateral lesioning of both olfactory bulbs. Somewhat surprisingly, pups with both types of lesion and with as little as 20% of the bulb remaining intact performed almost as well as control pups on all tasks tested. Thus, they continued to respond readily to the nipple-search pheromone and were well able to survive the rigours of normal nursing, they could readily learn to associate novel odours with nursing and, perhaps most importantly, continued to respond appropriately to artificial odour stimuli learned prior to lesioning (Hudson and Distel, 1987). Taken together, these findings suggest that in the newborn rabbit small and apparently arbitrary regions of the olfactory bulb are capable of a high degree of odour processing, and that no specific area of the bulb or epithelium is essential for mediating responsiveness to odours associated with suckling.

PARTIAL BULBECTOMY

Species	Task	N	Recovery (days)	Test	Post-operative experience	Stimuli
rabbit pup	recognition	8	1	nipple searching	—	pheromone
(Hudson & Distel, 1987)	"	4	"	"	—	citral
	"	8	"	"	—	Chanel No 5
	learning	8	4	"	3 learning trials	citral
rat pup	recognition	2	1	nipple attaching	—	maternal odor
(Risser & Slotnick, 1987)	"	5	5	"	suckling	"
adult rat	recognition	6	20	go / no go (water)	—	amyl acetate
(Slotnick et al., 1987)	threshold			"	post-recovery training	proprionic acid
	learning			"	"	butanol geraniol

Table 1: Studies reporting the effects of partial bulbectomy on odour-guided behaviour in vertebrates.

Nor is such robustness of peripheral olfactory function peculiar to the newborn rabbit or immature animals in general (Table 1). Rat pups are also dependent on odour cues for suckling, and whereas bulbectomy completely disrupts this, here too, small remnants of bulbar tissue appear sufficient to sustain the suckling response and ensure survival (Risser and Slotnick, 1987). Even adult rats subjected to partial bulbectomy maintain their responsiveness to odours learned prior to lesioning, and readily learn to detect novel odours post-operatively with very similar threshold scores to sham-lesioned controls (Slotnick et al., 1987). Thus, from the slowly accumulating behavioural evidence we may tentativly conclude that at the level of the peripheral olfactory system odour perception and coding appear to be based on some form of distributed information processing characterized by a considerable degree of regional autonomy. While local specialization cannot be excluded, the findings of these studies suggest that small and arbitrary regions of the bulb - and by extension, of the epithelium, must contain sufficient number and variety of neural elements for the processing of even quite complex odours such as perfumes.

DISRUPTION OF TEMPORAL CONTINUITY

In the experiments outlined above, part of the connectivity between epithelium and bulb was always left intact. However, as the sensory neurons of the olfactory epithelium are renewed throughout life and will be replaced following destruction or axotomy, one may ask to what extent odour coding and processing are affected by reinnervation of the bulb following complete nerve transsection - a question which has been extensively investigated in the goldfish (Table 2).

AXOTOMY

Species	Task	N	Recovery (days)	Test	Experience until recovery	Stimuli (s+)	Stimuli (s-)
pigeon	learning	4	211-231	shock avoidance	training	amyl acetate	
(Oley et al., 1975)	"	4	16-82	"	retraining	"	
goldfish	learning	7	90	food location	training	cumarine	
(Zippel et al., 1970)	discrimination	6	10	"	testing	cumarine	amyl acetate
	"	6	"	"	"	amyl acetate	cumarine
(Zippel et al., 1988)	recognition	10	6-7	"	testing	5 amino acids	
	discrimination	10	10	"	—	4 amino acids	1 amino acid
(anterior bulb-X)	recognition	10	8	"	testing	tubifex extract	
"	"	14	11	"	"	amyl acetate	
"	discrimination	10	10	"	—	4 amino acids	1 amino acid

Table 2: Studies reporting recovery of odour-guided behaviour in vertebrates following olfactory nerve transection.

In this species, bilateral transection of the primary olfactory nerves results in the immediate failure of even well trained fish to respond to odourants presented in concentrations below the taste threshold. However, within as little as 9 tó 10 days, the first clear histological signs of reinnervation are evident, the fish again start responding appropriately to single odourants learned prior to lesioning, and may even continue to discriminate between previously learned competing stimuli. If more time is allowed, the threshold performance of these fish becomes indistinguishable from that of unoperated controls (Zippel et al., 1970; 1988). A similar recovery of function has also been demonstrated in pigeons which, following regeneration of the transected olfactory nerves, can be trained to respond to an odour stimulus in a shock avoidance paradigm (Oley et al., 1975).

Given these findings, one may ask whether such behavioural recovery is also possible in mammals. To examine this question, all nerve strands leading to one olfactory bulb were transected intracranially in newborn rabbits from four litters (four operated and one control pup in each). After surgery, pups were well able to perceive the pheromone using the intact bulb and all could suckle in the nest. To test for functional recovery, the nares of either the intact or lesioned side were temporarily closed using tissue glue and the pups' response to a pheromone-producing doe recorded. Whereas no pup responded to pheromone with the lesioned side before post-surgery day 8, all animals raised to post-surgery day 14 did so, and in most animals functional recovery was already evident by about day 10. Even after removal of the intact olfactory bulb, 15-day-old animals showed vigorous nipple-search behaviour and were able to obtain normal amounts of milk in the nest using the reinnervated bulb only (Hudson and Distel, in prep.).

Animals were sacrificed 15 days after lesioning and the olfactory bulbs and septal epithelium for evidence of regeneration. To investigate the time course of reinnervation histologically, additional pups were sacrificed at various times following denervation. Five days after transsection, the bulb was characterized by an absence of connections and clear degeneration of all glomeruli. Nine days after surgery, however, small, irregularly formed glomeruli started to be visible on the ventrome-dial or ventrolateral surface of denervated bulbs. By day 15, the bulbar surface of all animals was almost completely reinnervated although the glomeruli were still smaller than those of the control bulb (Distel et al., in prep).

Faced by the small number and size of glomeruli sufficient for a functional recovery, the question then arises to what extent such recovery is dependent on the reestablishment of specific connections or classes of connections between the epithelium and bulb. While there has as yet been no direct investigation of this fundamental question, it would appear that the pattern of reinnervation need not be very specific. Thus, rabbit pups responded equally well to the pheromone regardless of whether reinnervation first occurred medially or laterally, and goldfish can respond to odour stimuli after 9 to 10 days even when axotomy is combined with removal of the rostral half of the olfactory bulbs (Zippel et al., 1988). Most remarkable of all are reports that mice and rats with most, if not all, of their olfactory bulbs removed, also show long term functional recovery in odour avoidance and food finding tasks (Wright and Harding, 1982; Amemori et al., 1988). While the exact nature and extent of the lesions inflicted may be debated, these studies nevertheless suggest that even the reinnervation of minimal bulbar remnants may be sufficient for the recovery of olfactory function.

RECOVERY: PRESERVATION OR RECODING?

Given these findings, one may further ask whether odour signals must be relearned following nerve transection or whether the coding is preserved. While in most of the cases reviewed above the animals were either trained post-operatively or had the opportunity to relearn the odour signals during repeated testing, there is evidence from the goldfish that recovery of specific olfactory function may occur independently of post-lesion experience.

Goldfish subjected to combined axotomy and rostral bulbectomy and then allowed to recover for 9 days without being tested, perform as well as controls in odour discrimination tasks learned prior to lesioning (Zippel et al., 1988). Further, in a particularly stringent reinvestigation of these findings, 12 fish trained to discriminate between the amino acids lysine and argenine were subjected to axotomy and during the recovery period were neither tested nor fed. As shown in Fig. 3, these fish appeared to retain their ability to discriminate between the two odours, even without the benefit of any form of task-related, post-operative experience.

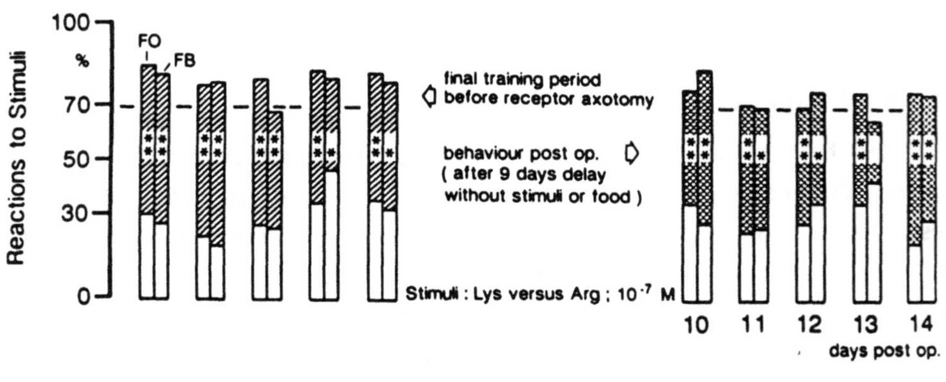

Fig. 3 Test performance before and 9 days after receptor axotomy in 12 goldfish trained to discriminate between the amino acids lysine and argenine. Mean values for funnel orientation (left bar) and funnel biting (right bar) are shown for the stimulus-free control condition (open bars) and in response to the training stimulus (total bars). Significance was tested by comparing values from the test sessions with values for the same funnel during the control registrations (Wilcoxon rank test; ** $p < 0.01$, * $p < 0.05$).

In rabbit pups, however, the situation appears somewhat different. To test whether the recovery of pheromone recognition is dependent on post-operative experience, four litters each containing four lesioned and one control pup, were hand-raised using a pheromone-free milk formula. Responsiveness to the pheromone was then tested 12 days after surgery following temporary closure of the nare on the intact side. While the four control pups demonstrated clear nipple-search behaviour and thus retention of the pheromonal signal, the lesioned pups failed to respond (Fig. 4).

All but five of the operated pups were then tested with the intact side open. Showing normal search behaviour they were then given to their mother to be nursed, while the others were again bottle-fed. On the following day, the pups suckled by their mother now responded vigorously to the pheromone with the reinnervated side alone, whereas their hand-raised litter mates did so only after the nare of the intact side was also opened. Histological examination indicated clear reinnervation of the denervated bulb in all cases, and no systematic difference in the pattern or extent of reinnervation between the different groups. Thus, these findings would suggest that following axotomy in rabbit pups, experience is necessary, or at least greatly facilitates, the recovery of specific olfactory function - even in the case of a pheromonal signal.

Fig. 4: Effect of axotomy and subsequent hand-raising on the response of 15 newborn rabbits to nipple-search pheromone. While hand-raised controls responded positively, lesioned pups only showed nipple-search behaviour after suckling experience.

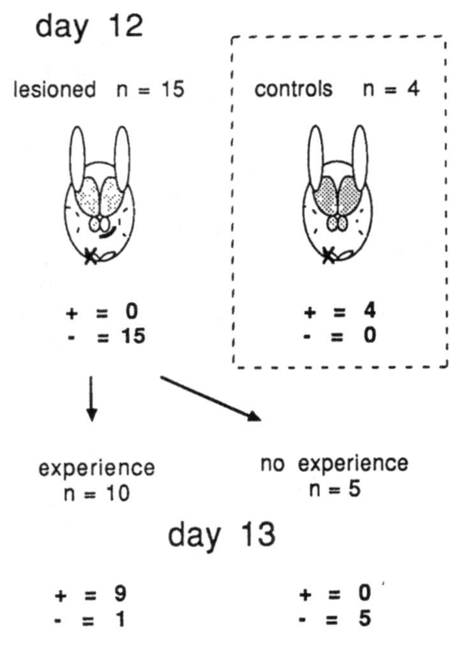

In the past there has been an understandable reluctance to use lesions to probe the mechanisms underlying olfactory function. As has been rightly pointed out, the effects of lesions are often non-specific and difficult to interpret (Alberts, 1974) - a criticism which, however, hardly applies to the positive results reported here. Confronted by the robustness of peripheral function and apparent redundancy in olfactory coding, one may well ask why in most vertebrates such large amounts of neural space are allocated to the processing of olfactory information. One answer most probably relates to the fact that animals generally have to discriminate between large numbers of subtly varying odours and be able to distinguish signals at low concentrations from complex background 'noise'. Thus it may be expected that presenting lesioned animals with tasks equivalent to the challenges of normal life should reveal deficits not evident from the relatively simple tests employed to date.

REFERENCES

Alberts JR (1974) Producing and interpreting experimental olfactory deficits. Physiol Behav 12: 657-670

Amemori T, Valuskova V, Zigova T, Galik J, Racekova E, Bures J (1988) Functional recovery after olfactory bulbectomy in rats: Effect of embryonal brain grafts. Physiol Bohemoslovaca 37: 385-394

Distel H, Hudson R (1984) Nipple-search performance by rabbit pups: Changes with age and time of day. Anim Behav 32: 501-507

Distel H, Stahl B, Hudson R (in prep) Reinnervation of the olfactory bulb following transection of the olfactory nerves in newborn rabbits.

Hudson R (1985) Do newborn rabbits learn the odor stimuli releasing nipple-search behavior? Dev Psychobiol 18: 575-585

Hudson R, Distel H (1986) Olfactory guidance of nipple-search behaviour in newborn rabbits. In: Breipohl W (ed) Ontogeny of Olfaction. Springer, Berlin Heidelberg New York, pp 243-254

Hudson R, Distel H (1987) Regional autonomy in the peripheral processing of odor signals in newborn rabbits. Brain Res 421: 85-94

Hudson R, Distel H (in prep) Recovery of behavioral function following transection of the olfactory nerves in newborn rabbits.

Kauer JS (1987) Coding in the olfactory system. In: Finger TF, Silver WL (eds) Neurobiology of Taste and Smell. Wiley & Sons, New York, pp 205-231

Oley N, DeHan RS, Tucker D, Smith JC, Graziadei PPC (1975) Recovery of structure and function following transection of the primary olfactory nerves in pigeons. J Comp Physiol Psychol 88:477-495

Risser JM, Slotnick BM (1987) Nipple attachment and survival in neonatal olfactory bulbectomized rats. Physiol Behav 40: 545-549

Shepherd GM (1985) The olfactory system: The uses of neural space for a non-spatial modality. In: Correia MJ, Derachio, AA (eds) Contemporary Sensory Neurobiology. Alan Riss Inc, New York, pp 99-114

Slotnick BM, Graham S, Laing DG, Bell GA (1987) Detection of proprionic acid vapor by rats with lesions of olfactory bulb areas associated with high 2-DG uptake. Brain Res 417:343-346

Stewart WB, Kauer JS, Shepherd GM (1979) Functional organisation of the rat olfactory bulb analysed by the 2-deoxyglucose method. J Comp Neurol 185:715-734.

Wright JW, Harding JW (1982) Recovery of olfactory function after bilateral bulbectomy. Science 216: 322-323

Zippel HP, Baumgarten Rv, Westerman RA (1970) Histologische, funktionelle und spezifische Regeneration nach Durchtrennung der Fila olfactoria beim Goldfisch (Carassius auratus). Z vgl Physiol 69: 79-98

Zippel HP, Breipohl W, Schoon H (1981) Functional and morphological changes in fish chemoreception systems following ablation of the olfactory bulbs. In: Flohr H, Precht W (eds) Lesion-Induced Neuronal Plasticity in sensory motor systems. Springer, Berlin Heidelberg New York, pp 377-394

Zippel HP, Meyer DL, Knaust M (1988) Peripheral and central post-lesion plasticity in the olfactory system of the goldfish: Behavior and morphology. In: Flohr H (ed) Post-Lesion Neural Plasticity. Springer, Berlin Heidelberg New York, pp 577-591

SCENT TRAILING BY TRACKING DOGS. WHAT IS THE PHYSIOLOGICAL BASIS FOR CONCENTRATION CODING ?

Kjell B. Døving

Department of Biology, P.O.Box 1051
University of Oslo
0316 Oslo 3, Norway

INTRODUCTION

The olfactory organs of mammals have several outstanding properties. Some of these are well known to us. Police dogs are able to sniff out drugs, and avalanche dogs rescue unfortunate skiers trapped under snow. In these acts the dogs demonstrate their ability to detect small quantities of odorants and discriminate different smells.

The qualitative aspect of olfactory function has been the main subject for very many physiological studies of the olfactory system. Several authors have discussed the physiological basis of the extreme sensitivity of the olfactory system. The large number of receptors, the specific sensitivity of single receptor cells, and the convergence of the many thousands primary receptor neurones to a single secondary neurone, are all part of this quantitative aspect of the olfactory function.

A phenomenon that has attracted the interest of Johan B. Steen at the Department of Biology, University of Oslo and Erik Wilsson, The Swedish Dog Training Center, Sollefteå, Sweden; concerns the ability of a dog to track the direction of the trail. A hunting dog can track the trail left by a prey and follow the trail to find the quarry. This feature of the dog's nose concerns an aspect not often mentioned in the physiological literature, but one which has been commonly observed by hunters watching the behaviour of the dog, and by ecologists. In the approach to this phenomenon I will first describe some properties of the trail and the complex chemistry that the track can present. Secondly, I shall mention some of the results of physiological studies which may be of interest in our attempts to understand the basis for the dog's outstanding

NATO ASI Series, Vol. H 39
Chemosensory Information Processing. Edited by D. Schild

performance. Hopefully this will excite new ideas that produce incentives for discussion and suggestions for new experiments.

THE TRAIL

Chemistry. A stationary human subject will leave a scent around him. The scent that a dog perceives is a composite of many odorants. These odorants come from the body, breath, hair and sweat. Gas-chromatographic studies of axillary sweat demonstrate up to 300 different volatile compounds (Sommerville and Gee 1988). These authors have also shown that different human races produce varying amounts of volatile substances. For example Caucasian Europeans, exude more smells than the Japanese do. The body odours will be mixed with any artificial odorants that the person is carrying. It has been suggested that "heavy" odorants form a small area of concentrated scent immediately around the person of origin. "Lighter particles" move further away, so that their concentration on the ground decrease, and they cover a large area. The lightest particles float away in the air, and their concentration decreases until they become imperceptible.

The track. As the subject moves through vegetation, for example, he forms a "track picture", of varying concentrations of scents. The footprints, in addition to being visible, will also comprise scents from crushed vegetation, footwear and odorants from the person's feet. The heaviest particles will form a scent track about 1.5 m wide. The second phenomenon has been elegantly demonstrated by letting dogs track subjects walking in the middle of a shallow stream of varying breadth (Clifford 1964). Finally there is a wide band of lighter odorants, extending to approximately 10 m on either side of the track itself. The dog is also able be perceive these odorants, as demonstrated by the way a dog follows a track across a road on a hot day. The tracking dog almost invariably has its "head down" as he approaches the road, then he lifts his head and continues with "head up" as he crosses the road, and then puts his "head down" again when he reaches the other side of the road. The reason for this difference in behaviour is to be found in the physical parameters of the environment. In daytime, a hot road produces rising air currents which carry any odorants with them, and these odorants are deposited on the vegetation verging the road.

Rain. It is said that the odorants forming a scent dissolve slowly in water. Clifford (1964) claims that a tracking dog following a track laid and followed during dry weather works at strength "A". If there is heavy rain when the track is laid, and dry weather when it is followed, the dog works at strength "A-". The heavier the rain the more difficult it seems to be for the dog to follow the track. However, if there is light rain or drizzle while the track is laid, and the weather is dry during the follow-up period, the dog works at strength "A+". A possible reason for the last phenomenon is that the odorants are entrained in the evaporating water. The strong aroma that we sense when we walk in a sunny forest after rain is another example of odorants being carried to our noses by minute drops of water.

We know from experience that as the air increases in humidity the odorants or scent particles become more "buoyant" and are dispersed further away from the track. This makes it increasingly difficult for a dog to track a scent successfully.

A dog cannot follow odorants deposited by the subject through water or a swamp. However, if there is abundant vegetation, a swimming dog can scent odorants left on vegetation.

Sun. Direct sunlight rapidly reduces the time a dog is able to follow a track. In the worst conditions a track can last as little as 15 min. In overcast conditions, or in deep vegetation, tracks can be perceived for up to 48 hours.

Wind. Wind will naturally disturb the tracking picture. A tracking dog will move outside the footprint to the fallen odorants carried aside by the wind. Distances as great as 18 m have been quoted.

DIRECTION DETECTION

It is a valuable asset for the survival of wild dogs and other hunting animals to be able to follow their quarry. When a dog first encounters a track he will stop and take a few sniffs at one spot, before soon moving on to another spot and sniffing again. Sniffing at only two different footprints may be enough to tell him in which direction to track. An experienced dog trainer can change the natural tendency of the dog to track in the direction in which the subject has been walking.

Track picture. The ability of the dog to take direction depends on how the track has been made. If the human subject walks backwards, the dog will still track in the direction in which the subject is walking, if so trained. However, the dog is fooled if the subject drags his feet along the ground. The dog is similarly fooled if the person's boots are dragged across a field (Steen and Wilsson,1986 and in preparation). Thus it seems that, if a dog is to be able to track direction, there has to be some space between the prints or, in other words, the track must not be continuous.

The concentration gradient. The differences in concentration between consecutive foot-steps are small. A person normally walks at a pace of about 90 steps a min., giving 5 400 steps in an hour or 10 800 steps in two hours. Roughly calculated, this means that after 2 hours a dog has to perceive a difference in concentration difference of 1/10 800 in two adjacent footprints.

The dog's ability to sense direction depends on the performance of his olfactory system. The dog has to perceive differences in concentration. A possible way to do this may be to compare the odour concentration of one footprint with that of the next. Thus the dog reacts to chemical gradients by klinotaxis, and discerns the axis of a chemical gradient by measuring differences in concentration through repeated sampling in time and space. This requires that the dog react and memorize the impression left by sniffing the first footprint and then compare this with the second footprint in terms of concentration of odorants.

PHYSIOLOGY

In order to elucidate the mechanisms by which odorants are discriminated, we studied the electrical activity of the relay neurones of the rat olfactory bulb when the animals were exposed to different concentration of different odorants (Døving 1987). The rats were anesthetized with nembutal, and a tracheostomy was performed to gain access to the choana. Suction through the left nostril was kept at 10 ml/s to make sure that the total olfactory cavity was exposed to air. The left olfactory bulb was exposed, in order to undertake extracellular recording of single neurone activity. The neurones were found to project to the lateral olfactory tract (LOT) by stimulating this structure electrically.

During continuous suction of air through the choana, various odorants were mixed in a turbulent flow of air and applied at increasing concentrations. The stimulation formed a ramp of increasing odorant strength. The odorant concentration was monitored using a flame-ionization device.

For a given odorant, the number of neurones excited, corresponded approximately to the number of cells that were suppressed. The effectiveness of the 21 odorants tested, in inducing changes in spike discharges, varied considerably. Isoamyl acetate was the most efficient. This substance excited about 30 percent of the neurones and inhibited the same percentage of cells.

The response profile of spike discharge in the relay neurones of the olfactory bulb to a presentation of a ramp stimulation could be classified into different types. In one type the odour stimulation led to an abrupt increase in spike frequency. The cell started to discharge impulses at a fixed concentration and remained active throughout the stimulation period. The spike frequency was at a maximum at the beginning of the stimulation period. The cell discharged impulses at about the same frequency at the beginning as at the end of the stimulation period, even though the concentration had increased ten-fold. This type of response profile was seen in 44 percent of the odour presentations causing to excitation.

The other response profile, found in 37 percent of the odour presentations producing excitation of the cell activity, was characterized by an increase in spike discharge following the increase in concentration.

For each response evoked in a cell by a stimulus, we related the start of impulse firing to the odorant concentration as monitored by the flame ionization detector. Thus, for each response to any odorant, it was possible to determine a threshold concentration at which the neurone changed its impulse activity. The threshold in response to amyl acetate was determined for 86 mitral cells projecting to the LOT. The thresholds varied in the concentration range from 5.62×10^{-8} to 2.69×10^{-5} M. Of these units, 73 had thresholds between 3×10^{-7} and 3×10^{-6} M.

DISCUSSION

As the tracking dog is exposed to progressively higher concentrations of one single odorant more relay neurones are recruited. The neurones are recruited from different areas of the olfactory bulb. It is possible to imagine that each concentration step makes up a special spatial pattern of active neurones.

Some neurones, namely those responding to increasing concentration by an increasing spike discharge, will be more affected, that is excited more extensively by the higher concentrations. I refer in this connection to the article on neurone doctrines of perception by Barlow (1972). One of these doctrines states that a neurone is excited in the presence of a trigger stimulus. Accordingly a high discharge frequency indicates, with great probability, the presence of a trigger stimulus. An increasing discharge frequency which follows the concentration of the odorant may also provide a basis for discrimination of intensity. For example, it is possible to imagine that the relay cells of the olfactory bulb connect to cells of the olfactory cortex in such a way that cortical neurones are to be excited only if the frequency of the synaptic input exceeds certain values. Thus increasing odour concentrations will recruit more neurones in the olfactory cortex.

One might suppose that the dog uses tropotaxis, which implies a simultaneous comparison of the odour concentrations reaching the right and left nostrils. Here the hunting dog could use the differences in the odour concentrations from the toe and the heel of a single footprint. There are several objections to this hypothesis, however. The quarry of a wild hunting dog, may be an antelope, walks on its toes, and there is thus little difference for the dog to detect. Secondly, the dog can be fooled into following the track in the reverse direction to the movement of the object. Steen and Wilsson made tracks with a bicycle on the wheel of which they had fastened two leather bands impregnated with sausage. In this case, the concentration of the object being tracked declines with time. The fresh track will naturally be strongest at the start and gradually diminish in concentration along the track. In such cases the faithful German Shepherd tracker dog always tracked in the direction of increasing concentration. Thus he was found happily waving his tail at the start of the track. The observation concerning bicycle tracks, together with the observation that the dog needs to compare the scents of two different footprints, suggests that this tropotaxic performance is not the preferred mechanism for detecting the direction of tracks.

It should be recalled that when the level of concentration of the track increases, this also increases the number of odorants that the dog perceives as he pursues and approaches his quarry. It can be assumed that this process also changes the quality of the scent. It is possible that the two processes, namely recruitment with increased odorant concentration and recruitment as a response to new odorants, cooperate, and together supply meaningful information to the animal.

REFERENCES

BARLOW, H.B. 1972. Single units and sensation: A neuron doctrine for perceptual psychology? Perception. 1:371-394.

CLIFFORD, R.J. 1964. Some notes and theories on 'scent'. Its formation, properties, and usage as derived from observations on and experience with tracker and patrol dogs in Malaya. Veterinary Annual. 10-21.

DØVING, K.B. 1987. Response properties of neurones in the rat olfactory bulb to various parameters of odour stimulation. Acta physiologica scandinavica. 130: 285-298.

SOMMERVILLE, B. and D. GEE. 1984. Research on body odours: New prospects for combating crime? International Criminal Police Review. July-August. 407:18-22.

STEEN, J.B. and E. WILSON. 1986. A sense of direction on the scent of trouble. New Scientist. March 13. p 26.

STEEN, J.B. and E. WILSON. How do dogs determine the direction of a track ? Appl. Animal Behav. Sci. Submitted.

EXTRACTING INFORMATION FROM SPIKE TRAINS OF OLFACTORY BULB NEURONS

John W. Scott

Department of Anatomy and Cell Biology
Emory University School of Medicine
Atlanta, Georgia 30322
USA

This conference is devoted primarily to the transduction of olfactory sensory signals and to the processing of those signals. This presentation will deal with the issue of how we can evaluate those signals when we observe them as trains of spikes in the vertebrate olfactory bulb or other central nervous structures.

It is perhaps first important to point out why it is still useful to collect data on sensory processing by single neuron recording, when there exist other techniques that allow observation of functional changes in the nervous system, several of which have been applied to the vertebrate olfactory system. Certainly some of these techniques, such as 2-deoxyglucose autoradiography, multiple electrode EEG recording, and voltage sensitive dye recording, allow an extensive, simultaneous sampling of the spatial distribution of activity in the nervous system that cannot be attained by single neuron recording. However, at present these techniques do not allow the spatial and temporal resolution afforded by single neuron methods. Thus they do not allow correlation of functional data with the particular cell types that can be studied by morphological techniques. In addition, it is likely that as sophistication in data gathering proceeds with the voltage sensitive dye methods, some of the same data analysis problems observed with single neuron recording methods will manifest themselves.

Single neuron recording methods have been used since the fifties to study the responses of single neurons in the olfactory bulb, and some smaller number of investigations have dealt with more central regions, both cortical and subcortical. An almost constant feature of these reports has been the appearance of temporally patterned responses to odor stimuli. The complexity of these temporal patterns has been the subject of many of those reports (Doving, 1987;Kauer, 1974; Mair, 1982;

NATO ASI Series, Vol. H 39
Chemosensory Information Processing. Edited by D. Schild

Macrides and Chorover, 1972). The origin of these complex spike trains probably lies in the inhibitory interactions involving the olfactory bulb granule and periglomerular cells. The existence of these intrabulbar inhibitory interactions has long been recognized from anatomical studies and from physiological and pharmacological studies of the olfactory bulb responses to electrical stimulation (Shepherd,1979). Strong evidence exists for the participation granule cells in odorant responses (Freeman,1975; Scott and Aaron,1977; Wellis and Scott,1989). Most evidence indicates that olfactory stimuli produce spatially localized responses in receptor populations and that, acting through a topographic mucoso-bulbar projection, these produce spatially localized activation of the second order cells (Kauer and Moulton,1974; Makay-Sim and Shaman,1984; Stewart et al,1979; Jourdan et al,1980; Saucier and Astic,1986). The spatial and temporal patterns of activation of the second order cells should produce equally complicated patterns of activation of inhibitory interneurons, which should, in turn, inhibit or disinhibit those second order neurons.

Presumably it is this recruitment of inhibitiory circuits that has lead some observers to describe "non-monotonic responses" evoked by a series of odorant concentrations and to describe a number of "types" of response by second order neurons (Duchamp-Viret,1988; Kauer,1974; Mair,1982). By this reasoning the different response types recognized some of these workers correspond to recordings made in regions that are excited to different degrees, on the edges of excited regions that may receive primarily inhibitory inputs or from neurons that are modulated by different types of local circuits. It may be more reasonable to view these "non-monotonic responses" and responses of different types as essentially continuous effects spreading across a sheet of neurons than in terms that do not encourage thought about continuous processes. This interpretation remains to be verified and, at least in the amphibian, it has been difficult to demonstrate surround inhibitory fields like those that I have suggested (Kauer and Moulton,1974).

One implication of this conception of olfactory bulb function is that the temporal patterns of response of olfactory bulb output cells and interneurons should change depending upon which population of receptor cells is activated. This should occur when odors that activate different populations of receptors are used, although the differences may depend upon the particular circuits acting on the particular cell and upon where the receptive field of that cell lies relative to the receptors activated by the particular odors tested. There have been some reports indicating that different odors produce different temporal patterns of response, but these reports have not shown sufficient data to demonstrate that these changes were independent of changes produced by variation in concentration (cf. Harrison and Scott,1986; Meredith,1986).

We have made both qualitative and quantitative studies of these response patterns in rat olfactory bulb studied under sodium pentobarbital anesthesia. We have used both intracellular and extracellular recording methods. Recorded neurons were identified by their pattern of antidromic activation (Schneider and Scott,1983; Scott,1981) and/or by intracellular marking with horseradish peroxidase. All reported intracellular recordings were conducted using sodium acetate electrodes to avoid reversing inhibitory postsynaptic potentials. Odor stimuli were generated by dilution of a stream of air saturated with odorant into a stream of filtered air (Harrison and Scott,1986). Our experiments used a cyclic controlled artificial sniff of 1800 msec duration (experiments varying the duration of this sniff show that the response patterns are essentially similar over a range of sniff periods down to as short as 600 msec). Usually the odors were presented in bouts of ten successive sniffs at each concentration separated by a period of at least 30 seconds between bouts. These repeated sniff periods gave a means of assessing the variability of the response. This technique has some limitations because there is often some adaptation between the first and second sniffs. Future experiments will utilize a computer controlled system that allows rapid interleafing of single sniffs at varying concentrations, a more ideal method that was not practical with our manual switching methods. Cells were often held long enough for several complete series of six or more concentrations of several odorants. Odors were always presented in an alternating order of intermediate, low, and high concentrations in order to reduce confounding of concentration and order effects. Data are presented as raster plots that are rearranged with responses to ascending concentrations below the spontaneous activity record. The raster plots provide a much greater temporal resolution and illustration of response variability than do post stimulus time histograms.

Figure 1 is from a recent paper (Wellis et al,in press) showing that reliable differences in response pattern can be evoked with different odor stimuli. This figure demonstrates that these differences in pattern are apparent at the lowest concentrations at which a response appears so that unlike the case in figure 2B, the differences in pattern are not explanable simply by the strength of the odor stimulus. Reliable differences like this had been documented in long term recordings from ten cells at the time of that paper. It is worthy of note that some cells showed more than two discriminable patterns and that a pattern produced by one odor in a particular cell could be produced by a different odor in a different cell. Odor pairs that evoked very distinct patterns in one cell could produce strikingly similar patterns in another cell. At present it seems likely that there is considerable variation in the degree to which individual mitral/tufted cells can discriminate between pairs of odors. To what extent this variation is related to a spatial gradient of input from different receptor populations or to differences in local circuit properties that may be related to differences in morphological type remains to be studied.

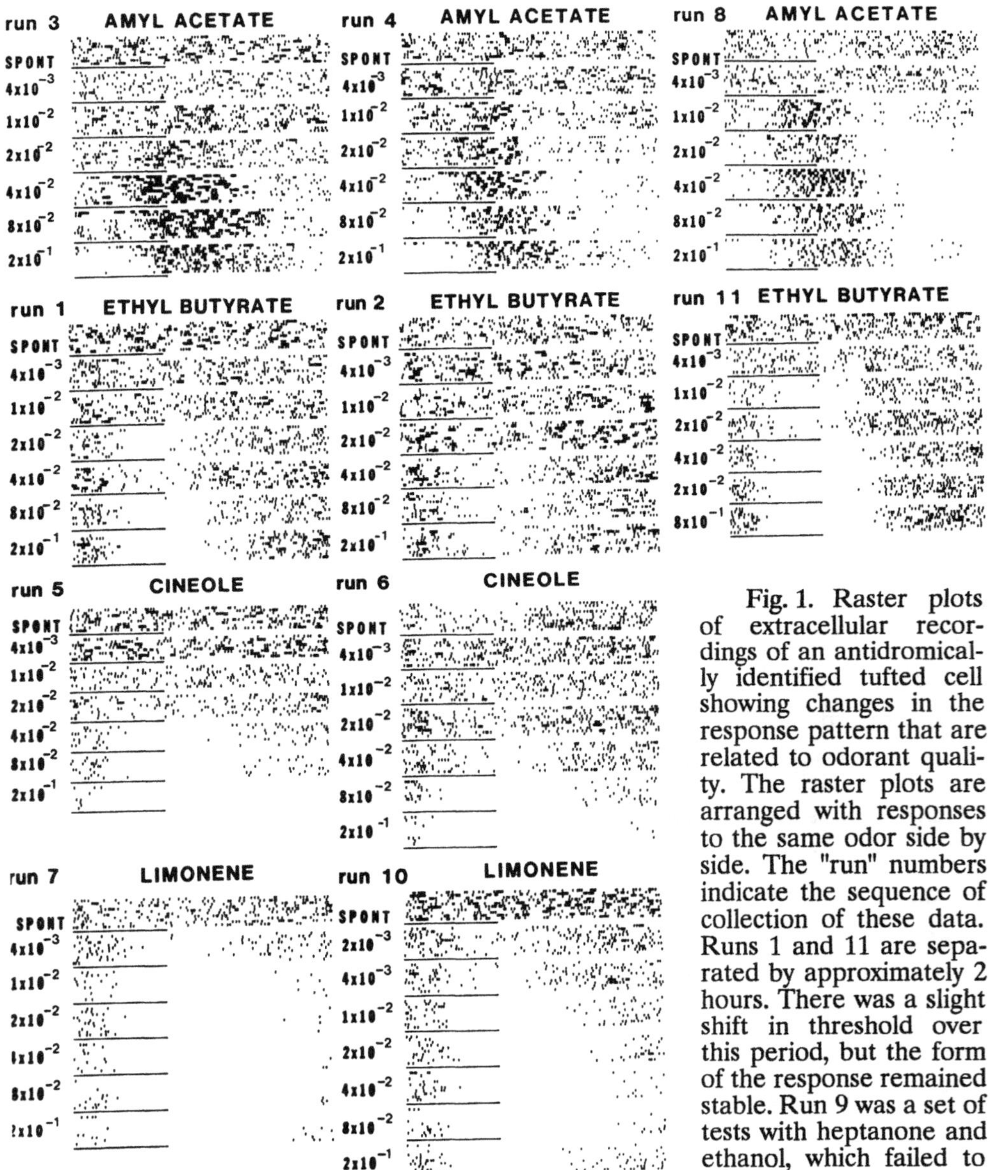

Fig. 1. Raster plots of extracellular recordings of an antidromically identified tufted cell showing changes in the response pattern that are related to odorant quality. The raster plots are arranged with responses to the same odor side by side. The "run" numbers indicate the sequence of collection of these data. Runs 1 and 11 are separated by approximately 2 hours. There was a slight shift in threshold over this period, but the form of the response remained stable. Run 9 was a set of tests with heptanone and ethanol, which failed to evoke responses. From Wellis et al.

Intracellular recordings in identified neurons of the olfactory bulb have demonstrated that many of the elements necessary for the conceptual model of responses molded by inhibitory circuits are present. We have observed both depolarizing and hyperpolarizing potentials associated with the changes in spike occurrence. Conductance measurements taken during these potentials suggest that excitatory, inhibitory, and disinhibitory mechanisms are all present in mitral/tufted cell responses (Wellis et al,in press). Recordings from marked granule cells indicate that many of these cells undergo spiking responses to odorants and thus are in a position to modify mitral/tufted cell behavior (Wellis and Scott,1989).

These observations have much in common with previous studies that categorized types of odor responses. For example, the two response forms shown in figure 2, one characterized almost exclusively by a period of spike suppression that increases in duration with increase in concentration and the other characterized by an excitatory burst that shifts in latency with increase in concentration, are very common in our data. The first bears a striking resemblance to Kauer's "S-type" response, although we have chosen not to emphasize that resemblance because of the differences in our experimental techniques and because we often see intermediate forms of response. Our results are definitely in accord with his original observation that a single neuron may respond to odors with at least two different temporal patterns, depending upon the odor. It will be interesting to determine whether there is a limited number of basic response forms that can be related to particular synaptic mechanisms. However, the

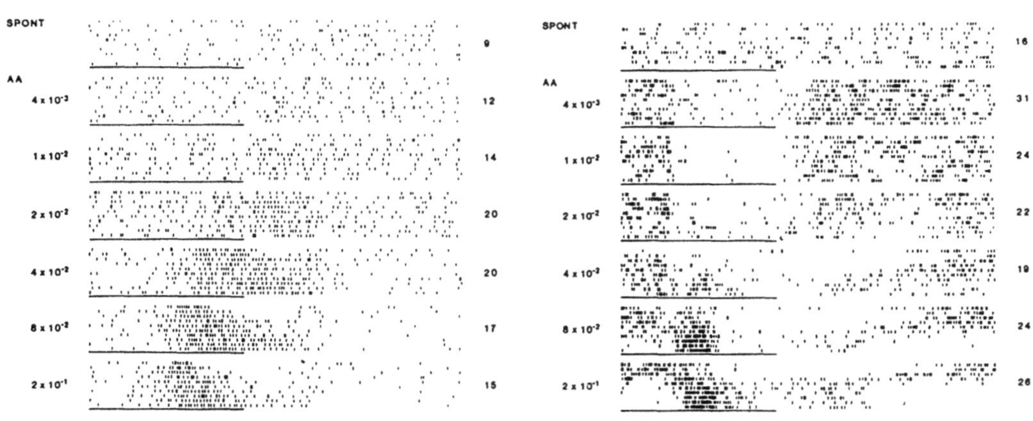

Fig. 2. Raster plots of two cells showing response patterns that are difficult to quantify by simple counting methods. The cell in panel A is a presumed interneuron, while the cell in panel B is an antidromically identified tufted cell. The numbers at the right of each raster give the average spike count over the 10 artificial sniff cycles. As would be expected from the changes in pattern over concentration, these counts are not systematically predictable by concentration.

presence of intermediate forms of response is clearly in favor of the idea that the mechanisms controlling the response pattern are continuous (cf. Hamilton and Kauer, in press).

While these changes in response pattern are convenient for demonstrating differences in the way olfactory bulb neurons process information about odor quality, the strong temporal patterning of these odor responses has always presented a problem for quantitation of odor responses (Duchamp-Viret,1988; Meredith and Moulton,1978; Onoda et al,1984; Scott,1977). Estimation of the magnitude of the odor response of central neurons has been an important problem in a number of investigations. For example, recent investigations have attempted to relate the positions of neurons to the size of response to odor stimulation (Meredith,1986; Wilson and Leon,1987). Others have attempted to evaluate changes in the degree of response of olfactory bulb neurons as the population of stimulated receptors is changed (Duchamp-Viret,1988; Kauer and Moulton,1974) or to compare the efficiency of processing at different levels of the nervous system (Duchamp and Sicard,1984; Tanabe et al,1975). In our own laboratory, we would like to test for functional differences in processing of odor related information by mitral versus tufted cells that might parallel the differences that we have observed in the responses of these cells to electrical stimulation of the olfactory nerve (Schneider and Scott,1983; Wellis et al,in press). All of these types of studies require an efficient estimate of the information contained in the spike train.

Recently, we have renewed our efforts to find a method for describing olfactory bulb odor responses that would use as much information as possible from the spike pattern and would preserve the idea of continuity of response form. This preservation of response form continuity has practical as well as theoretical bases. In the evaluation of response magnitude, it is not feasible to a priori select the portion of the response that is carrying information. To do so would bias the measurement. We have chosen to try to evaluate the total spike train to avoid this problem.

The essential features of our approach have been to construct a response measure that tracks the changes in response pattern as stimulus concentration is changed and to express those changes in terms related to intrinsic response variability. Note that this approach implies that it is not possible to assess the response strength on the basis of studies at a single concentration or with single presentations of the stimulus. The first conclusion derives not only from technical features of our analysis but from observations of the data indicating that the form of the response may change as a function of stimulus intensity. The second conclusion is based on variations from cell to cell in background activity and response range that preclude use of any absolute standard for measurement. Therefore measurements were based on changes related to internal variability.

In practice, we use the following steps: (1) a measure of response similarity applied to a closely spaced series of stimulus concentrations, (2) estimate of variability in that measure, (3) validation of the measure by testing for systematic relationship with stimulus concentration, and (4) construction of stimulus-response curves based on increments in the similarity measure. The first step, that of the measure of response similarity, must be one that makes efficient use of the temporal distribution of spikes in the response and it should be easy to compute. However, the third step, that of validation of the response measure is the most important, since it is necessary to demonstrate that the measure detects something systematically related to variation in stimulus concentration.

MEASUREMENT OF RESPONSE SIMILARITY

Since we had concluded from many observations that it was not possible to adequately quantify many odor response without taking into account complex changes in response pattern, we sought a measure that would be sensitive to a wide variety of differences in pattern. It was clear from data such as those in figures 1 and 2 that counts of numbers of spikes or computation of mean frequency do not always change monotonically with concentration in all responses. By contrast there are clearly systematic changes in these patterns represented in some cases by shifts in response latency (as in figure 2a) or by gradual appearance of additional excitatory or inhibitory components (as in figure 2b). There are several possible approaches to quantifying these differences. We chose one that had some physiological appeal to us under the assumption that similar patterns should have similar summation effects on the neural postsynaptic membrane. Therefore to compare two responses, each consisting of a train of spikes, we represented each spike by a pair of exponential curves and added the area under these curves. These sums were largest when the spike trains representing the two responses were identical and became smaller for responses pairs that had very different spike trains. It is obvious that the quality of the results will depend upon the decay constant chosen for the exponential curve. We conducted these calculations with a range of decay constants (.002 msec to .2 msec) and has usually used a composite of the results based on a weighted sum of the results from the different calculations. For convenience we have called this method of response evaluation the coincidence index. It differs in significant details from an earlier form of analysis we employed (Harrison and Scott,1986).

Since our data have consisted of series of several artificial inspirations or "sniffs" we could can responses to similar odor concentrations to get estimates of response variability. Therefore any comparisons of responses to different concentrations could be expressed in terms of standard scores, that is a mean of between concentration comparisons minus within concentration comparisons divided by the standard deviation. This standard score is small when there is little difference between the responses at the two concentrations being compared and becomes large if the two concentrations evoke responses that are reliabily very different. Note that this expression assumes that a spike train carries more significance to the nervous system to the degree that it represents a large change relative to the inherent variability of the system.

VALIDATION OF RESPONSE MEASURE

It is desirable to test whether the measurements made of changes in response pattern change systematically with concentration. Some of the differences between sets of spike trains will be due to noise, or to uncontrolled changes in the state of the neuron or of the experimental animal. In addition, a highly derived measure, such as this, should always be regarded with suspicion until validated in some way. Our method of validation is to test whether the observed values can be fit to a sigmoid stimulus-response function. Ideally such a function should be adequate to deal with the fact that some of the stimulus concentrations we presented are below threshold for some neurons and that some concentrations are above the saturation point of the stimulus response function for some neurons. We have compared out observations to such curves using nonlinear regression. In this regression analysis we used two other sets of parameters (1) an exponent representing a power function for the comparisons between any pair of concentrations and (2) a decrementing function that relates to that fact that the similarity measure is often more sensitive to small changes in pattern than to very large pattern changes, a result of the exponential form of the matching function. Thus the nonlinear regression estimates five parameters, while a typical set of data involves at least six concentrations plus a set of spontaneous activity, generating 21 comparisons between concentrations. Concentrations of odorants are expressed as proportions of dilution of saturated odorant in clean air, typically 0.001 to 0.2. The spontaneous activity records are represented as very low dilutions (.0001) to avoid problems with expression on a logarithmic scale.

With this regression analysis, we can determine two facts of interest: the coefficient of determination that indicates the proportion of the variance in the response measure that is related systematically to the differences in stimulus concentration and the curve representing the best fit through the data points. This best fit curve reduces the effect of any unusual event that may have occurred during collection of data for a particular odor concentration because it is based on all comparisons with all concentrations. The values from this fitted curve can then be used to construct a stimulus response curve that begins at zero for the spontaneous activity and increases by the calculated amount for each increase in concentration.

Some of these analyses and curves are represented in the next figure. They demonstrate that substantial portions of the variance can be accounted for in this regression analysis. We have also tested this system of analysis by simply mixing up the data to determine whether the analysis is finding patterns that are unrelated to the systematic changes that the eye can detect related to changes in odor concentration. We find that interchanging pairs of responses elicited by subthreshold concentrations or pairs of responses that were elicited by odors at the point of saturation of the response has no effect on the coefficient of determination. On the other hand, interchanging pairs of responses that look very different to the eye, even if they are evoked by concentration pairs that are very close together, can reduce the coefficient of determination by up to 50%.

Fig. 3. (next page). A: The relationship between the standard score values of the coincidence index and the concentration differences for odor concentration pairs (standardized difference curves). The data are from figure 1. The numbers along the curves correspond to the lowest concentration for the pair: 0 indicates comparisons involving no odor stimulation; 7 indicates comparisons involving 2.3 x 10^{-3}; 2 indicates comparisons involving 1.8 x 10^{-1}; etc. (expressed as proportions of saturated odorant in the stimulus mixture). These might have been plotted as several curves for each series (for example for the amyl acetate records there would have been a curve for comparisons with no odorant and five others representing comparisons with 4.6 x 10^{-3}, 1.1 x 10^{-2}, etc.). These have been collapsed onto a single curve by normalizing the, to the comparison with no odorant. The coefficient of determination (R2) indicates the proportion of variance in the values of the coincidence index that is accounted for by knowledge of concentration differences.

B: Cumulative stimulus-response curves. These were constructed by summing the standard score values for adjacent comparisons from records with no odor stimulus up to the highest concentration. These adjacent comparisons, marked with arrows in A, are the comparisons of each stimulus concentration with the next highest comparison in the series. Note that the form of the curves in parts A and B are different. The standardized difference curves of part A are distorted by the fact that the coincidence index algorithm has greater sensitivity to small differences than to big differences. This point can be appreciated in the amyl acetate curve where the standardized difference curve of panel A saturates at high concentrations.

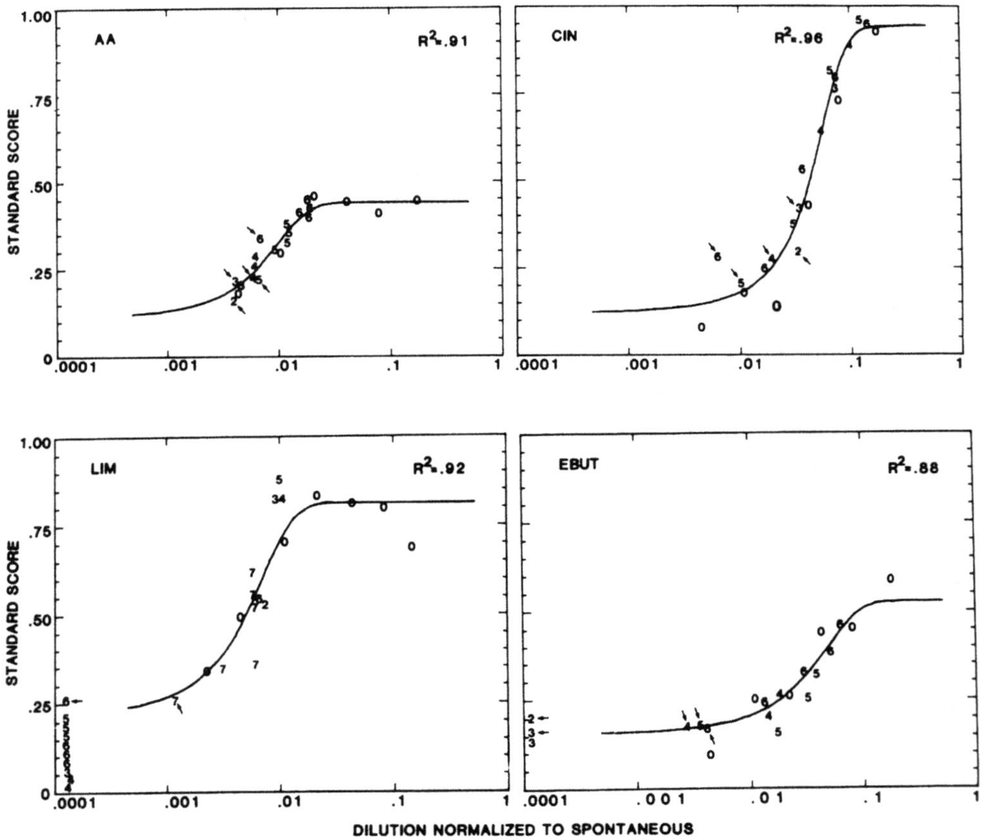
AA
R^2=.91
CIN
R^2=.96
LIM
R^2=.92
EBUT
R^2=.88
STANDARD SCORE
1.00
.75
.50
.25
0
.0001
.001
.01
.1
1
DILUTION NORMALIZED TO SPONTANEOUS

A

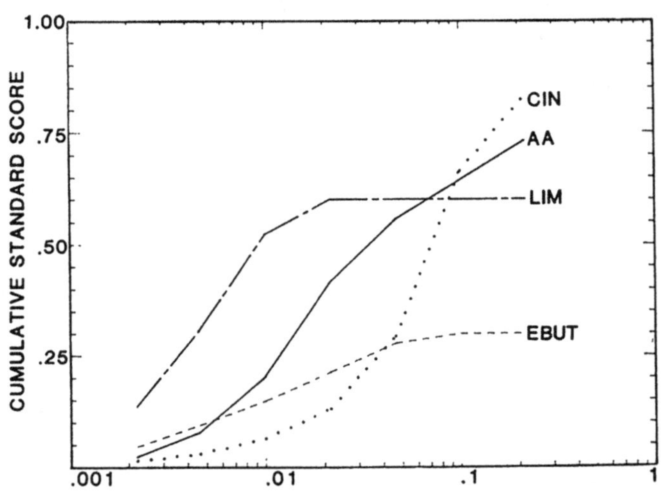
CIN
AA
LIM
EBUT
CUMULATIVE STANDARD SCORE
1.00
.75
.50
.25
.001
.01
.1
1
DILUTION

B

We have compared these results with other methods of analysis, including another method for measuring differences in pattern by performing a principal components analysis of digitally smoothed functions based on the spike trains[1]. In cases where the response has a strongly patterned nature that changes with concentration, our similarity measure or the differences based on the principal components analysis were generally superior in terms of the coefficient of determination criterion. These methods never gave poorer fits than the simpler count or frequency based methods. Table 1 presents an example of this comparison for the neuron of figure 2. We interpret this table as indicating a general superiority of methods that compare the form of the response over those that are based on simple counts, because there are multiple patterns of response that could be represented by the same spike count or mean frequency. Our own similarity method, which we have termed the coincidence index, is not necessarily consistently more efficient than a method such as that based on the principal conponents analysis, which should in fact extract all of the variance for a particular spike pattern, assuming a particular smoothing function. The coincidence index was computationally easier and seems to us to be justifiable on some physiological terms. That is, the differences that we describe are based on the temporal disparity between spikes and those temporal disparities are expressed in an exponential fashion that may be related to temporal properties of the membrane. It is interesting in this context that some highly patterned responses have much better coefficients of determination for some decay constants than for others (see figure 2). One possible source of this variation could be sets of terminals terminating at different distances from the spike trigger zone so that the dendritic internal resistance imposes different time constants on them.

It should be possible to extend this quantitative analysis of intensity relationships to study differences induced by varying odor quality. It is possible, for example to reconstruct the curves for the ethyl butyrate and limonene responses of the cell of figure 2 by computing the differences between those responses and the cineole response. That means the the difference between responses of similar form but different threshold can be easily measured. Whether the analysis could be appropriately extended to responses of different form (such as those of cineole and amyl acetate of figure 2) is not yet clear.

1) In the case of the principal components analysis the distance between each response was determined from the first six components. These distances could be determined as though they were represented in six dimensional space (ie. the square root of the sum of the differences between the coordinates of the two responses for each dimension). The values in table 1 were obtained by entering differences calculated between responses at different concentrations into the same analysis used for our own similarity measure.

However, analysis of the shape of the stimulus-response function already gives some considerable information about quality coding. This approach can certainly be used to test whether the stimulus response profile for a set of odors differs between neurons of different morphological type or of different position within the bulb (ie; do the profiles change systematically with the difference between cells).

	Amyl Acetate	Ethyl Butyrate	Cineole	Limonene
Mean frequency	.20	.14	.63	.48
Mean spike count	.34	.57	.57	.94
Coincidence index	.77	.70	.90	.89

Table 1. Coefficients of determination for three methods of analysis of the data of figure 1. These are mean values for the total series. Note that the coincidence index is more strongly related to concentration differences for this series than are the methods based on count or frequency. The exception is the limonene record where a very strong concentration dependent suppression imposes a relationship with the spike count.

REFERENCES

Doving KB(1987)Response properties of neurones in the rat olfactory bulb to various parameters of odour stimulation.Acta Physiol Scand 130:285-298

Duchamp A, Sicard G(1984)Influence of stimulus intensity on odour discrimination by olfactory bulb neurons as compared with receptor cells.Chem Senses 8:355-366

Duchamp-Viret P(1988)Le codage intensitif du stimulus olfactif. Thesis L'Universite Claude-Bernard Lyon

Freeman WJ (1975)Mass action in the nervous system.Academic Press, New York

Hamilton KA, Kauer JS(in press)Patterns of intracellular potentials in salamander mitral/tufted`cells in response to odor stimulation.J Neurophysiol

Harrison TA,Scott JW(1986)Olfactory bulb responses to odor stimulation:analysis of response pattern and intensity relationships.J Neurophysiol 56:1571-1589

Jourdan F,Duveau A,Astic L,Holley A (1980)Spatial distribution of [14C]2-deoxyglucose uptake in the olfactory bulbs of rats stimulated with two different odours.Brain Res 188:139-154

Kauer J(1974)Response patterns of amphibian olfactory bulb neurones to odour stimulation.J Physiol Lond 243:695-715

Kauer J(1987)Coding in the olfactory system.In Finger TE,Silver WL (eds)Neurobiology of taste and smell.Wiley,New York,p 205
Kauer J,Moulton DG(1974)Responses of olfactory bulb neurones to odour stimulation of small nasal areas in the salamander.J Physiol Lond 243:717-737
Mackay-Sim A,Shaman P(1984)Topographic coding of odorant quality is maintained at different concentrations in the salamander olfactory epithelium.Brain Res 297:207-216
Macrides F,Chorover SL(1972)Olfactory bulb units:activity correlated with inhalation cycles and odor quality.Science 175:84-87
Mair RG(1982)Response properties of rat olfactory bulb neurones.J Physiol Lond 326:341-359
Meredith M(1986)Patterned response to odor in mammalian olfactory bulb:the influence of intensity.J Neurophysiol 56:572-597
Meredith M,Moulton DG(1978)Patterned response to odor in single neurons of goldfish olfactory bulb: influence of odor quality and other stimulus parameters.J Gen Physiol 71:615-643
Onoda N,Imamura K,Obata E,Iino M.(1984)Response selectivity of neocortical neurons to specific odors in the rabbit.J Neurophysiol 52:638-652
Saucier D, Astic L(1986)Analysis of the topographical organization of olfactory epithelium projection in the rat.Brain Res 16:455-462
Schneider SP,Scott JW(1983)Orthodromic response properties of rat olfactory bulb mitral and tufted cells correlate with their projection patterns.J Neurophysiol 50:358-378
Scott JW(1977)A measure of extracellular unit responses to repeated stimulation applied to observations of the time course of olfactory responses. Brain Res 132:247-258
Scott JW(1981)Electrophysiological identification of mitral and tufted cells and distribution of their axons in the olfactory system of the rat.J Neurophysiol 45:918-931
Scott and Aaron(1977)Averaged odor-induced waves in the olfactory bulb of the rat.Exp Neurol 55:654-665
Scott JW, Harrison TA(1987)The olfactory bulb:anatomy and physiology.In Finger TE,Silver WL (eds)Neurobiology of taste and smell.Wiley,New York,p 151
Shepherd GM(1979)The synaptic organization of the nervous system,2nd edn.Oxford University Press,Oxford,UK
Stewart WB,Kauer JS,Shepherd GM(1979)Functional organization of rat olfactory bulb analyzed by the 2-deoxyglucose method.J Comp Neurol 185:715-734
Tanabe T,Iino M,Takagi SF(1975)Discrimination of odors in olfactory bulb, pyriform amygdaloid areas, and orbitofrontal cortex of the monkey.J Neurophysiol 38:1284-1296
Wellis DW, Scott JW(1989)Odor induced responses of rat olfactory bulb granule cells.Xth International Symposium on Olfaction and Taste
Wilson DA, Leon M(1987)Evidence of lateral synaptic interactions in olfactory bulb output cell responses to odors.Brain Res 417:175-180

OPTICAL RECORDING OF NEURONAL ACTIVITY: PARALLEL *VERSUS* SERIAL METHODS

P. Saggau

Dept. of Physiology, University of Munich
Pettenkoferstraße 12
D-8000 München 2, F.R.G.

I. INTRODUCTION

Optical recording of neural activity using voltage-sensitive dyes has become a powerful tool in the investigation of complex neuronal networks, from invertebrate ganglia to intact primate cortex. By this method, complementary to classical electrophysiological techniques, optical parameters that are related to membrane potential are monitored, thus enabling non-invasive, quasi-intracellular recording from multiple sites (for reviews see: Blasdel, 1989; Cohen et al., 1989; Cohen and Lesher, 1986; Grinvald, 1985; Lieke et al., 1989; Salzberg, 1983, 1989).

I.A. Fundamentals of optical recording

The general principle is illustrated in Fig. 1. Assume a neuronal population (in this case a ganglion) that has been incubated with a voltage-sensitive dye - these dyes bind to neural membranes and are sensitive to transmembrane potential. When this preparation is illuminated monochromatically, extrinsic optical signals can be detected that are proportional to the *change* in membrane potential (ΔMP). Depending on the type of dye used, these signals will be either a change of fluorescence ($\Delta F \propto \Delta MP$) or transmission ($\Delta T \propto \Delta MP$).

Due to the size of the objects to be recorded from (usually in the micrometer range) this technique uses light microscopy to collect information, thus theoretical limitations of resolution are dependent only on diffraction effects. Both lateral

NATO ASI Series, Vol. H 39
Chemosensory Information Processing. Edited by D. Schild

resolution ($\Delta x, \Delta y$) and axial resolution (Δz, depth of focus) are related to the numerical aperture (N.A.) of the lens.

$$\Delta x, \Delta y \propto \text{N.A.}^{-1}, \qquad \Delta z \propto \text{N.A.}^{-2}$$

Another important factor in optical recording is the signal-to-noise ratio (S/N) that can be achieved. Assuming the ideal situation of a system limited by shot noise only, then S/N is proportional to the fractional change of the luminous flux ($\Delta I/I$) and the square root of the energy density of the detected photons ($I/\Delta f$, where Δf is the bandwidth of the detection system).

$$S/N \propto \Delta I/I * (I/\Delta f)^{1/2}$$

Since the light-collecting efficiency of the lens is a function of N.A. ($I \propto \text{N.A.}^2$), it follows that good results are likely to be obtained when high N.A. lenses are used in conjunction with an optically optimized (thin, non-scattering) preparation (e.g. acutely isolated or cultured neurones).

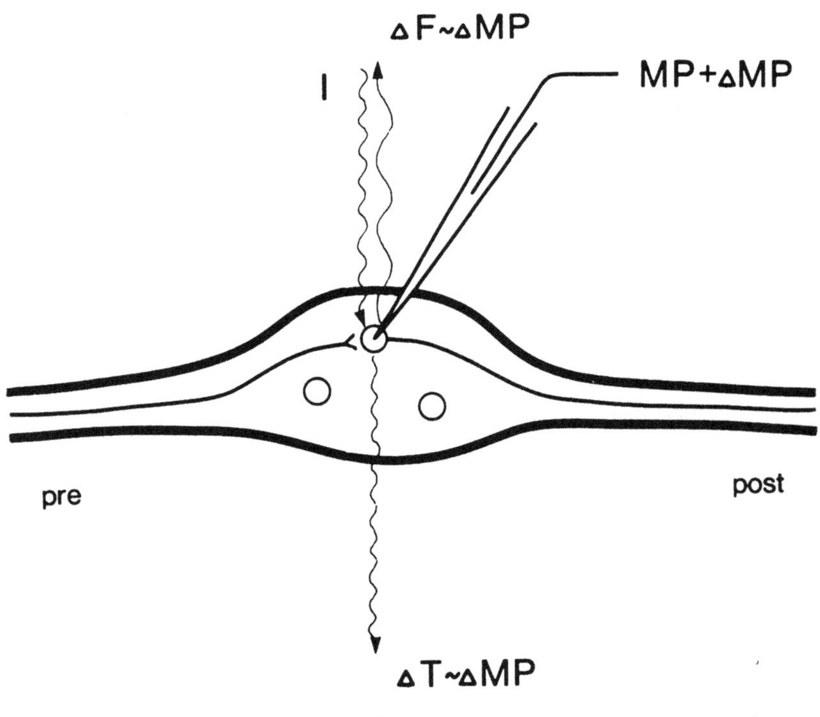

Fig. 1. Principle of optical recording. The classical method uses an intracellular microelectrode to measure the membrane potential, consisting of a static (MP) as well as a dynamic (ΔMP) component. The optical method, in contrast, employs monochromatic illumination of the indicator-stained neural structures to be investigated, detecting potential-correlated changes in optical transmission ($\Delta T \propto \Delta MP$) or fluorescence ($\Delta F \propto \Delta MP$).

I.B. Specifications for recording of neural activity

The application of optical principles to the recording from multiple sites within neural networks has specific requirements of stability and resolution.

Stability

Stability is related to the reduction of extraneous noise sources that prevent shot noise limitation. Such non-photonic fluctuations (D) of the detected light that are not correlated to neural activity degrade S/N directly.

$$S/N \propto \Delta I/I * 1/D$$

Taking into account the low fractional change of optical parameters exhibited by voltage-sensitive probes (e.g. $\Delta I/I \approx 10^{-3}$ for a change of membrane potential of 100 mV), and attempting to achieve a S/N that allows recording without a need for averaging (i.e. S/N = 10), then the necessary stability is high ($1/D = 10^4$). This level of stability is required from the source of illumination, through the optical system and the preparation, to the detector.

Resolution in time

Resolution in time is determined by the bandwidth of the neural signals to be recorded. For example, if synaptic or action potentials are to be detected, the frequency of all individual measurements and thus, in multi-site recordings, the frame frequency, has to be several thousands per second ($f_{frame} \geq 1$ kHz).

Resolution in space

In practise, resolution in space has to deal with the number of detected points per frame (n), rather than the theoretical diffraction limitation (about 1 μm). The speed of the recording system is thus formed by the product of the number of points per frame and the number of frames per second. This equals the bandwidth of the detection unit ($\Delta f = n * f_{frame}$), influencing S/N (see I.A).

II. PARALLEL AND SERIAL PRINCIPLES

Optical recording from multiple sites can be achieved by a variety of different approaches. The general elements of such recording systems (see Fig. 2) are discussed here with respect to their parallel or serial principles.

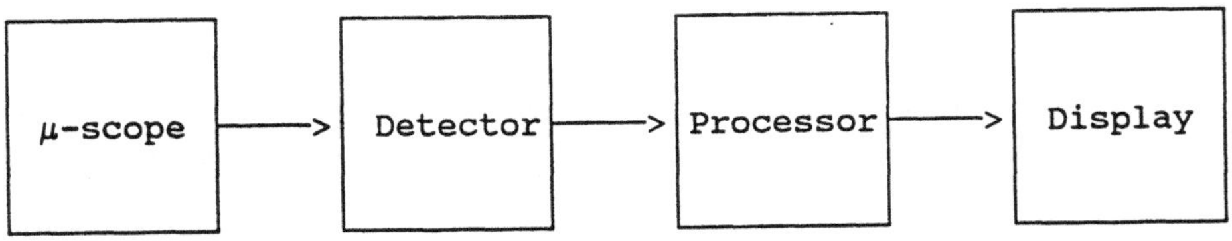

Fig. 2. Elements of optical recording systems

II.A. Microscopes

The difference between a parallel and a serial type microscope is best illustrated by comparing the principles of conventional microscopy with scanning and confocal microscopy.

Conventional microscopy

In conventional microscopy a light source illuminates homogeneously the object under investigation via the condenser lens (Fig. 3.A). An objective lens collects the light emerging from the object and parallely forms an image. This real image can be visualized by spatially resolving detectors like the retina of the eye or a camera target, or it can be stored on photographic media.

Scanning microscopy

Classical scanning microscopy (Young and Roberts, 1951) makes use of a serially scanning point illumination of the object. This can be achieved either by a stationary point source in connection with a scanning object, or a scanning point source and a stationary object. Obviously, only the latter method may be applied to optical recording from neural preparations. As Fig. 3.B indicates, the point source is imaged on the object via an objective lens. A second lens collects the light on a spatially non-resolving detector without forming an image. This microscope gives no improvement in resolution compared to the conventional type (for review see: Wilson and Sheppard, 1984).

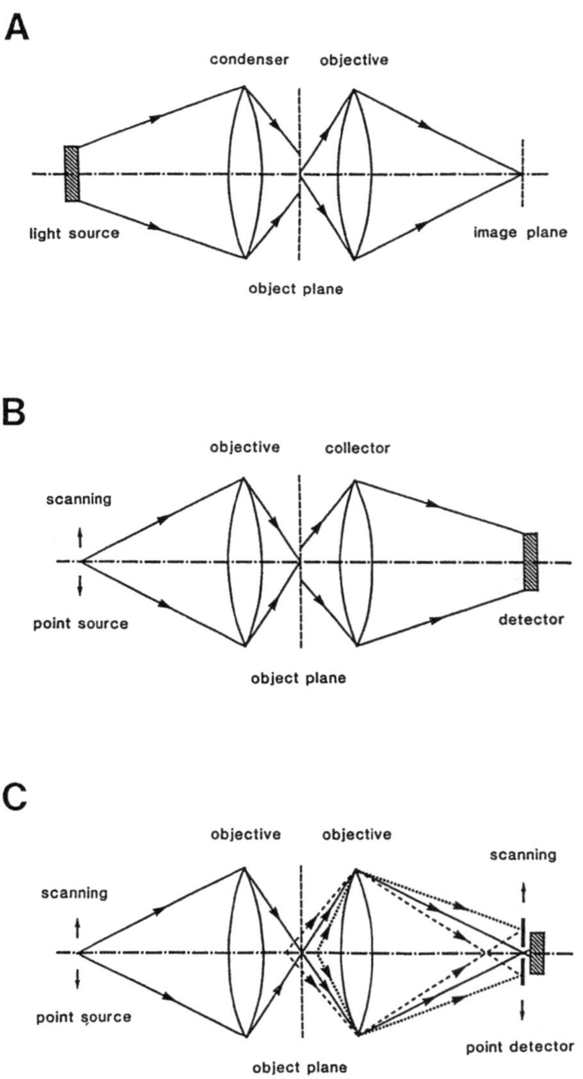

Fig. 3. Comparison of microscopic principles

A: Conventional microscope
B: Scanning microscope
C: Confocal microscope (Non-solid lines represent light from off the object plane.)

Confocal microscopy

Another serial microscopic principle exhibiting superior optical performance was first pointed out by Minsky (Minsky, 1957) and experimentally demonstrated by Petran (Egger and Petran, 1967).

Whereas in classical scanning microscopy a point source is used, in confocal microscopy both point source and point detector are used in conjunction. The basic principle is outlined in Fig. 3.C. A point aperture prevents scattered in-focus light from reaching the detector, resulting in an improvement of lateral resolution. In addition, any point of the object not situated exactly in the object plane is imaged by the objective lens to a corresponding defocused spot in the detector plane. The pinhole prevents most of this out-of-focus light from reaching the detector, thus reducing the practical depth of focus drastically (for review see: Wilson and Sheppard, 1984).

II.B. Detectors

Depending on the type of microscope used (see II.A), spatially resolving or non-resolving detectors are required. According to readout (but not to detection principle) they are classified as parallel or serial.

Video cameras

With a conventional microscope, the most obvious type of detector would be the target of a video camera, such as CCDs or vidicons. These devices represent spatially-resolvent serial readout detectors. Yet in many applications their low dynamic range ($< 10^3$), together with the restrictions imposed by low frame frequency, make them unsuitable.

Detector arrays

One solution to the problems mentioned above is the use of a photodiode-matrix detector. Silicon photodiodes (representing the elements of such an array) cover the required dynamic range and can be read out parallelly. Usually the number of array elements, and thus their resolution in space, is limited to several hundreds.

Single Detectors

In scanning microscopy single photodiodes or photomultiplier tubes are used. These devices are spatially non-resolvent serial readout detectors, achieving both the required readout speed and dynamic range.

II.C. Processors

Optical data from multiple-site recording can be processed in both parallel and serial ways. This holds for the analog preprocessing, the analog-to-digital conversion, as well as for the digital computation.

Analog preprocessor

When a serial-readout detector (video camera or single photodetector) is employed, single-channel, high-speed processing is essential ($\Delta f = n * f_{frame}$, see I.B). In contrast, a detector of parallel-readout type requires low-speed ($\Delta f = f_{frame}$), multi-channel (n) processing. In both cases amplification and filtering are the primary aims of preprocessing.

Analog-to-digital converter

Before being processed by a computer the data have to be digitized. Parallelly-preprocessed data have to be converted either in a parallel way or, after time-multiplexing them, a single high-speed converter can be used in the same way as for serial signals. While speed of conversion is dependent on the parallel or serial representation of data, accuracy is determined only by the dynamic range of the signals to be converted. Because of the low fractional change of the optical parameters (10^{-3}, see I.B) accuracy has to be high. If only optical changes (AC-coupled signals) are of interest, then 10^{-2} (8 bit A/D resolution) is sufficient, while DC-recordings require 10^{-5} (16 bit).

Digital computer

Even modern minicomputers are, with very few exceptions, serial machines that can only process one piece of data at a time. Since both temporal and spatial resolution determine the bandwidth of the signals to be processed (and thus the speed of the computer to be used), improvements of resolution quickly lead to the need for supercomputers. Such machines with very high instruction rates (> 100 MIPS, mega instructions per second) are extremely costly when organized serially. Even less expensive array processors, which are employed as additional processing units that can deal with blocks of data (to increase the effective instruction rate substantially), do not pass this principle bottle-neck by being organized serially as well. Parallelly processing computers can deal with multiple data simultaneously and can be quite affordable when organized as a network of less expensive microcomputers.

II.D. Displays

The data display has to provide the experimenter with synoptic information of the preparation under investigation. Thus, in multi-site recording only systems that allow parallel display of data may be used (e.g. high-resolution graphics to monitor many traces of local neural activity superimposed on a video micrograph).

III. REAL RECORDING SYSTEMS AND THEIR APPLICATIONS

Experimentally applied optical multi-site recording systems, containing parallel as well as serial elements are described in this section. The structure of each of these systems is indicated in Fig. 4.

III.A. Photodiode-matrix system

The most widely used recording system is based on conventional microscopy (CON), combined with a photodiode-matrix (PDM). The photocurrent emerging from each PDM element are parallelly conducted to individual analog preprocessors (PROC), each consisting of a current-to-voltage converter, filter (lowpass filtering, AC-coupling) and amplifier. A time-multiplexing unit (MUX) codes the parallel to serial signals; thus, a single analog-to-digital (A/D) converter is required. The digital data are then fed into a minicomputer (COMP) where they are serially processed with specific algorithms (e.g. signal averaging, bleaching correction). Signals are then passed to a display unit (DIS) where they can be parallely visualized.

Originally introduced for simultaneously monitoring the activity of many neurones in an isolated invertebrate ganglion (Grinvald et al., 1981), the photodiode-matrix system has since become widely used to investigate neural networks. The applications range from simple invertebrate *in vitro* preparations to mammalian brain *in situ.* The studies include: excitation-secretion coupling in Xenopus neurohypophysis (Salzberg, 1983), behaviour-correlated activity in intact invertebrate preparations (London et al., 1987), spread of activation in mammalian brain slices (Grinvald et al., 1982; Saggau and ten Bruggencate, 1988), long-term potentiation of synaptic transmission in the hippocampus (Saggau and ten Bruggencate, 1988), sensory-evoked local activity in intact vertebrate and mammalian brain (Grinvald et al., 1984; Orbach et al., 1985). Specific applications of the photodiode-matrix system to the olfactory system have been reported in Salamander (Orbach and Cohen, 1983; Kauer et al., 1987) and the rat (Cattarelli and Cohen, in press). For further results from the olfactory system, obtained with a photodiode-matix system, see the contributions from both Cattarelli and Senseman in this volume.

III.B. Video system

The most obvious recording combination, as mentioned above, is based on conventional microscopy with a video camera (VID). Since such detectors are serial devices, only single-channel analog preprocessing is required which is directly followed by A/D conversion. For digital processing and displaying see III.A.

Because of the low frame frequency and dynamic range of video systems, their experimental application has been limited to the few experimental situations where time- resolution is not critical. The first reported use of such a system was made on the intact primate cortex (Blasdel and Salama, 1986), where stationary spatial patterns of neural activity were shown to be dependent on spatial patterns of visual stimuli. Video techniques have also been successfully applied to the olfactory system of the Salamander (Kauer, 1988). For further data on the olfactory system, recorded with a video system, refer to the contributions from both Kauer and Senseman in this volume.

III.C. Laser scanning system

This strictly serial recording system employs classical scanning micoscopy (SCAN) and a single photodiode (PD) or photomultiplier tube. The photocurrent is passed to a single analog preprocessor, A/D converted, and finally digitally processed and displayed as before.

There are no reports of a successful application of such a laser scanning system to a neural network, but this system has been used with voltage-sensitive dyes to monitor the spread of excitation from a non-neural network of excitable cells: the heart (Dillon and Morad, 1981).

III.D. Confocal system

This serial recording system differs from the classical scanning technique only by virtue of the confocal microscope optics (see II.D).

In contrast to the many useful applications of confocal microscopy to investigate the morphology of neural populations, attempts to apply this technique to optical recording of neural activity have not been very successful (for review see Fine et al., 1988). The reason for this is the poor stability ($D < 10^2$, see I.B) and the extremely low temporal resolution of commercially available confocal microscopes ($f_{frame} > 1$ s).

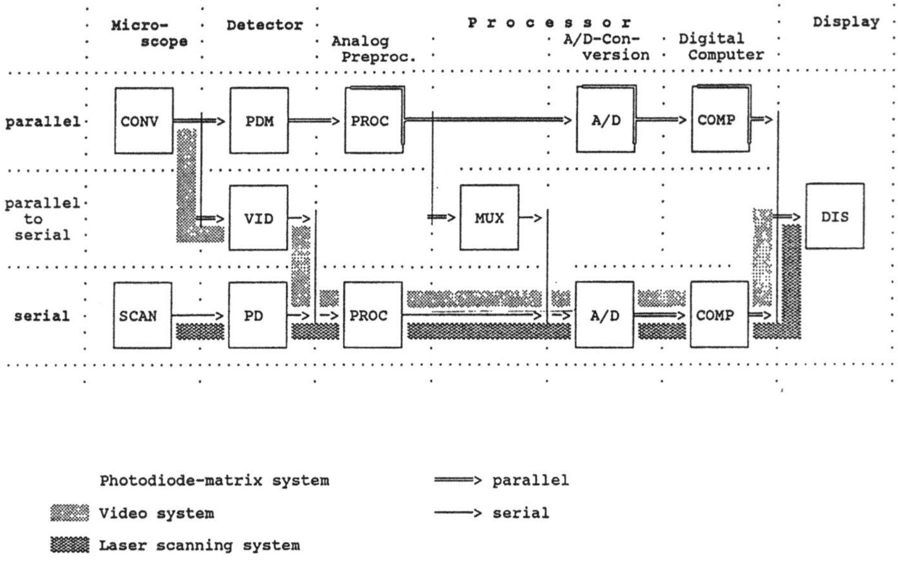

Fig. 4. Structure of real optical recording systems

IV. PRACTICAL LIMITATIONS AND FUTURE PROSPECTS

There are practical contraints that prevent optical recording methods from reaching the theoretical limitations of spatial resolution which are imposed by diffraction effects (see I.A). On the one hand, the spatio-temporal resolution of currently available detectors, as well as the speed of affordable digital computers, are low. On the other hand, most preparations of interest exhibit strong intrinsic optical scattering, resulting in a dramatic reduction of spatial resolution if not viewed confocally. This optical distortion holds especially for mammalian neural tissue which demonstrates its scattering properties by looking whitish even on macroscopic scales, and it is generally true with thick preparations like intact brain structures.

Future prospects include spatially-resolvent detectors exhibiting video resolution, high frame frequencies and sufficient dynamic range. Such devices that are currently being commercially developed follow both parallel and serial principles. Nevertheless, the potential user of such high-resolution detectors has to keep in mind that S/N strongly depends on the number of photons that reach a single detector element per frame (see I.A). As a consequence, more intense illumination has to be applied, which can be practically obtained with lasers only. However, this light source introduces speckle noise (caused by the coherence of laser light) if used for conventional (homogeneous) illumination. The most promising way out of this trap is the use of fluorescence techniques in combination with a point illumination. Both these requirements are met in laser scanning fluorescence microscopy - a serial principle which consequently makes spatially-resolvent detectors at first view redundant (F. Rucker, M.Sc. Thesis, unpublished; Saggau et al., 1988).

To overcome scattering problems, confocal microscopy seems to be the most promising method, yet standard serial techniques do not apply. As can be seen in Fig. 3.C, the detector pinhole has to scan synchronously with the point source. In commercially available systems this requirement is fulfilled either by virtually scanning pinholes (rotating coupled disk pairs) or by reflective systems that scan and descan the beam with the same mirrors to make it pass a stationary pinhole (rotating polygonal or oscillating galvanometer mirrors). These mechanical systems cause stability problems even at slow scanning speed, and are totally impractical at the required frame frequencies (range of kHz, see I.B.). One way out of this problem is the application of an acousto-optical, diffractive system (modulated gratings, for review see Milton et al.,

1983) combined with a spatially resolvent, randomly accessible detector. Such a parallel detector, when used in a "1 of n mode" (e.g. a PDM connected to a synchronized fast switching device), can serve as an electronically controlled "flying pinhole" (Kagerbauer; Kaufmann, M.Sc. Theses, unpublished; Saggau et al., 1988). This seems to be a promising solution, since it has been shown (Wilson and Carlini, 1988), that even large pinhole diameters mimicked by such a detector (e.g. d = 150 μm, N.A. = 0.5) can give a favourably small depth of focus ($\Delta z < 10$ μm). When this can be achieved, three-dimensional optical recording of neural activity by applying the method of optical sectioning should be feasible (Hiendl, M.Sc. Thesis, unpublished), which up to now has been applied only in anatomical studies. In testing a commercial confocal microscope with high spatial resolution (n = 512 x 512) and low resolution in time ($f_{frame} < 1$ Hz), we found unblurred imaging possible at levels between 100 - 200 μm below the surface. The preparations employed were hippocampal brain slices of different thicknesses, stained with a potential- sensitive fluorescent dye. This is in accordance with the findings of another group (Fine et al., 1988).

Future computers for optical recording systems will be organized parallely, preferably in networks of autonomous microprocessors. Depending on their structure, such transputer networks (Holzmann, M.Sc. Thesis, unpublished) can be perfectly suited to deal with parallel incoming data that have be to treated independently (e.g. temporal averaging) as well as non-dependently (e.g. spatial averaging, image processing).

Purely parallel and serial optical recording systems are depicted in Fig. 5. The ideal parallel system would thus consist of a conventional microscope in connection with a photodiode-matrix, followed by one "local" low-speed processor per detector element, consisting each of an analog preprocessor, an A/D converter and a microcomputer. The corresponding serial system would contain scanning microscope, single photodiode and single high-speed processor. The future systems will go on being compounds of parallel and serial elements, depending on the specific preparation and experimental situation, but mostly they will be based on confocal principles.

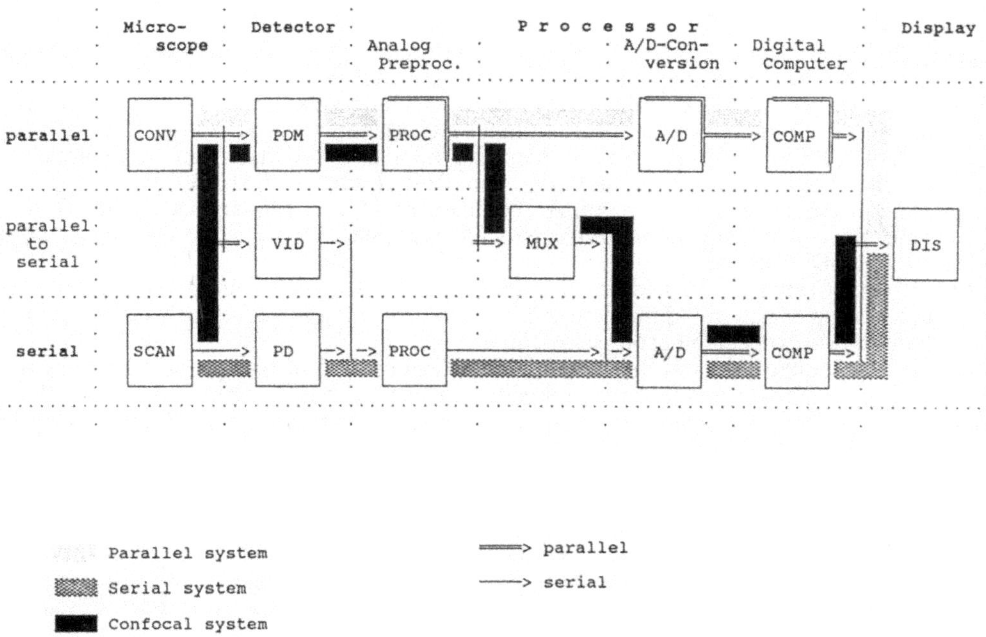

Fig 5. Structure of future optical recording systems

ACKNOWLEDGEMENTS - I want to thank my present and former student coworkers R. Hiendl, G. Holzmann, R. Kagerbauer, R. Kaufmann and F. Rucker for brain-, hard- and software emerging from their theses. This work was supported by the DFG, SFB- 220, and the Heidenhain-Foundation.

V. REFERENCES

Blasdel, G.G. (1989) Visualization of neuronal activity in monkey striate cortex. Ann. Rev. Physiol., 51, 561-581.

Blasdel, G.G. and Salama G. (1986) Voltage-sensitive dyes reveal a modular organization in the monkey striate cortex. Nature, 218, 578-585.

Cattarelli, M. and Cohen, L.B. (1989) Optical recording of the *in vivo* piriform cortex responses to electrical stimulation of the lateral olfactory tract in the rat. Chem. Senses, in press.

Cohen, L.B., Höpp, H.-P., Wu, J.-Y., Xiao, Ch., London, J. and Zecevic, D. (1989) Optical measurements of action potential activity in invertebrate ganglia. Ann. Rev. Physiol., 51, 527-541.

Cohen, L.B. and Lesher S. (1986) Optical monitoring of membrane potential; methods of multisite optical measurements. In: Optical Methods in Cell Physiology. Eds: P. De Weer & B.M. Salzberg, pp.71-99, Wiley, New York.

Dillon, S. and Morad, M. (1981) A new laser scanning system for measuring action potential propagation in the heart. Science, 214, 453-456.

Egger, M.D. and Petran, M. (1967) New reflected-light microscope for viewing unstained brain and ganglion cells. Science, 157, 305-307.

Fine, A., Amos, W.B., Durbin, R.M. and McNaughton, P.A. (1988) Confocal microscopy: applications to neurobiology. Trends Neurosci., 11, 346-351.

Grinvald, A. (1985) Real-time optical mapping of neural activity: from single growth cones to the intact mammalian brain. Ann. Rev. Neurosci., 8, 263-305.

Grinvald, A., Anglister, L., Freeman, A., Hildesheim, R. and Manker, A. (1984) Real-time optical imaging of naturally evoked electrical activity in intact frog brain. Nature, 308, 848-850.

Grinvald, A., Cohen, L.B., Lesher, S. and Boyle, M.B. (1981) Simultaneous optical monitoring of activity of many neurons in invertebrate ganglia using a 124-element photodiode array. J. Neurophysiol., 45, 829-839.

Grinvald, A., Manker, A. and Segal, M. (1982) Visualization of the spread of activity in rat hippocampal slices by voltage sensitive optical probes. J. Physiol., 333, 269-291.

Kauer, J.S. (1988) Real-time imaging of evoked activity in local circuits of the salamander olfactory bulb. Nature, 331, 188-168.

Kauer, J.S., Senseman, D.M. and Cohen, L.B. (1987) Odor-elicited activity monitored simultaneously from 124 regions of Salamander olfactory bulb using a voltage-sensitive dye. Brain Res., 418, 255-261.

Lieke, E.E, Frostig, R.D, Arieli, A., Ts'o, D.Y., Hildesheim, R. and Grinvald, A. (1989) Optical imaging of cortical activity: Real-time imaging using extrinsic dye-signals and high resolution imaging based on slow instrinsic-signals. Ann. Rev. Physiol., 51, 543-559.

London, J.A., Zecevic, D. and Cohen, L.B. (1987) Simultaneous optical recording of activity from many neurons during feeding in Navanax. J. Neurosci., 7, 649-661.

Milton, G., Ireland, C.L.M. and Ley, J.M. (1983) Electro-optic and Acousto-optic Scanning and Deflection. Optical Engineering Vol.3. Marcel Dekker, New York and Basel.

Minsky, M. (1957) Microscopy apparatus. U.S. Patent 3013467.

Orbach, S.H. and Cohen, L.B. (1983) Optical monitoring of activity from many areas of the *in vitro* and *in vivo* Salamander olfactory bulb. A new method for studying functional organization in the vertebrate central nervous system. J. Neurosci., 3, 2251-2262.

Orbach, S.H., Cohen, L.B. and Grinvald. A. (1985) Optical mapping of electrical activity in the rat somatosensory and visual cortex. J. Neurosci., 5, 1886-1895.

Saggau, P. and ten Bruggencate, G. (1988) Topology related real-time monitoring of neural activity in hippocampal brain slices by noninvasive optical recording - a step towards functional aspects of long-term potentiation. In: Synaptic Plasticity in the Hippocampus. Eds: H.L. Haas and G. Buzsaki, pp.159-162. Springer, Berlin Heidelberg.

Saggau, P., Hiendl, R., Rucker F. and Wenzel, G. (1988) Fundamentals of laser scanning microscopy in optical recording of neural activity. In: Optical Methods in Neurobiology. Eds: H.-R. Lüscher and M.G. Rioult, pp.46-60. Technical Workshop I, 11th Annual Meeting of ENA, Bern, Switzerland.

Salzberg, B.M. (1983) Optical recording of electrical activity in neurons using molecular probes. In: Current Methods in Cellular Neurobiology. Vol.III: Electrophysiological and optical recording techniques. Eds: J.L. Barker and J.F. McKelvy, pp.139-180. Wiley, New York.

Salzberg, B.M. (1989) Optical recording of voltage changes in nerve terminals and fine neuronal processes. Ann. Rev. Physiol., 51, 543-559.

Wilson, T. and Carlini, A.R. (1988) Three-dimensional imaging in confocal imaging systems with finite sized detectors. J. Microscopy, 149, 51-66.

Wilson, T. and Sheppard, C. (1984) Theory and Practice of Scanning Optical Microscopy. Academic Press, London

Young, J.Z. and Roberts, F. (1951) A flying-spot microscope. Nature, 167, 231.

TEMPORAL PATTERNS OF MEMBRANE POTENTIAL IN THE OLFACTORY BULB OBSERVED WITH INTRACELLULAR RECORDING AND VOLTAGE-SENSITIVE DYE IMAGING: EARLY HYPERPOLARIZATION.

J. S. Kauer, K.A. Hamilton, S.R. Neff, A.R. Cinelli

New England Medical Center/Tufts Medical School,
Departments of Neurosurgery and Anatomy & Cell Biology,
and the Section Of Neuroscience,
750 Washington St., Boston, MA. 02111
USA

INTRODUCTION

Adrian's (1951, 1956) early work was instrumental in suggesting that changes in the time course of spikes and membrane potentials (temporal patterning) and differences in the degree of activation across cells (spatial patterning) could play roles in encoding characteristics of odorous stimuli. In light of the relatively broad-spectrum responses observed at receptor (Getchell and Shepherd, 1978a,b; Revial et al., 1982, 1983) and olfactory bulb (Kauer, 1974; Meredith and Moulton, 1978; Mair, 1982) levels, it continues to be generally believed that many neurons at each stage in the olfactory pathway are activated by odor stimulation and it is, therefore, this aggregate activity that carries the neural representation of the quality and intensity of the stimulus.

Precise characterizations of odor-elicited, spatial and temporal patterns and the relationships of these patterns to the stimuli have, however, been difficult to define. Extracellular (Kauer, 1974; Meredith and Moulton, 1978; Mair, 1982) and intracellular (Hamilton and Kauer, 1985, 1989) studies using odor stimulation have revealed what appear to be a limited number of characteristic temporal patterns in several different species, suggesting that the time course of spikes and membrane potentials per se do not have the complexity needed to encode large numbers of odor qualities. 2-deoxglucose mapping has shown widespread distribution of activity after odor stimulation (Sharp et al., 1975; Stewart et al., 1979; Jourdan et al., 1980), but this

NATO ASI Series, Vol. H 39
Chemosensory Information Processing. Edited by D. Schild

method has still not allowed specific hypotheses to be made about how an odorous stimulus activates the circuitry of the olfactory epithelium/bulb to generate a particular response pattern over the normal time course of a stimulus presentation. In part this has been due to the technical difficulties of recording, with high temporal and spatial resolution, from the large numbers of elements that make up the neural network. This may also have been due to the difficulty of relating the complexities of widely distributed spatial and temporal neuronal responses to what are, presumably, unitary stimuli, i.e. odorant molecules of a single type. Indeed, analysis of other sensory systems have not provided insight and guidance on what distributed information processing should look like.

In the present paper we have observed the presence of a period of hyperpolarization which precedes the onset of the well-described sequence of excitation and inhibition (Shepherd, 1963; Mori et al., 1981a,b,c) in mitral/tufted cells of the salamander olfactory bulb. We suggest that this inhibition, which is difficult to see under circumstances in which recordings are usually made, may serve to provide a background of reduced neuronal noise against which activity carrying information characteristic an odorant stimulus may stand in contrast. By focussing on the patterns of activity across cells in particular time domains during the evolution of responses, we hope to be able to improve our chances for observing critical aspects of odor-elicited patterns.

METHODS

Tiger salamanders (Ambystoma tigrinum) were anaesthetized and prepared as previously described (Kauer, 1974). Intracellular recordings were made using standard methods (Hamilton and Kauer, 1988). Electrical stimulation was performed with concentric stainless steel electrodes placed either on the cranial side of the olfactory nerve (stimulating electrode position 'on.' in Fig. 1A) as it entered the olfactory bulb, or on a peripheral olfactory nerve fascicle (stimulating electrode position 'm.' or 'l.' in Fig. 1A) of the dorsal nasal mucosa. Some of the electrical stimulation experiments were made in water-phase animals in which the olfactory nerves were particularly long (~1 cm). The results from the water-phase animals were confirmed in land-phase phase specimens which were used in all the other experiments.

Olfactory stimulation was carried out using controlled and monitored odorant delivery as previously described (Kauer and Shepherd, 1975).

Real-time imaging of olfactory bulb activity by measuring voltage-sensitive dye fluorescence after electrical and olfactory stimulation was carried out as described in

Kauer, 1988. The setup of the stimulating electrodes, light source and microscope objective is shown in Fig. 1A. Fig. 1B shows a schematic drawing of the dorsal view of the planar layers of the olfactory bulb of the salamander (Herrick, 1948) as they are observed in each frame of a dye imaging run (see Fig. 4A,B).

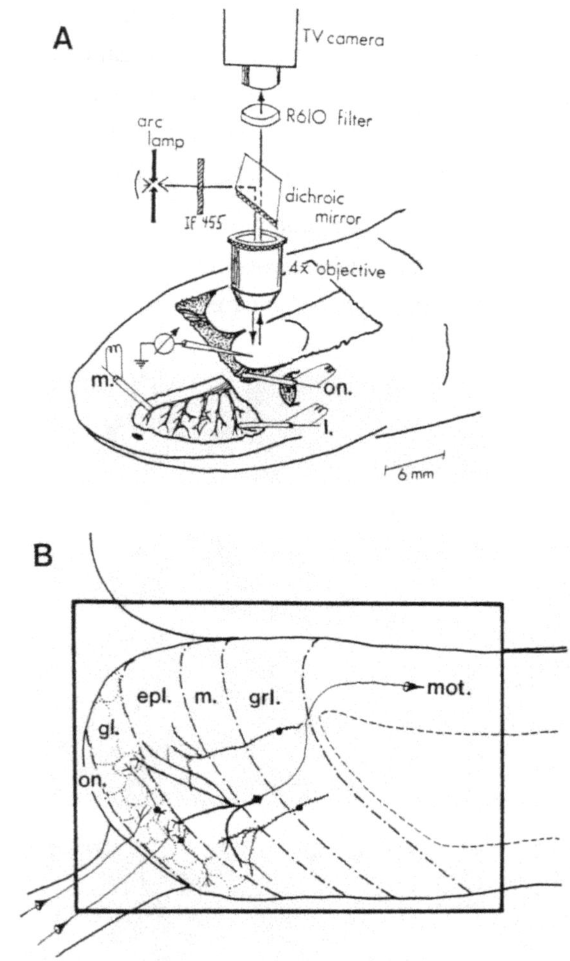

Figure 1. A: Diagram of experimental setup for voltage-sensitive dye recording. The exposures of the olfactory epithelium and bulb are also used in the intracellular recording experiments. Sites of electrical stimulation of olfactory nerves are marked 'm.'- medial fascicles; 'l.'- lateral fascicles; 'on.'- olfactory nerve bundle as it enters the olfactory bulb. R610 and IF455 are emission and excitation filters.

B: Schematic drawing of the dorsal view of the salamander olfactory bulb showing cell layers. Dark rectangle delineates the view seen in each frame of the voltage-sensitive dye experiments (see Fig. 4). on. = olfactory nerve layer; gl. = glomerular layer; epl. = external plexiform layer; m. = mitral body layer; grl. = granule cell layer; mot. = medial olfactory tract.

RESULTS

Fig. 2A illustrates a set of intracellular records of spikes elicited from a mitral/tufted cell by three intensities of brief orthodromic shocks to an olfactory nerve fascicle at the site labelled 'm' in Fig. 1A. For the shortest latency spike, the stimulus intensity was 720 μA, (substantially above spike threshold), for the middle-latency spike the intensity was 670 μA, and for the longèst latency spike the intensity was 600 μA. In these records, one can see that for each intensity the spike arises from the baseline after the occurance of a brief period of hyperpolarization. A longer period of hyperpolarization, the duration of which was dependent on stimulus intensity, followed the spike. Notice that the period of preceding hyperpolarization became more difficult to see as stimulus intensity was increased. This early period of hyperpolarization was completely obscured at intensities several-fold above threshold, the level usually used for these kinds of experiments (see Hamilton and Kauer, 1988)

Fig. 2B, on a faster time scale than Fig. 2A, shows another example from a different cell. In this case a brief period of depolarization may precede the hyperpolarization. This was seen in a number of cases.

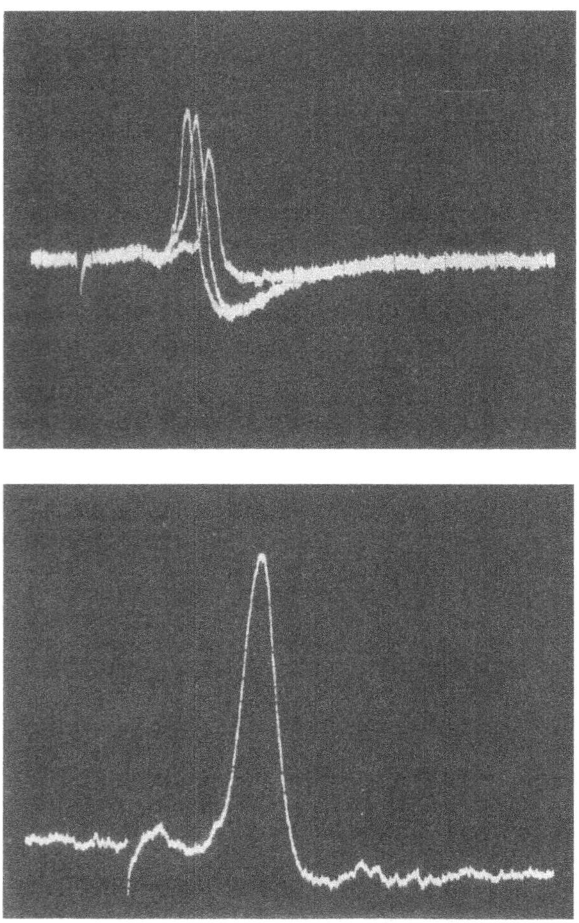

Figure 2. A: Three superimposed intracellular records from a water-phase salamander mitral/tufted cell after electrical orthodromic stimulation of a peripheral olfactory nerve fascicle. Record showing the longest latency spike was obtained with a stimulus of 600 μA, middle-latency record with a stimulus at 670 μA, and shortest latency record at 720 μA. Notice the hyperpolarizing period preceeding spikes. Note also the decrease in latency of spikes and increase in after-hyperpolarization with increase in stimulus intensity.

B: Another cell in another animal. Note preceding hyperpolarization.

Time scales = 20 msec in both A. and B.

As shown in Fig. 3, brief, initial hyperpolarizations are also frequently observed in intracellular recordings made from mitral/tufted cells after square-pulse odor stimulation (see also Hamilton and Kauer, 1985, 1989). In a-d of Fig. 3, the top trace is the intracellular record and the lower trace is the electro-olfactorgram (EOG), plotted negative up. In Fig. 3 e, the third trace is the output of the CO_2 monitor which delineates the time course of the odor stimulus. Fig. 3 a-d are increasing stimulus intensities in dilutions of saturated vapor of cineole and Fig. 3e is a stimulus of amyl acetate presented at 10^{-1} of saturation. In the classification system we have developed for describing odor responses of mitral/tufted cells in the salamander (Kauer, 1974; Hamilton and Kauer, 1989), the records in Fig. 3 a-d would be considered E_{2-3}-type responses and the record in Fig. 3 e an S-type response.

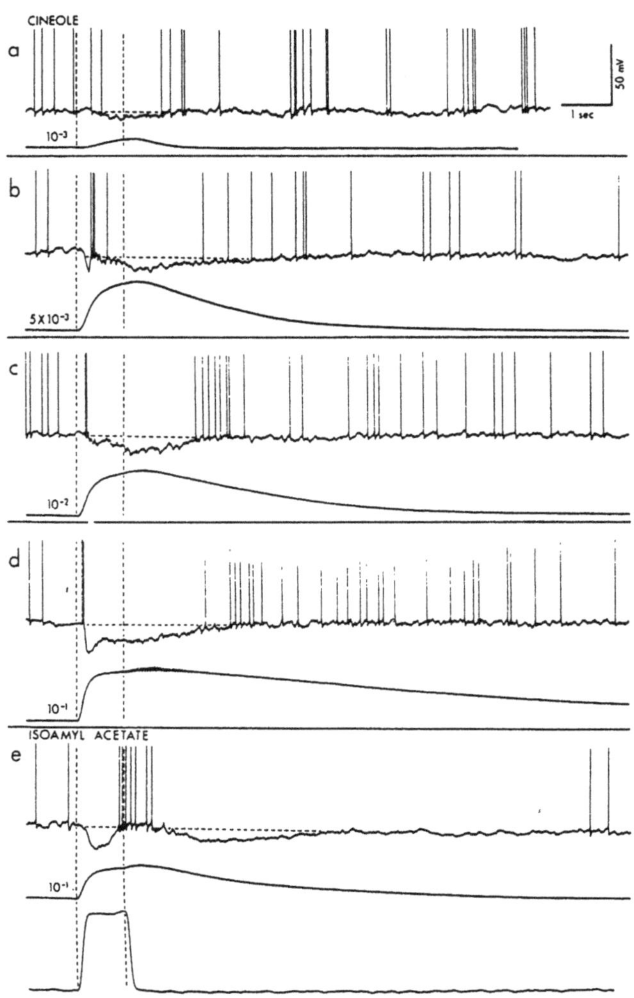

Figure 3. Intracellular records from a land-phase salamander mitral/tufted cell during controlled odor stimulation.
a-d: Top trace in each panel is intracellular record, bottom trace is EOG, plotted negative up. Stimulus is cineole at concentrations in percent saturation as indicated. This is an E_{2-3}-type response.
e: Top and middle traces as above, bottom trace is odor stimulus monitor. Stimulus is amyl acetate at 10^{-1} of saturation. This is an S-type response.

One can see, as the intensity of the stimulus is increased (Fig. 3a-d), the temporal pattern of the E-type response changes dramatically from a slight, uniform hyperpolarization in Fig. 3a, to a brief hyperpolarization, followed by depolarization, spikes, and a longer-lasting hyperpolarization in Fig. 3b. In Figs. 3c and d the patterns consist of a brief burst of spikes and long-lasting hyperpolarization with little or no preceding hyperpolarization (see Hamilton and Kauer, 1985 for detailed description). The initial sequence in Fig. 3b is similar to that seen with electrical stimulation in Fig. 2, although the response pattern is elaborated over a longer period of time. We think that the critical factor for seeing these patterns is the use of either moderate intensity or low intensity odor stimulation applied peripherally. Stimulation by either of these methods seems to result in less synchronous activation of the bulbar circuitry than the use of higher intensities applied to large areas of the mucosa, and therefore provide opportunities for synaptic interactions to occur sequentially over longer time courses.

Fig. 3e shows a longer-lasting, preceding hyperpolarization, which lasts for much of the odor pulse stimulus with the burst of spikes occuring near the end of the stimulus pulse, typical of S-type responses.

After observing these periods of preceding hyperpolarization in the intracellular recordings we looked to see if periods of inhibition with similar time courses appeared in our recordings of ensemble activity made by video imaging of voltage-sensitive dye fluorescence. Fig. 4A shows the results of electrically stimulating a peripheral olfactory nerve fascicle approximately at site 'm' in Fig. 1A. Although the details of the fluorescent patterns are difficult to see, the gray background in this figure consists of a sequence of 16 sequential frames (33 ms in duration) each of which represents the change in fluorescence of the voltage-sensitive dye across the view of the bulb shown in Fig. 1B (see Kauer, 1988 for details). The electrical stimulus, just above threshold, was applied at the beginning of frame 2 (small arrow on pink graph). The plot maps the change in fluorescence at the sites circled in each frame (approximately in the glomerular layer). The vertical lines on the graph occur at 100 ms intervals. From this graph, it can be seen that a period of hyperpolarization (large arrow) precedes a large depolarization and the beginning of a long lasting hyperpolarization.

We carried out similar experiments using odor stimulation, as shown in Fig. 4B. This experiment was performed as described above for electrical stimulation, except that the nasal cavity was opened and an odor stimulus (amyl acetate 10^{-2} of saturation; see Kauer and Moulton, 1974 for details) was applied to the ventral olfactory epithelium. In this experiment, 64 sequential frames were taken, only 16 of which are shown here. The vertical lines on the graph represent 100 ms intervals. The colors code for changes in fluorescence (voltage) in the depolarizing and hyperpolarizing directions (green = one standard deviation (S.D.) above baseline, red = two S.D.'s above baseline, blue = one or more S.D.'s below baseline). As in Fig. 4A, the graph plots the activity in each frame at the sites indicated by the circles. Here one can also see a sequence of hyperpolarization, depolarization, and hyperpolarization. The sequence of activity is more spread out in time with an odorous stimulus because odors presumably activate olfactory receptors less synchronously than do electrical shocks.

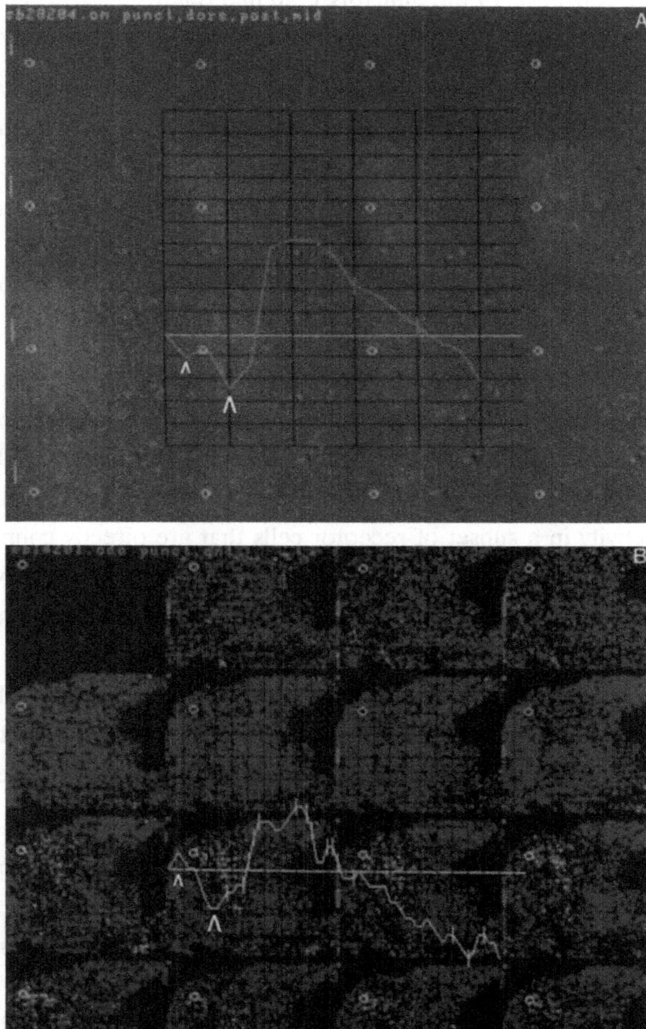

Figure 4. A: 16 frame, voltage-sensitive dye recording from land-phase salamander after orthodromic electrical stimulus applied to an olfactory nerve fascicle approximately at site labelled 'm' in Fig. 1A. Each frame represents changes in voltage in the region depicted in Fig. 1B. Circles mark spot on each frame (approximately the glomerular layer) from which the values were obtained for the graphical plot shown in the center of the figure. Vertical black lines mark 100 ms intervals. Small arrow indicates when stimulus was applied. Large arrow indicates preceding hyperpolarization. B: 64 frame, voltage-sensitive dye recording after odor stimulation of ventral olfactory epithelium. Only 16 of the 64 frames are shown. Colors encode changes in fluorescence reflecting differences in voltage. Green = one S.D. in the depolarizing direction above the baseline; red = two S.D.'s above baseline; blue = one or more S.D.'s below baseline in the hyperpolarizing direction. The points plotted in the graph were taken from the sites marked by circles (approximately in the glomerular layer). Vertical lines mark 100 ms intervals.

DISCUSSION

With low intensity, peripherally placed, odor and electrical stimulation, a period of hyperpolarization can be observed in single unit and ensemble recordings from the olfactory bulb that precedes the first period of spike activity elicited in mitral/tufted cells. One of the critical requirements for observing this hyperpolarization appears to be the activation of the olfactory bulb in a relatively asynchronous manner, thus permitting observation of the various components of the synaptic activity.

While electrical stimulation of the olfactory nerve is a useful method for synchronously activating the bulbar circuitry to dissect its function, it does not mimic the natural activation sequence generated by odorant stimulation. By recording intracellularly from mitral/tufted cells during controlled odor stimulation (Hamilton and Kauer, 1985, 1989), we think we may have revealed events occuring early in the response that could carry information specifically related to odor type.

We suggest that the early hyperpolarization seen in the present experiments is a manifestation of inhibition generated by inhibitory periglomerular cells (see Halasz and Shepherd, 1983), at least some of which appear to be GABAergic in the salamander (Hamilton et al., 1987). We further suggest that, because this hyperpolarization is generated in the absence of preceding spikes in the mitral/tufted cells, it is initiated by activity in a subset of receptor cells that are directly connected to periglomerular cells (see Kauer, 1987). Since this period occurs before the activation of mitral/tufted cells it may be that this subpopulation of receptor cells has axons that are larger in diameter and thus have higher conduction velocities than those that mediate excitation of mitral/tufted primary dendrites. This initial period of inhibition of the output mitral/tufted cells may serve to reduce the backgound activity upon which patterns of excitation are differentially distributed depending on the characteristics of the odor stimulus.

As previously discussed (Kauer, 1987), this hypothesis takes into account the patterns of connectivity that exist between the olfactory epithelium and bulb (Kauer, 1981; Jastreboff et al., 1984), the distribution of sensitivities to odor across the epithelium observed in EOG studies (McKay-Sim et al., 1982), the characteristics of mitral/tufted receptive fields using punctate odor stimulation (Kauer and Moulton, 1974), and the present findings.

We think the use of intracellular recording from single neuronal elements in a synaptic network, in combination with ensemble recording using imaging of voltage-sensitive dye fluorescence, provides a unique opportunity for studying sensory coding that involves distributed information processing mechanisms.

REFERENCES

Adrian ED (1951) Olfactory discrimination. L'Anee Psychol 50:107-113

Adrian ED (1956) The action of the mammalian olfactory organ. J Laryngol Otol 70:1-14

Getchell TV, Shepherd GM (1978a) Responses of olfactory receptor cells to step pulses of odour at different concentrations in the salamander. J Physiol 282:521-540

Getchell TV, Shepherd GM (1978b) Adaptive properties of olfactory receptors analyzed with odour pulses of varying durations. J Physiol 282:541-560

Halasz N, Shepherd GM (1983) Neurochemistry of the vertebrate olfactory bulb. Neurosci 10:579-620

Hamilton KA, Kauer JS (1985) Intracellular potentials of salamander mitral/tufted neurons in response to odor stimulation. Br Res 338:181-185

Hamilton KA, Kauer JS (1988) Responses of mitral/tufted cells to orthodromic and antidromic electrical stimulation in the olfactory bulb of the tiger salamander. J Neurophysiol 59:1736-1755

Hamilton KA, Kauer JS (1989) Responses of mitral/tufted cells to odor stimulation in the olfactory bulb of the tiger salamander. J Neurophysiol (in press)

Hamilton KA, Kream RM, Kauer JS (1987) Neurochemical analysis of the salamander olfactory bulb. Chem Senses 12:663 (abst)

Herrick CJ (1948) The Brain of the Tiger Salamander, Univ. of Chicago Press, Chicago

Jastreboff PJ, Pedersen PE, Greer CA, Stewart WB, Kauer JS, Benson TE, Shepherd GM (1984) Specific olfactory receptor populations projecting to identified glomeruli in the rat olfactory bulb. PNAS 81:5250-5254

Jourdan F, Duveau A, Astic L, Holley A (1980) Spatial patterns of 2-deoxyglucose uptake in the olfactory bulb of rats stimulated with two different odors. Br Res 188:139-154

Kauer JS (1974) Response patterns of amphibian olfactory bulb neurones to odour stimulation. J Physiol 243:695-716

Kauer JS (1981) Olfactory receptor cell staining using horseradish peroxidase. Anat Rec 200:331-336

Kauer JS (1987) Coding in the olfactory system. In: Finger TE, Silver W (eds) Neurobiology of Taste and Smell. Wiley and Son, New York, pp 205-231

Kauer JS (1988) Real-time imaging of evoked activity in local circuits of the salamander olfactory bulb. Nature 331:166-168

Kauer JS, Moulton DG (1974) Responses of olfactory bulb neurones to odour stimulation of small nasal areas in the salamander. J Physiol 243:717-737

Kauer JS, Shepherd GM (1975) Olfactory stimulation with controlled and monitored step pulses of odor. Br Res 85:108-113

MacKay-Sim A, Shaman P, Moulton DG (1982) Topographic coding of olfactory quality: Odorant specific patterns of epithelial responsivity in the salamander. J Neurophys 48:584-596

Mair RG (1982) Response properties of rat olfactory bulb neurones. J Physiol 326:341-359

Meredith M, Moulton DG (1978) Patterned response to odor in single neurones of goldfish olfactory bulb: influence of odor quality and other stimulus parameters. J Gen Physiol 71:615-643

Mori K, Nowycky MC, Shepherd GM (1981a) Electrophysiological analysis of mitral cells in the isolated turtle olfactory bulb. J Physiol 314:281-294

Mori K, Nowycky MC, Shepherd GM (1981b) Analysis of synaptic potentials in mitral cells in isolated turtle olfactory bulb. J Physiol 314:295-309

Mori K, Nowycky MC, Shepherd GM (1981c) Analysis of a long-duration potentials in mitral cells of the isolated turtle olfactory bulb. J Physiol 314:311-320

Revial MF, Sicard A, Duchamp A, Holley A (1982) New studies on odour discrimination in the frog's olfactory receptor cells. I. Experimental results. Chem Senses 7:175-190

Revial MF, Sicard A, Duchamp A, Holley A (1983) New studies on odour discrimination in the frog's olfactory receptor cells. II. Mathematical analysis of electrophysiological responses. Chem Senses 8:179-190

Sharp FR, Kauer JS, Shepherd GM (1975) Local sites of activity-related glucose metabolism in rat olfactory bulb during odor stimulation. Br Res 98:596-600

Shepherd GM (1963) Neuronal systems controlling mitral cell excitability. J Physiol 168:101-117

Stewart WB, Kauer JS, Shepherd GM (1979) Functional organization of rat olfactory bulb analyzed by the 2-deoxyglucose method. J Comp Neurol 185:715-734.

OPTICAL MAPPING OF THE OLFACTORY SYSTEM ACTIVITY USING VOLTAGE-SENSITIVE DYES

Martine Cattarelli

Physiologie neurosensorielle
CNRS - Université Claude Bernard, LYON
69622 Villeurbanne cedex
France

INTRODUCTION

What can optical recordings of neuronal activity using voltage-sensitive dyes tell us about the olfactory nervous system ?

Since a century, different methodological approaches have been extensively used to understand the organization and function of the central nervous system. Anatomical studies have pointed out the basic tenet of the nervous system organization, i.e., the interconnections of neurons into complex circuits and networks. Any sensory information processing depends on the activity of millions of neurons, which are largely interconnected. Thus, each individual neuron activity will be modulated by the network activity, and its response properties will be influenced by other brain activities. For example, olfactory system studies have demonstrated that secondary order neuron responsiveness differ according to the nature of the odorant and the internal motivational state of the organism (Pager, 1976; Cattarelli, 1983). Until now, electrophysiology, histochemistry, biochemistry, psychophysics and behavior study have been the main tools used to examine the functional processing of information in the central nervous system. In a general way, the methods used are related to the conceptual frameworks underlying the neurosciences. Thus several approaches have been considered (Varela, 1989). For example, knowledge of individual neuron properties would elucidate the whole brain functioning. Neuronal networks were compared to hardware in a computer; their functions would be inferred from their connectivity. In this case, each recorded unit activity would inform about the whole

NATO ASI Series, Vol. H 39
Chemosensory Information Processing. Edited by D. Schild

system function according to the pattern of connections in which it is involved. But it turns out that neuronal networks are impressively complex: they exhibit elaborated properties of their own which do not directly depend on the individual properties of the constituting neurons. Moreover, the activity of a particular network also depends on the activity of other neurons which belong to other neuronal networks. It is the case for the olfactory system activity which is modulated by at least three factors i) the internal state of satiety (Pager, 1976) ii) the motivational state induced by the odorants themselves (Cattarelli, 1983) and iii) the vigilance level of the animal (Cattarelli, 1979; Gervais, 1983). Thus, the functioning of a network brain area cannot be directly inferred from its anatomy and connectivity only, but it rather emerges from the network itself which is involved in the whole brain. Therefore, data recorded from isolated neurons appear as elements which participate in the network area activity but are not the direct image itself. As underlined by Grinvald et al. (1988), no single network is actually fully understood. Particularly in the brain of vertebrates, the large redundancy observed among neuronal networks suggests that a specific function will be rather controlled by the activity level of the network than by the activity of any given nervous cell.

In parallel with the conceptual evolution in neurosciences, needs for new research tools for investigating the activity of large neuronal assemblies appeared three decades ago (Grinvald et al., 1988). In this purpose different methods have been developed. Real-time recordings of unit activity are allowed by microelectrodes, but they are limited to a small number of neurons, usually less than five, at the same time. Conversely, metabolic mapping of brain activity, using the 2-deoxyglucose method (Sokoloff et al., 1977) allows a fine localization of all the areas activated during a given test. But this method has a very slow temporal resolution since the exposure to the experimental conditions is about 45 minutes. Moreover, only one stimulation condition can be tested per animal. Other methods such as electroencephalography, positron emission tomography, nuclear magnetic resonance imaging, blood flow radioactive imaging ... offer either a good temporal (real-time) or spatial (three-dimensional) resolution but also have limitations. However, a new possibility is offered now by the optical method of neuronal activity recording using voltage-sensitive dyes. This method was developed at Yale University by Cohen and his colleagues (Cohen et al., 1971; Cohen and Salzberg, 1978; Cohen and Lesher, 1986; Grinvald, 1985; Grinvald et al., 1986, 1988; Orbach et al., 1985; Ross et al., 1977). This method, as the others, presents advantages and limitations which will be discussed hereafter.

PRINCIPLE, LIMITATIONS AND ADVANTAGES OF THE OPTICAL RECORDING USING VOLTAGE-SENSITIVE DYES

The optical method of neuronal activity recording is based on measurements of changes in fluorescence or light absorption of preparations vitally stained with voltage-sensitive dyes. The dye molecules bind to the cell membranes and become molecular transducers which translate variations of membrane potential into optical signals. These optical signals are nearly concomitant (in the microsecond range) and linearly proportional to the electrical activity of the stained cells. The variations of light (i.e. optical signals) are detected by an array of photodiodes (more than 100), located in the microscope image plane. The optical signals are converted into electrical signals by the photodiodes. After appropriate amplification and filtering, the analog signals are digitized, then stored in a computer RAM. Each photodiode signal can be sampled at a high given rhythm, up to 7 KHz in our device (Vigouroux et al., 1988). The duration of the recording depends on the sampling rate and the number of samples recorded during each experiment. Then the activity recorded by each photodiode of the array is visualized. According to the preparation studied and the magnification of the microscope, the photodiode array can image either fine neuronal processes or a brain area of several square millimeters. Thus direct recordings of synaptic potentials in fine neuronal processes as well as spatio-temporal patterns of neuronal population activity can be performed (See: Cohen and Salzberg, 1978 and Grinvald et al., 1988; as reviews).

If the principle of the optical method seems rather simple, some problems are nevertheless encountered (for extensive discussions see: Cohen and Salzberg, 1978; Cohen and Lesher, 1986; Grinvald et al., 1988). Briefly it can be mentioned that the small size of optical signals obliges to reduce as much as possible the different sources of noise, i.e. light shot noise, dark noise of diodes, vibrations from any elements of the device and particularly from the preparation itself (respiration and heartbeat). The use of voltage-sensitive dyes in presence of intense light can also induce pharmacological effects and/or photodynamic damage of the preparation and they have to be evaluated for each new experiment in order to limit their influence. Lastly, in *in vivo* brain studies, only exposed surfaces of the brain can be optically imaged up to a depth of 0.5 to about 2 mm (Lieke et al., 1989). Since vertebrate neurons are rather small, optically recorded activities result from the summation of all potential variations occurring in the area imaged on each diode. Optical signals could result from EPSP, IPSP and potential variations of soma and axon membranes. As the membrane area of mammalian dendrites is larger than that of soma and axon, the contribution of synaptic events

(EPSP and IPSP) should be important (Grinvald et al., 1988). The eventual contribution of glial cells to the optical signals remains unknown but it might be of special concern at the slower portions of the signals.

Nevertheless, optical recording of neuronal activity presents many advantages which make it a useful tool for brain functioning studies. They are summarized hereafter:

- Simultaneous monitoring of neuronal activity at many (more than one hundred) juxtaposed sites becomes possible.
- Time resolution is in the submillisecond range.
- Spatial resolution is reasonable enough since it is compatible with cortical columns for example (Orbach et al., 1985).
- Recordings are made in real-time on living animals.
- A same animal can be used as its own control and can be submitted to several successive stimulations.
- Optical signals reflect the electrical events (i.e. intracellular membrane potential changes) occuring in their origin sites looked at by each photodiode, in contrast to electrical field potential recordings.

Previous experiments have demonstrated the usefulness of the optical monitoring for mapping the activity of either brain slices (Grinvald et al., 1982; Saggau et al., 1986) or of some whole brain areas *in vitro* (Orbach and Cohen, 1983; Kauer et al., 1987) and *in vivo* (Kauer et al., 1987; Kauer, 1988; Cattarelli and Cohen, 1989) in Vertebrates. Particularly, somatosensory and visual systems have been studied in this way (Orbach et al., 1985). It can be noted that the spatial resolution of this method has been improved by using a CCD camera instead of a photodiode array (Blasdel and Salama, 1986; Kauer, 1988).

THE OLFACTORY NERVOUS SYSTEM: ANATOMICAL AND FUNCTIONAL DATA

In neurosciences, according to a basic hypothesis, sensory information should be encoded in spatio-temporel patterns of neural activity (Freeman, 1983). Moreover, in all biological systems, selectivity should be enhanced at the successive neural network levels following the receptor cell level (Schild, 1988).

The olfactory nervous system can be described as a neural network, but every olfactory area also acts as a neuronal network. In mammals, several studies have demonstrated that secondary neurons of the olfactory bulb directly project onto various cortical areas of the basal forebrain via the lateral olfactory tract. This primary olfactory cortex includes the anterior olfactory nucleus, the piriform cortex, the olfactory tubercle, the periamygdaloid cortex, the nucleus of the lateral olfactory tract, the lateral entorhinal area and the ventral part of the tenia tecta (Heimer, 1968; Price, 1973; Ojima et al., 1984). Thus, in the olfactory nervous system, a striking feature is that the third sensory neuron of the chain is not a thalamic neuron, as usually observed in the other sensory systems, but a cortical cell. Such an organization certainly implies some specific consequences in the olfactory information processing. The main projection area of the olfactory bulb and the largest area of the olfactory cortex is the piriform cortex, a paleocortical structure (Shepherd, 1974). It is likely that this structure plays an important role in the olfactory information processing and integration. As a paleocortex, it offers advantages, such a three histological well-defined layer structure and a synaptic proximity to the receptor cells, allowing the study of neuronal integration and coding in a sensory cortex. Moreover, the piriform cortex receives inputs from the ipsi- and the contra-lateral olfactory bulb and from the contralateral piriform cortex, via the anterior and posterior branches of the anterior commissure. Beside its interconnections with all the primary olfactory cortical areas, the piriform cortex output neurons project to the neocortex either directly or after relays in limbic, thalamic or hypothalamic structures (Krettek and Price, 1974; Leonard, 1969; Powell et al., 1965; Price and Slotnick, 1983; Scott and Leonard, 1971; Switzer et al., 1985). Lastly the piriform cortex is the origin of centrifugal fibers which reach the ipsilateral olfactory bulb, either via the lateral olfactory tract (Willey et al., 1975), or via the anterior branch of the anterior commissure (Carson, 1984; Davis and Macrides, 1981).

The organization of the projections, both at the olfactory bulb and at the piriform cortex levels, is characterized by convergence and divergence phenomena (Holley and

Mac Leod, 1977; Mori, 1987). The olfactory receptor cell axons converge with a factor of 1000 onto the glomeruli of the olfactory bulb. But it is noticeable that one glomerulus receives inputs from receptor cells largely distributed among the sensory epithelium (Astic et al., 1987). Similarly, at the piriform cortex level, each pyramidal cell receives synapses from several olfactory bulb output neurons. These latter make synapses with several pyramidal cells, via numerous collaterals of their axons (Stevens, 1969; Ojima et al., 1984). Moreover, the pyramidal cell axons emit collaterals which establish synaptic contacts with neighboring or distant pyramidal cells (Haberly and Bower, 1984). These association fibers contribute to a large distribution of the olfactory messages through the whole piriform cortex. Anatomical convergence and divergence have also to be considered from a functional point of view. In the olfactory bulb and the piriform cortex, convergence of the afferent fibers induces an integration of the inputs. But the inputs are not simply relayed at these two levels, they are modulated and controlled by interneurons through feedback and lateral inhibition (Holley and Mac Leod, 1977; Mori, 1987; Haberly and Bower, 1984). Moreover, at the piriform level, a feedforward inhibitory mechanism has been described and also possible loops of reactivation of pyramidal cells via association fibers (Haberly and Shepherd, 1973; Haberly and Bower, 1984; Rodriguez and Haberly, 1989; Satou et al., 1983; Tseng and Haberly, 1988). Such a complexity of the anatomo-functional organization suggests that both the olfactory bulb and piriform cortex are largely involved in the information processing as integrative centers. So it can be asked what is the coding of the odorants at both these two levels.

Odorous stimuli have no spatial characteristics per se, and the stimulation source location is not encoded by the spatial distribution of neuroreceptors on the olfactory epithelium. Thus, spatial patterns of activity of second and third order olfactory neurons are disposable to encode any other stimulus particularities, such as quality or intensity (Kauer, 1987a). Electrophysiological studies of neuroreceptors and physicochemical studies of odorants have pointed out that spatial patterns of olfactory structure activity encode molecular properties of the odorant - adsorption properties principally - (Chanel, 1987). They also pointed out that the selectivity of the olfactory bulb neurons is higher than that of the receptor cells (Duchamp, 1982; Duchamp and Sicard, 1984). Thus, the spatial pattern of bulbar activity is involved in the quality coding of odorants. Therefore, in every olfactory neuron, the characteristics of the burst of activity (spike frequency) can be used for coding slight variations of stimulus intensity. It must be remembered that large variations of intensity are coded via recruitment of neurons with another range of sensitivity (Duchamp-Viret, 1988). Thus, the intensity is temporally and spatially coded as it has been also observed in the visual and auditory systems.

At the piriform cortex level, only few studies have been carried out. They established that the piriform cell selectivity is nearly similar in the rat (Haberly, 1969; Giachetti and Mac Leod, 1975) or higher in the monkey (Tanabe et al., 1975) and in the rabbit (Onoda and Imamura, 1988) than that observed at the olfactory bulb level. However, nearly no information is available about the spatio-temporal dimension of olfactory coding. Successive waves of reactivation could occur at the pyramidal cell level. Their timing as well as that of the inhibitory events which could reduce their effectiveness should be crucial for the piriform cortex functioning. such a sequence of synaptic events could occur during volleys of activity induced by odorants (Rodriguez and Haberly, 1989). An "ensemble coding" (Freeman, 1979; Freeman and Schneider, 1982), i.e. involvement of the whole neuronal population, has been proposed at the piriform cortex level (Haberly and Bower, 1984).

Metabolic mapping of activity using 2-deoxyglucose has shown different but overlapping patterns of activity related to different odorants in the olfactory bulb (Jourdan et al., 1980; Sharp et al., 1985).In the piriform cortex (Cattarelli et al., 1988), the spread of the increased metabolic activity varied with the odorant tested but, here also, overlapping patterns of 2-DG uptake were observed. One particular feature was that no local and well-defined foci of 2-DG uptake could be noted in contrast to what found in the olfactory bulb. This data suggests a redistribution of the olfactory bulb output activity throughout the piriform cortex. Moreover, the redistribution could have been enhanced by the reactivation loops mediated by association fibers, during the 45 minutes of the experiment. These data are in agreement with the "ensemble coding" hypothesis above mentioned.

All these briefly summarized data underline that the olfactory central areas could be good candidates for an investigation of the spatio-temporal dimensions of sensory coding. As a matter of fact, it could be asked if a whole structure is involved in the coding or if it is subdivided into functional sub-units able to encode the stimuli and specifically connected to different areas of the olfactory network. It can be hypothesized that optical recordings of neuronal activity could be useful for understanding the olfactory system functioning.

OPTICAL RECORDING OF THE OLFACTORY AREAS: FIRST DATA

The first olfactory structure optically recorded was the salamander olfactory bulb. Orbach and Cohen (1983) demonstrated that optical signals can be recorded from this structure in response to electrical stimulation of the primary olfactory nerves. Different waveforms could be observed, even if the authors could not precisely correlate them with the antero-posterior successive bulbar layers observed. The main data of this first experiment indicated that optical recording can be used to provide a real-time map of brain activity and allow the study of the functional organization of the olfactory system.

Then olfactory stimulations have been tested in the same preparation (Kauer and al., 1984; Cattarelli and al., 1986; Kauer and al., 1987). Optical signals were recorded on the salamander olfactory bulb in response to odorant stimuli. Amplitudes, waveforms and latencies of the signals varied across the bulbar surface in relation to the bulbar layers and the odorants tested. The amplitudes of the signals seemed to be correlated with the concentration of the odorants. Moreover, Kauer and al. (1987) noted variations of amplitudes of the signals within a defined layer, suggesting that the various parallel circuits of the olfactory bulb were heterogeneously activated. These data were the first demonstration in real-time that different spatio-temporal patterns of activity are induced by different odorants at the olfactory bulb level. Senseman and al. (1987, 1988) using *in vitro* salamander olfactory bulb preparation observed a widespread, but not spatially uniform activity in the different layers of the olfactory bulb in response to electrical stimulations of the olfactory mucosa. They confirmed that the receptors located in one small area of the mucosa project widely onto the olfactory bulb.

More recently, Kauer (1987b, 1988) and Kauer and Neff (1988) using a video camera, instead of a photodiode array, improved the visualization of the spatial patterns of olfactory bulb activity evoked by olfactory nerve electrical stimulations or by odorants. They observed the spread of activity through the different bulbar layers and even to the contra-lateral olfactory bulb, with a longer latency. They described non-homogeneous patterns of responses indicating a differential involvement of the local circuits in the olfactory bulb. They underlined also a spatial and temporal overlapping of the bulbar cell activation according to different stimulations. (For more details see Kauer or Senseman papers in this volume).

Data related to the ionic mechanisms involved in the olfactory bulb activity have been reported by Cinelli and Salzberg (1988). On salamander olfactory bulb slices, they observed that fast and slow optical signals at the mitral-granule dendrite level were not inhibited by TTX. Optical signals from mitral cells seemed correlated to the Ca^{++} concentration, particularly the long lasting excitation of mitral cell dendrites. Cinelli and Salzberg (1987) reported in olfactory bulb slices of mouse, fast and slow depolarizing optical responses after LOT electrical stimulations. The fast component might arise from potentials propagating in the LOT and the mitral cells. It could spread in the glomerular and external plexiform layers as the intensity of the stimulation was increased. This result might indicate an action potential genesis in the mitral or granule dendritic tufts. The slow component was sensitive to ionic concentration of bath solutions and to GABA. It should result from synaptic activity, electrotonic propagation of action potentials into dendrites or glial cell depolarization.

Thus, optical recordings have been useful to demonstrate in real-time, different spatio-temporal patterns of activity according to the stimulus tested. Moreover, they allow to study the ionic mechanisms underlying the olfactory bulb cell functioning.

At the piriform cortex level, optical signals have been recorded in response to electrical stimulations of the rat lateral olfactory tract (Cattarelli and Cohen, 1989). Using mammals as experimental subjects oblige to take into account the brain movement artefact induced to the heartbeat and the respiration. As described by Orbach et al. (1985), such an artefact is reduced by triggering two recording periods from the R-wave of the electrocardiogram and substracting them: LOT stimulation was delivered during only one of the two recording periods. The noise due to respiratory movements was eliminated by turning off (for the 1.5 s required for each recording) the computer controlled respirator used to artificially ventilate the curarized rat. Usually, amplitude and duration of the optical signals vary in parallel with the stimulation. A decrease of these parameters of the optical signals could be sometimes noted for large and long stimuli. The latencies of the 124 optical signals recorded increased with increasing distance from the LOT to the cortical area. Identical stimulations induced a spatially similar distribution of the optical responses. When different groups of axons were stimulated (by using different pairs of electrodes) the patterns of responses overlapped largely. Nevertheless, some differences were noted: some pattern distributions were more restricted that others, even if the distance between the electrodes was similar. Moreover, beside the differences in the spatial distribution of activity patterns, differences in the amplitude and the waveforms of the responses were also observed, according to the pair of stimulating electrodes used. Thus, different internal piriform circuits have certainly been involved during the various stimulations.

From these first data, it can be hypothesized that a temporal distribution of the inputs occurs at the piriform cortex level since latencies of the responses differ.

This temporal distribution could depend on the propagation delays in the lateral olfactory tract fibers themselves. Besides it could also depend on the delays due to intrinsic loops of activation via association fibers. Moreover, the temporal distribution of the inputs could play an important role in the piriform cortex response to bilateral stimulation. This question will be studied soon. It is the first time that mapping of the activity on about 10 mm^2 of the piriform cortex was carried out. The responses were widely distributed on the piriform cortex surface, which is in agreement with the anatomical data. As the lateral olfactory tract axons and their collaterals are largely dispatched onto the piriform cortex, a redistribution and a large overlap of the olfactory inputs are allowed (Price, 1973). Moreover, it has been shown in the salamander that the bulbar activity elicited by a stimulation is, itself, not spatially confined. Nevertheless, some differences in the maps of piriform responses were noted, either in their spatial location or in their amplitudes and waveforms, according to the different pairs of stimulation electrodes used. Thus, it can be hypothesized that optical method is useful to visualize spatial patterns of activity. These latter could be induced by the different projections of the axons involved in the different stimulations (Scott et al., 1980) or/and by the different intrinsic association circuits activated. In the next future, the influence of spatially well-defined stimulations delivered in the olfactory bulb will be studied as well as the effect of olfactory stimulations with odorants which could be pure chemicals or complex mixtures having a significant biological meaning.

These first results about the piriform cortex permit also to point out some particular problems:

- Some voltage-sensitive dyes which have been successfully used in other preparations, even on the rat somatosensory cortex (Orbach et al., 1985), appeared toxic on the piriform cortex and induced a strong artery constriction. Thus it is always necessary to verify the influence of the dye used on a new biological preparation.
- A decrease in the amplitude of the electrical evoked potentials was noted after staining of the piriform cortex. Litaudon and Cattarelli (1989) have shown that the electrical evoked potentials were no longer altered if the dye RH 795 was diluted in artificial cerebrospinal fluid instead of isotonic sodium chloride.

Lastly, it can be mentioned that optical recording is a useful tool for mapping the olfactory projections from one structure to another. As a matter of fact, Cinelli and Salzberg (1987) observed optical signals in sulcal cortex in response to electrical stimulation of the piriform cortex in mouse brain slices.

CONCLUSIONS

Considering both the olfactory network organization and the first promising data obtained using optical recordings, it can be predicted that this method will be valuable for studying the mechanisms underlying the encoding of quality and intensity of odors. These mechanisms could include the topographical organization of the projections from one area to another and the anatomo-functional interactions of internal circuits in a given area (slices studies). They also include the spatio-temporal characteristics of the olfactory responses themselves, the role of the different messages conveyed by different pathways to a given structure, and the specific ionic phenomena underlying the cell activity at the membrane level. Of course, association of this optical method with other technical approaches (2-DG, HRP, electrophysiology ...) will improve our understanding of the data observed and finally of the functioning of the olfactory nervous system.

ACKNOWLEDGEMENT - We thank Dr A. Grinvald and R. Hidelsheim for useful advice on RH 795 and for having kindly given the dye to us.

REFERENCES

Astic L, Saucier D, Holley A (1987) Topographical relationships between olfactory receptor cells and glomerular foci in the rat olfactory bulb. Brain Res 424 : 144-152

Blasdel GG, Salama G (1986) Voltage-sensitive dyes reveal a modular organization in monkey striate cortex. Nature 321 : 579-585

Carson KA (1984) Quantitative localization of neurons projecting to the Mouse main olfactory bulb. Brain Res Bull 12 : 629-634

Cattarelli M, Chanel J (1979) Influence of some biologically meaningful odorants on the vigilance states of the Rat. Physiol Behav 23 : 831-838

Cattarelli M (1983) Le tri des informations à l'entrée du système olfactif. Commande centrale de la réactivité bulbaire et signification biologique des odeurs. Thèse de Doctorat d'Etat ès-Sciences, Université Claude-Bernard - Lyon I

Cattarelli M, Senseman D, London J, Kauer JS (1986) Optical recordings of salamander olfactory bulb responses to odorants. Proc Internat Union Physiol Sci 16 : 326

Cattarelli M, Astic L, Kauer JS (1988) Metabolic mapping of 2- deoxyglucose uptake in the rat piriform cortex using computerized image processing. Brain Research 442 : 180-184

Cattarelli M, Cohen LB (1989) Optical recording of the *in vivo* piriform cortex responses to electrical stimulation of the lateral olfactory tract in the rat. Chemical Senses 14 : to be published

Chanel J (1987) The olfactory system as a molecular descriptor. NIPS 2 :203-208

Cinelli AR, Salzberg BM (1987) Optical recording in slices of mammalian olfactory structures : extrinsic signals from olfactory bulb and pyriform and sulcal cortices of the mouse. Soc Neurosci Abstr 13 : 1411

Cinelli AR, Salzberg BM (1988) Early and late responses in slices of salamander olfactory bulb : optical recording of electrical events that depend upon Ca^{++} Soc Neurosci Abstr 14 : 1187

Cohen LB, Davila HV, Waggoner AS (1971) Changes in axon fluorescence (abstr.). Biol Bull Woods Hole 141 : 382

Cohen LB, Salzberg BM (1978) Optical measurement of membrane potential. Rev Physiol Biochem Pharmacol 83 : 35-88

Cohen LB, Lesher (S) (1986) Optical monitoring of membrane potential : methods of multisite optical measurement. Soc Gen Physiol Ser 40 : 71-79

Davis BJ, Macrides F (1981) The organization of centrifugal projections from the anterior olfactory nucleus, ventral hippocampal rudiment, and piriform cortex to the main olfactory bulb in the Hamster : an autoradiographic study. J comp Neurol 203 : 475-493

Duchamp A (1982) Electrophysiological responses of olfactory bulb neurons to odour stimuli in the Frog. A comparison with receptor cells. Chemical Senses 7 : 191-210

Duchamp A, Sicard G (1984) Odour discrimination by olfactory bulb neurons : statistical analysis of electrophysiological responses and comparison with odour discrimination by receptor cells. Chemical Senses 9 : 1-14

Duchamp-Viret P (1988) Le codage intensitif du stimulus olfactif, rôle des neuro-récepteurs et des deutoneurones, étude électrophysiologie chez la grenouille. Thèse de doctorat de neurosciences Université Claude-Bernard - Lyon I

Freeman WJ (1979) EEG analysis gives model of neuronal template-matching mechanism for sensory search with olfactory bulb. Biol Cybern 35 : 221-234

Freeman WJ, Schneider W (1982) Changes in spatial patterns of rabbit olfactory EEG with conditioning to odors. Psychophysiology 19 : 44-56

Freeman WJ (1983) The physiological basis of mental images. Biological Psychiatry 18 : 1107-1124

Gervais R (1983) Caractéristiques fonctionnelles de l'entrée olfactive pendant la veille et le sommeil. Etude électrophysiologique, neurochimique et comportementale chez le Rat. Thèse de Doctorat d'Etat ès-Sciences, Université Claude Bernard - Lyon I

Giachetti, I, Mac Leod P (1975) Cortical neuron responses to odours in the Rat. In : D.A. Denton & J.P. Coghlan Eds. Olfaction and Taste V, Academic Press New York p: 303-307

Grinvald A, Manker A, Segal M (1982) Visualization of the spread of electrical activity in rat hippocampal slices by voltage-sensitive optical probes. J Physiol (Lond) 333: 269-291

Grinvald A (1985) Real-time optical mapping of neuronal activity : from single growth cones to the intact mammalian brain. Annu Rev Neurosci 8 : 263-305

Grinvald A, Segal M, Kuhnt U, Hildesheim R, Manker A, Anglister L, Freeman JA (1986) Real-time optical mapping of neuronal activity in vertebrate CNS *in vitro* and *in vivo* Soc Gen Physiol Ser 40 : 165-197

Grinvald A, Frostig RD, Lieke E, Hildesheim R (1988) Optical imaging of neuronal activity. Physiol Rev 68 : 1285-1366

Haberly LB (1969) Single unit responses to odor in the prepyriform cortex of the Rat. Brain Res 12 : 481-484

Haberly LB, Shepherd GM (1973) Current-density analysis of summed evoked potentials in Opossum prepyriform cortex. J Neurophysiol 36 : 789-802
Haberly LB (1983) Structure of the piriform cortex of the opossum. I. Description of neuron types with Golgi methods. J Comp Neurol 213 : 163-187
Haberly LB, Bower JM (1984) Analysis of association fiber system in piriform cortex with intracellular recording and staining techniques. J Neurophysiol 51 : 90-112
Heimer L (1968) Synaptic distribution of centripetal and centrifugal nerve fibres in the olfactory system of the rat. An experimental anatomical study. J Anat (Lond) 103 : 413-432
Holley A, Mac Leod P (1977) Transduction et codage des informations olfactives chez les Vertébrés. J Physiol (Paris) 73 : 725-828
Jourdan F, Duveau A, Astic L, Holley A (1980) Spatial distribution of [14C]2-deoxyglucose uptake in the olfactory bulbs of rats stimulated with two different odours. Brain Research 188 : 139-154
Kauer JS, Senseman D, Cohen LB (1984) Voltage-sensitive dye recording from the olfactory system of the tiger salamander. Soc Neurosci Abstr 10 : 846
Kauer JS (1987a) Coding in the olfactory system. In : Finger TE (ed) Neurobiology of taste and smell, John Wiley & Sons, p 205-231
Kauer JS (1987b) Real-time video imaging of electrical events in the olfactory bulb using a potentiometric dye. Soc Neurosci Abstr 13 : 821
Kauer JS, Senseman DM, Cohen LB (1987) Odor-elicited activity monitored simultaneously from 124 regions of the salamander olfactory bulb using a voltage-sensitive dye. Brain Research 418 : 255-261
Kauer JS (1988) Real-time imaging of evoked activity in local circuits of the salamander olfactory bulb. Nature 331 : 166-178
Kauer JS, Neff SR (1988) High resolution video imaging of odor responses in the salamander olfactory system. Soc Neurosci Abstr 14 : 1187
Krettek JE, Price JL (1974) Projections from the amygdala to the perirhinal and entorhinal cortices and the subiculum. Brain Res 71 : 150-154
Leonard CM (1969) The prefrontal cortex of the Rat. I- Cortical projection of the medio-dorsal nucleus; II- Efferent connections. Brain Res 12 : 321-343
Lieke EE, Frostig RD, Arieli A, Ts'o DY, Hildesheim R, Grinvald A (1989) Optical imaging of cortical activity : real-time imaging using extrinsic dye-signals and high resolution imaging based on slow intrinsic signals. Annu Rev Physiol 51 : 543-559
Litaudon P, Cattarelli M (1989) Information transmission in the rat piriform cortex after staining with voltage-sensitive dyes : evoked potential recordings in response to lateral olfactory tract stimulation. Chemical Senses 14 : 217
Luskin MB, Price JL (1982) The distribution of axon collaterals from the olfactory bulb and the nucleus of the horizontal limb of the diagonal band to the olfactory cortex, demonstrated by double retrograde labeling techniques. J Comp Neurol 209 : 249-263
Mori K (1987) Membrane and synaptic properties of identified neurons in the olfactory bulb. Prog Neurobiol 29 : 275-320
Ojima H., Mori K, Kishi K (1984) The trajectory of mitral cell axons in the rabbit olfactory cortex revealed by intracellular HRP injection. J Comp Neurol 230 : 77-87
Onoda N, Imamura K (1988) Central processings of odor discrimination in non-primate mammals. Chemical Senses 13 : 312
Orbach HS, Cohen LB (1983) Optical monitoring of activity from many areas of the *in vitro* and *in vivo* salamander olfactory bulb : a new method for studying functional organization in the vertebrate central nervous system. J Neurosci 3 : 2251-2262
Orbach HS, Cohen LB, Grinvald A (1985) Optical mapping of electrical activity in rat somatosensory and visual cortex. J Neurosci 5 : 1886-1895
Pager J (1976) Contrôle centrifuge des réponses électriques du bulbe olfactif : étude chez le Rat en fonction de l'état nutritionnel. Thèse de Doctorat d'Etat ès-Sciences Université Claude-Bernard - Lyon I

Price JL (1973) An autoradiographic study of complementary laminar patterns of termination of afferent fibers to the olfactory cortex. J Comp Neurol 150 : 87-108

Price JL, Slotnick BM (1983) Dual olfactory representation in the Rat thalamus : an anatomical and electrophysiological study. J Comp Neurol 215 : 63-77

Powell TPS, Cowan WM, Raisman G (1965) The central olfactory connections. J Anat 99 : 791-813

Rodriguez R, Haberly LB (1989) Analysis of synaptic events in the opossum piriform cortex with improved current source-density techniques. J Neurophysiol 61 : 702-718

Ross WN, Salzberg BM, Cohen LB, Grinvald A, Davila HV, Waggoner AS, Chang CH (1977) Changes in absorption, fluorescence, dichroism and birefringence in stained axons : optical measurement of membrane potential. J Membr Biol 33 : 141-183

Saggau P, Galvan M, ten Bruggencate G (1986) Long-term potentiation in guinea pig hippocampal slices monitored by optical recording of neuronal activity. Neurosci Lett 69 : 53-58

Satou M, Mori K, Tazawa Y, Takagi S (1983) Monosynaptic and disynaptic activation of pyriform cortex neurons by synchronous lateral olfactory tract volleys in the Rabbit. Exp Neurol 81 : 571-585

Schild D (1988) Principles of odor coding and a neural network for odor discrimination. Biophys J 54 : 1001-1011

Scott J, Leonard C (1971) The olfactory connections of the lateral hypothalamus in the Rat, Mouse and Hamster. J Comp Neurol 141 : 331-344

Scott JW, McBride RL, Schneider SP (1980) The organization of projections from the olfactory bulb to the piriform cortex and olfactory tubercle in the rat. J Comp Neurol 194 : 519-534.

Senseman DM, Garcia CL, Wayner MJ (1987) Improved visualization of the salamander olfactory bulb for optical recording. Soc Neurosci Abstr 13 : 1411

Senseman DM, Muljadi P, Nash PL, Wayner MJ (1988) Spatio-temporal activation of cortical circuits : visualization with MTV/2. Soc Neurosci Abstr 14 : 158

Sharp FR, Kauer JS, Shepherd GM (1975) Local sites of activity-related glucose metabolism in rat olfactory bulb during olfactory stimulation. Brain Research 98 : 596-600

Shepherd GM (1974) The synaptic organization of the brain. Oxford University Press, New York

Sokoloff L, Reivich M, Kennedy C, Des Rosiers MH, Patlak CS, Pettigrew KO, Sakudara O, Shinohara M (1977) The 14C-deoxyglucose method for the measurement of local cerebral utilization : theory, procedure and normal values in the conscious and anaesthetized albino rat. J Neurochem 28 : 897-916

Stevens CF (1969) Structure of cat frontal olfactory cortex. J Neurophysiol 32 : 184-192

Switzer RC, de Olmos J, Heimer L (1985) Olfactory system. In : Paxinos G (ed) The Rat nervous system Vol.1 Forebrain and Midbrain Academic Press Sydney p :1-36

Tanabe T, IINO M, Takagi SF (1975) Discrimination of odors in olfactory bulb, pyriform-amygdaloid areas, and orbitofrontal cortex of the monkey. J. Neurophysiol 38 : 1284-1296

Tseng GF, Haberly LB (1988) Characterization of synaptically mediated fast and slow inhibitory processes in piriform cortex in an *in vitro* slice preparation. J. Neurophysiol 59 : 1352-1376

Varela FJ (1989) Connaître les sciences cognitives, tendances et perspectives. Seuil Paris

Vigouroux M., Brevet O, Cattarelli M (1988) Neuronal activity monitoring using voltage-sensitive dyes : fast image-recording device adapted to a PC/AT microcomputer. In : Optical methods in neurobiology, ENA technical Workshop, University of Bern Switzerland p : 172-176

Willey TJ, Maeda G, Rafuse D (1975) Antidromic units in the prepyriform cortex driven by olfactory peduncular volleys Brain Res 92 : 132-136

ANIMATED PSEUDOCOLOR ACTIVITY MAPS (PAM's): SCIENTIFIC VISUALIZATION OF BRAIN ELECTRICAL ACTIVITY

David M. Senseman[1],
Sylvia Vasquez[1],
Patrick L. Nash[2]

1) Brain Research Laboratory
Division of Life Sciences
The University of Texas at San Antonio
San Antonio, Texas, USA 78285

2) Division of Earth & Physical Sciences
The University of Texas at San Antonio
San Antonio, Texas, USA, 78285

INTRODUCTION

Major advances in neuroscience have often followed directly from the application of new and more powerful methodological approaches to the study of brain structure and function (Clarke & Jacyna, 1987). Within the last decade a new technique has been developed that allows both brain structure and function to be studied in a closely integrated and highly complimentary fashion. This technique is multiple-site optical recording of membrane potential, or more simply, optical recording. Optical recording is based upon the ability of certain vital dyes (potentiometric probes) to optically signal changes in intracellular membrane potential. By viewing brain tissue stained with a voltage-sensitive dye with a suitable light detector system, changes in neuronal activity can be monitored simultaneously from a 100 or more contiguous anatomical regions (*cf.* Grinvald *et al.*, 1988).

It is the ability of optical recording to provide a global view of evoked neuronal activity that makes it an attractive technique for attacking a central and long-standing problem of olfaction, *i.e.*, how does the vertebrate central nervous system extract odor quality information from afferent olfactory input? Because primary olfactory receptor neurons in vertebrates exhibit relatively little odor specificity (O'Connell & Mozell,

NATO ASI Series, Vol. H 39
Chemosensory Information Processing. Edited by D. Schild

1969; Gesteland, 1971; Getchell, 1974; Holley *et al.*, 1974; Getchell & Shepherd, 1978), it has been argued for some time that odor quality information must be encoded within the overall spatial and temporal patterning (STP) of the centrally evoked olfactory response (Adrian, 1950, 1951; Moulton, 1967, 1976; Moulton & Tucker, 1964; Moulton & Beidler, 1967). We will refer to this idea as the STP hypothesis (Spatial Temporal Patterning hypothesis) of odor quality discrimination. If the STP hypothesis is correct, how then does the encoding and decoding work?

The idea that optical recording might be useful for examining the STP hypothesis was first suggested by Orbach & Cohen (1983) in their pioneering application of this technique to study of evoked electrical activity in the intact vertebrate central nervous system. In their study, Orbach & Cohen used a 12 x 12 silicon photodiode array to monitor neuronal activity from 124 sites of the salamander olfactory bulb following electrical stimulation of the olfactory nerve . More recently, Kauer *et al.* (1987) used the same recording system to show that bulbar activity evoked by controlled-odor stimulation of the olfactory mucosa could also be monitored using voltage-sensitive dyes.

In both of these studies, much of the experimental data was presented in a 'page display' format in which the 124 optical signals where shown as a series of analog traces arranged in a 12 x 12 grid. The location of each optical trace in the grid corresponded to the location of the diode element within the 12 x 12 detector array that recorded the optical trace. Whereas the page display format has proven useful for comparing differences in signal amplitude and waveforms recorded from different anatomical loci (e.g. Orbach & Cohen, 1983; Salzberg, *et al.*, 1983; Senseman *et al.*, 1987 ; Kauer *et al.*, 1987), it has proven less useful for revealing the dynamic characteristics of the evoked response.

In this paper we show how diode array-based optical recording can be combined with image processing and computer animation techniques to visualize the spatio-temporal patterns of evoked electrical activity in the intact brain. We suggest that these visualization techniques may be of general utility for examining a number of problems related to the central processing of sensory information including the STP hypothesis.

MATERIALS AND METHODS

Preparation

Adult, land phase tiger salamanders (*Ambystoma tigrinum*) purchased from a commercial supplier (Amphibians of North America, Nashville, TN) served as our experimental model. Animals were maintained at 10 $^{\circ}$C in plastic containers lined with moist paper towels prior to use. For the experiments presented below, the *in vitro* brain preparation described previously by Orbach and Cohen (1983) was employed. Composition of the Ringer's solution (mM) was as follows: NaCl 104, KCl 2.1, $CaCl_2$ 3.6, $MgCl_2$ 0.7, $NaHCO_3$ 26, D-glucose 5.0, adjusted to pH 7.3.

Optical Recording

Because the small salamander forebrain is relatively translucent, comparatively large optical signals can be recorded using voltage-sensitive dyes that alter, in a rapid and linear fashion, their absorbance as well as their fluorescence in response to changes in membrane potential (Orbach & Cohen 1983; Kauer *et al.*, 1987; Kauer, 1988). In most experiments the 'absorbance' potentiometric probes WW375, NK2367 and RH 155 (Nippon Kankoh-Shikiso Kenykyusho Co., Okayama, Japan) were used since these probes generated optical signals with consistently better signal-to-noise ratios than those produced by the fluorescent probe RH 414. Dye concentrations of the staining solutions ranged from 0.3 to 0.5 mg/ml (dissolved in Ringer's) with a staining period between 30 and 60 min.

A large binocular microscope (Zeiss UEM) served as the optical platform. Brightfield illumination of the preparation was accomplished using a 12 volt, 100 watt tungsten-halogen lamp 'over-driven' to 9 amps by a regulated DC power supply (Lambda model LK 351-PM). The light was collimated and passed through heat and bandpass interference filters before being brought to a focus on the preparation via the microscope's substage condenser. A Wild 7X (0.2 NA) objective lens then projected a real image of the preparation on to the surface of a 12 x 12 silicon photodiode array (MD 144-O, Centronic Electro-Optics, Inc, Newbury Park, CA) which was mounted on the microscope's trinocular tube (Fig. 1). At an objective magnification of 7X, each element in the photodiode array monitored electrical activity from a 183μ x 183μ 'patch' of brain tissue. Additional details concerning the parallel amplification of the diode photocurrents, analog-to-digital conversion of these amplified signals as well as the digital signal processing of the optical data using a minicomputer-based data acquisition system can be found in Cohen & Lesher (1986).

Video Imaging

Conventional procedures were followed for the acquisition, processing and storage of high-resolution video images of the experimental preparation (512 x 512 pixels, 256 grey levels). Video images were obtained by replacing the photodiode array on the microscope's trinocular tube with a TV camera optimized for video microscopy (Hammamatsu model C2400). To increase the camera's field of view, we substituted a microscope objective with a lower magnification (2.5X) than the one use for optical recording. Improved video contrast was achieved by illuminating the preparation with a pair of fiber optical light guides, positioned tangentially to the microscope's optical axis, to produce a 'pseudo darkfield' image. The TV camera's video output was fed to a Panasonic (Model TQ-2025F) Optical Disc Memory Recorder (ODMR) which provided archival storage of the image data. Selected video images were subsequently digitized ('frame grabbed') with a commercially available image processor system (Imaging Technology Inc., Series 151) and the digitized image file transferred to a graphics workstation (Fig. 1).

Registration of the Optical Signals with the Video Image

The following procedure was used to achieve registration of the optical signals recorded with the photodiode array with the video image of the preparation. After the final recording trial, the photodiode array housing was removed from the trinocular tube and replaced with a ground glass viewing screen with an inscribed outline of the diode array. The preparation appeared on the viewing screen as a brightfield image superimposed on the array outline. From this image a drawing was prepared showing the location of the array outline relative to the tissue's surface landmarks. Care was taken to insure that the preparation was not moved within the image field during this procedure. After the drawing was completed, the viewing screen was replaced with the TV camera and a video image of the preparation recorded as described in the preceding section.

An interactive computer graphics program was then used to map the x and y coordinates of diode array outline to the video image. We displayed the video image on the screen of the graphics workstation with a superimposed outline of the photodiode array. The array outline was then sized and translated relative to the video image using

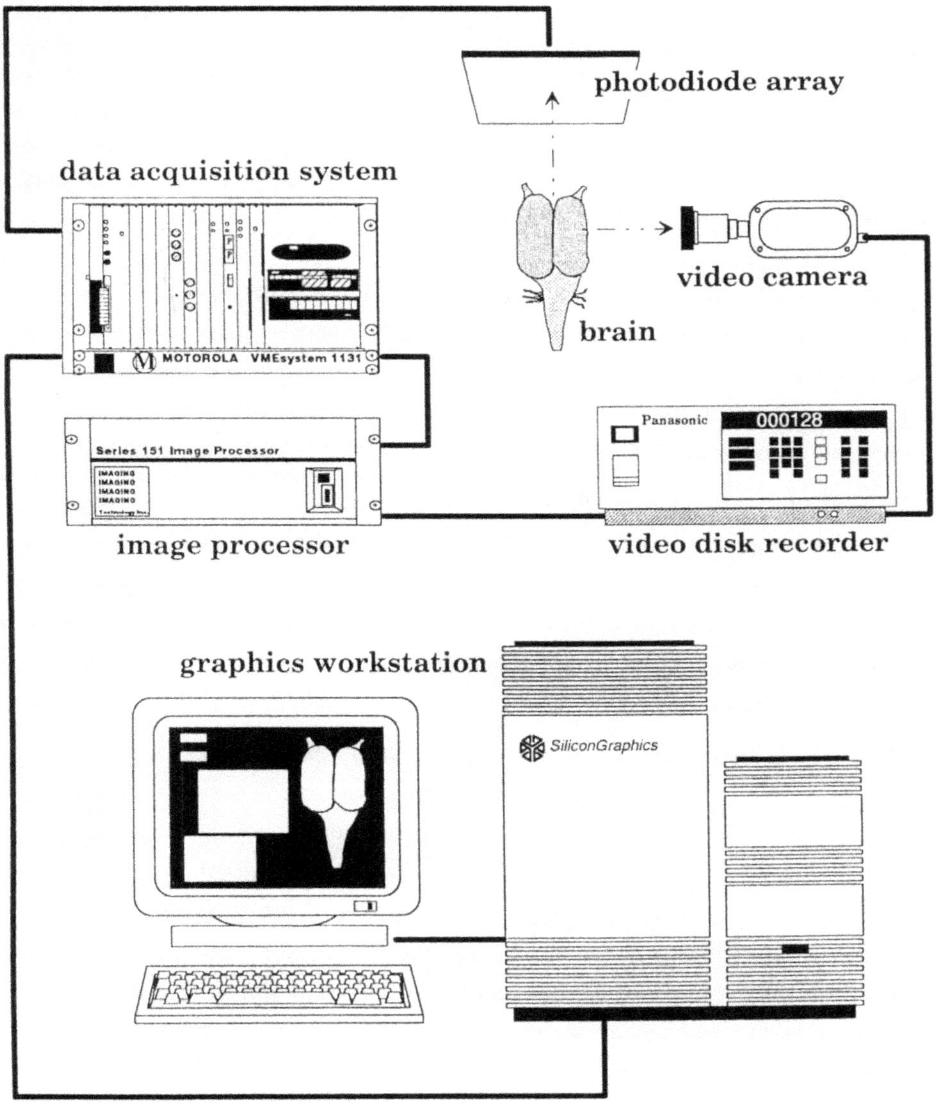

Figure 1. Schematic drawing of the experimental apparatus. A 12 x 12 silicon photodiode array was used for high-speed optical recording (1ms/frame) of voltage-sensitive signals from the experimental preparation (brain). Photocurrents from the 124 central array elements were passed in parallel to an amplifier bank (not shown) and then to a 128 channel digital-to-analog converter system (not shown) before processing on a 32-bit Motorola MC68030-based microcomputer system (data acquisition system). A Hammamatsu C2400 video camera system was used at the end of an experiment to record a high-resolution video image of the experimental preparation. The video image was stored on a Panasonic video disc recorder prior to digitization and processing on a Imaging Technology Series 151 image processor system. A Silicon Graphics Inc. 4D/80GT color graphics workstation produced the animated PAM displays from the combined optical recording data and the enhanced image data.

a 'mouse' input device to match the prepared drawing. Once the correct array size and location were determined, coordinates for the diode array were written as header information to the image data file.

Mapping Optical Signals to a Pseudo-color Scale

We used the following procedure to map the instantaneous voltage of each optical signal to a pseudo-color scale was as follows. The analog traces from each of the array's 124 active diode channels were first displayed on the screen of the graphics workstation in a 'page-display' format. Diode signals contaminated with excessive noise were selected using a mouse input device and were excluded from further analysis.

Optical signals relatively free from noise contamination were then subjected to a two-step normalization procedure. In the first step, each data point in an optical trace was divided by the resting DC light level reaching the diode element generating the trace. This procedure is necessary to reduce artifactual differences in signal amplitude between recording elements that arise from non-uniform illumination of the image field (e.g. Senseman *et al.*, 1987).

A second normalization process was then performed to express the data points in each trace as a percentage of the overall maximum positive signal value recorded during the trial. Using a simple 'post office' function, each data point was assigned one of 14 colors based on its percentage value relative to the maximum signal. Data points within ± 10% of baseline were assigned dark green, larger 'positive signals' (i.e. the optical direction indicating membrane depolarization) were assigned progressively shorter wavelengths (i.e. 'warmer' colors) with the largest positive signals (≥80% of maximum) assigned dark red. 'Negative signals' (i.e. the optical direction indicating membrane hyperpolarization) were assigned progressively 'cooler' colors along the blue-to-violet axis of the visible light spectrum.

Animation of the PAM Display

Once a color value was assigned to each data point in the 124 optical traces, a Pseudocolor Activity Map (PAM) was generated by drawing a 12 x 12 array of colored polygons superimposed on the grey-scale video image. To give the illusion of transparency and to prevent complete obscuring of the underlying video image, polygons were 'filled' with a stipple pattern. By assigning different 'z-values' (depth values) to the image background, the illuminated brain tissue and the colored polygons, it was possible to prevent display of complete polygons, or irregular portions of polygons, that extended beyond the tissue surface.

We used standard 'double buffering' techniques for animation of the sequential PAM's in which the graphics hardware generated the next PAM image on a hidden 'back' buffer while the current PAM frame was being displayed from the visible 'front' buffer. By varying the time interval between swapping of the front and back display buffers, PAM 'movies' can viewed at various speeds up to maximum of 60 frames/sec.

RESULTS

Optical Recording from the Dorsal Bulbar Surface

Figure 2 shows a 'page display' of the optical signals recorded from the dorsal surface of the salamander olfactory bulb following a brief (0.7 msec) shock delivered to to the transected olfactory nerve. In order to compare directly our results with those of Orbach & Cohen, we attempted to match, as closely as possible, the orientation of the preparation as well as the stimulation and optical recording parameters they used for the experiment presented in Figure 2 of their 1983 paper.

While in general, the optical signals are similar in amplitude and waveform for the two experiments, some differences can be seen. In their experiment, Orbach & Cohen could distinguish three rapid peaks and a slow response in a few selected traces while other traces showed only a subset of these components. The earliest fast peak, which propagated caudally at ~0.2 m/sec, was presumed to originate from action potentials in the olfactory nerve axons. A similar peak (marked *P1)* can be seen in our recording and by analogy we assume this peak represents the incoming afferent volley.

A direct correlation between the second rapid peak seen by Orbach & Cohen and our data is somewhat more problematical. This second peak was distinguished by the fact that it propagated in a rostral rather than a caudal direction. Orbach & Cohen suggested that this peak was also generated by propagated activity in the olfactory nerve axons. They suggested that the slight delay in the second peak, as well as its rostral propagation, could be explained by the anatomy of the incoming fiber pathway. Part of the incoming afferent fibers follows an '>'-shaped pathway - traveling first in a caudal direction before turning back to travel in a dorsorostral direction. Because of this flexure in the fiber pathway, the second peak could have been generated by afferent ac-

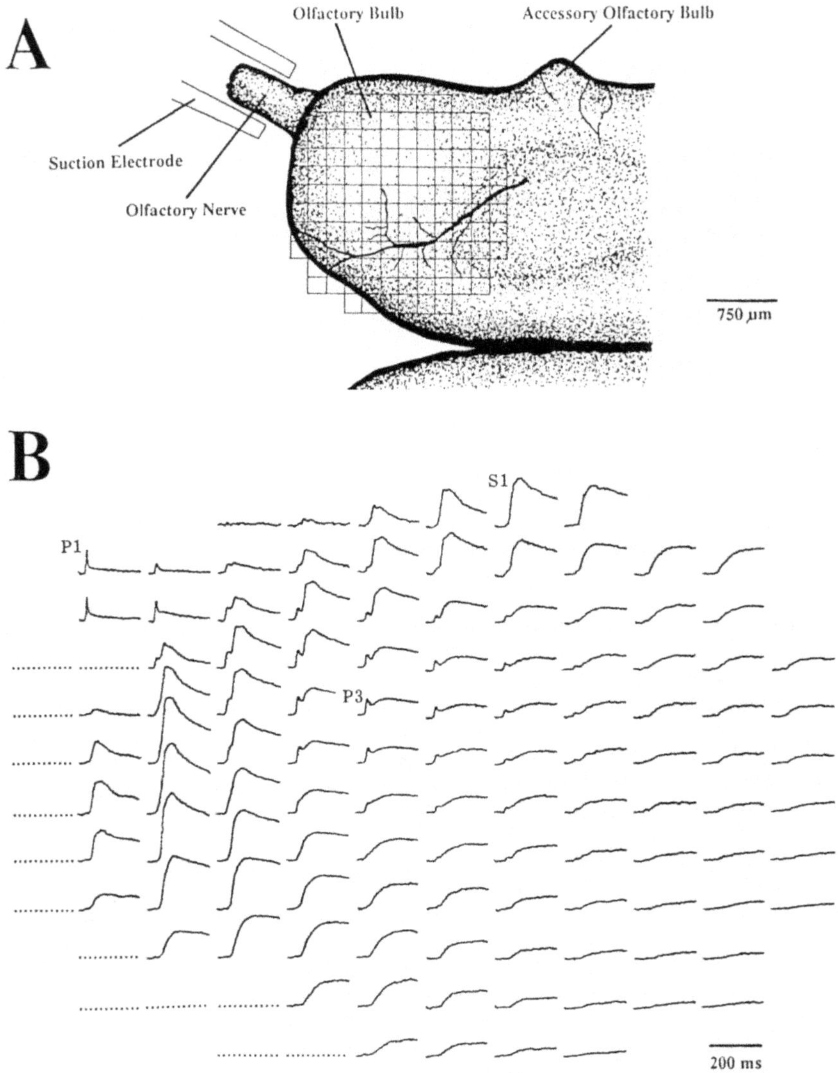

Figure 2. Page display of optical signals recorded from the dorsal surface of the salamander olfactory bulb. Part A: Drawing of the rostral 2/3's of a right cerebral hemisphere, viewed from its dorsal aspect. The rectangular grid indicates the brain region monitored by each of the 124 elements in the photodiode array. Each array element monitored approximately a 183μ x 183μ tissue area. A plastic suction electrode was used to stimulate the transected olfactory nerve. Part B: Optical signals recorded in response to brief electrical stimulation of the transected olfactory nerve (0.7 msec, 50 volts). The brain had been stained 60 min with the voltage-sensitive dye RH155 (0.6 mg/ml). Signals were recorded under brightfield illumination using a 750 $\pm$ 30nm bandpass interference filter and a Schott RG9 heat filter. Dotted-lines indicate optical traces that have been omitted due to low signal-to-noise ratios. Examples of signal waveforms corresponding to Orbach & Cohen's earliest rapid peak (P1), third rapid peak (P3) and slow signal (S1) are indicated.

tivity moving in this dorsorostral limb that had previously passed through the recording site as a *P1* peak moving in a rostral direction.

Our failure to observe this second peak in the data shown in Figure 2 is not too surprising since Orbach & Cohen reported seeing it in only a few preparations. What Orbach & Cohen labelled as their third rapid peak can, however, be clearly seen in our data (*P3*). This peak is thought to be generated, at least in part, by postsynaptic activity in mitral, tufted and periglomerular cells (Orbach & Cohen, 1983).

The slower signals observed by Orbach & Cohen can also be seen in our data (*S1* in Fig. 2). Whereas Orbach & Cohen's slow signals generally returned to base-line within 100 msec of the response onset, the slow signals shown in Figure 2 are much broader, continuing for several hundred milliseconds. Orbach & Cohen admitted that the time course of their slow signals was likely to have been distorted by the relatively fast 100-msec coupling time constant of their photodiode amplifiers. Consequently, we have extended the time constant of our amplifiers to 5 sec so that the broader *S1* signals shown in Figure 2 are a more faithful representation of the underlying neuronal activity. While the exact cellular basis of these slower signals is still largely unknown, there is evidence that they are related to postsynaptic neuronal activity (Orbach & Cohen, 1983).

We used the optical signals presented in Figure 2 to generate an animated PAM display of the evoked response. The first 37 msec of this PAM series are shown in Figure 3. These data were acquired at 1000 frames/sec (1 msec/frame) but only odd-numbered frames are shown for compactness. Frames acquired prior to the onset of stimulation (not shown) were uniformly dark green indicating ongoing neural activity was below threshold for optical detection. At 1 msec following nerve stimulation, the first indication of the incipient response could be observed in the olfactory bulb at the entrance of the olfactory nerve. From 1 to 10 msec, there is a continuous increase in the area affected with the spread of activity occurring diagonally in a rostromedial fashion.

For approximately the next 10 msec (15 to 25 msec post-stimulation), growth of the response slows and appears to stabilize momentarily. When viewing the animated PAM display on the graphics workstation, the subjective impression is that the response 'hovers' in the central bulbar region. A possible mechanism to account for this momentary stabilization will be presented below.

At 27 msec into the response, the momentary stability is replaced by increased activity which spreads first to the lateral bulbar surface (at the top of the image) followed by spread in a caudal direction (to the right of the image). Concomitant with this rostrolateral spread, activity can also be seen to to spread rostrally and medially.

Between 31 to 37 msec post-stimulation, the rate of expansion again begins to slow while the level of activity, as indicated by the 'red-shift', begins to rise. This general increase in neural activity continues for the next 30 msec reaching a maximum approximately 60 msec after nerve stimulation (data not shown). A complete account of the evoked response, which lasted for more than 800 msec, is beyond the scope of the current report.

Optical Recording from the Lateral Bulbar Surface

A significantly different pattern of evoked neural activity was observed if optical recordings were made from the lateral bulbar surface. Figure 4 shows a 'page display' of optical signals recorded from a lateral perspective. The most notable difference is the increased number of optical traces exhibiting the early rapid *P1* peak. From the anatomical drawing of the preparation shown in Part A of Figure 4, it can be seen that diode elements generating the largest *P1* waveforms monitored activity from the implantation cone, a central continuation of the olfactory nerve containing unmyelinated axons of the primary olfactory receptors (Herrick, 1948). *P1*-like waveforms, with somewhat smaller amplitudes, can also be seen in the most rostral bulbar regions. In more caudal bulbar regions, dorsal and ventral to the implantation cone, optical responses were similar in amplitude and waveform to the slow signals (*S1*) shown in Figure 2.

Figure 3. PAM display of the analog optical data presented in Figure 2. The first 37 msec of the response presented in Fig. 2 is shown as a 19-frame PAM sequence. Normalized optical signals, mapped to a 14-level pseudocolor scale, are displayed as colored polygons superimposed on a grey-scale video image of the preparation. Dark green area (polygons) indicate optical signals within ± 10% of base-line activity. Each color to the right of dark green on the pseudocolor scale ('warmer colors') indicates a 10% increase in signal amplitude in the direction of net membrane depolarization. The rightmost color value, dark red, indicates the amplitude of the optical traces was within 80% to 100% of the maximum signal amplitude recorded during the trial. Each color to the left of dark green on the pseudocolor scale ('cooler colors') indicates a 10% increase in signal amplitude in the direction of net membrane hyperpolarization. (No brain areas with a net membrane hyperpolarization were observed in this trial). Frame at the extreme upper left shows registration of the diode array outline with the video image. Time of recording since stimulus onset is indicated at the lower left of each frame. See text for a description of the spatio-temporal spread of activity.

Figure 4. Page display of optical signals recorded from the lateral surface of the salamander olfactory bulb. Part A: Drawing of the rostral 2/3's of a right cerebral hemisphere, viewed from its lateral aspect. The rectangular grid indicates the brain region monitored by each of the 124 elements in the photodiode array. The plastic suction electrode used to stimulate the transected olfactory nerve can be seen at left. Each array element monitored approximately a 183μ x 183μ tissue area. Part B: Optical signals recorded in response to brief electrical stimulation of the transected olfactory

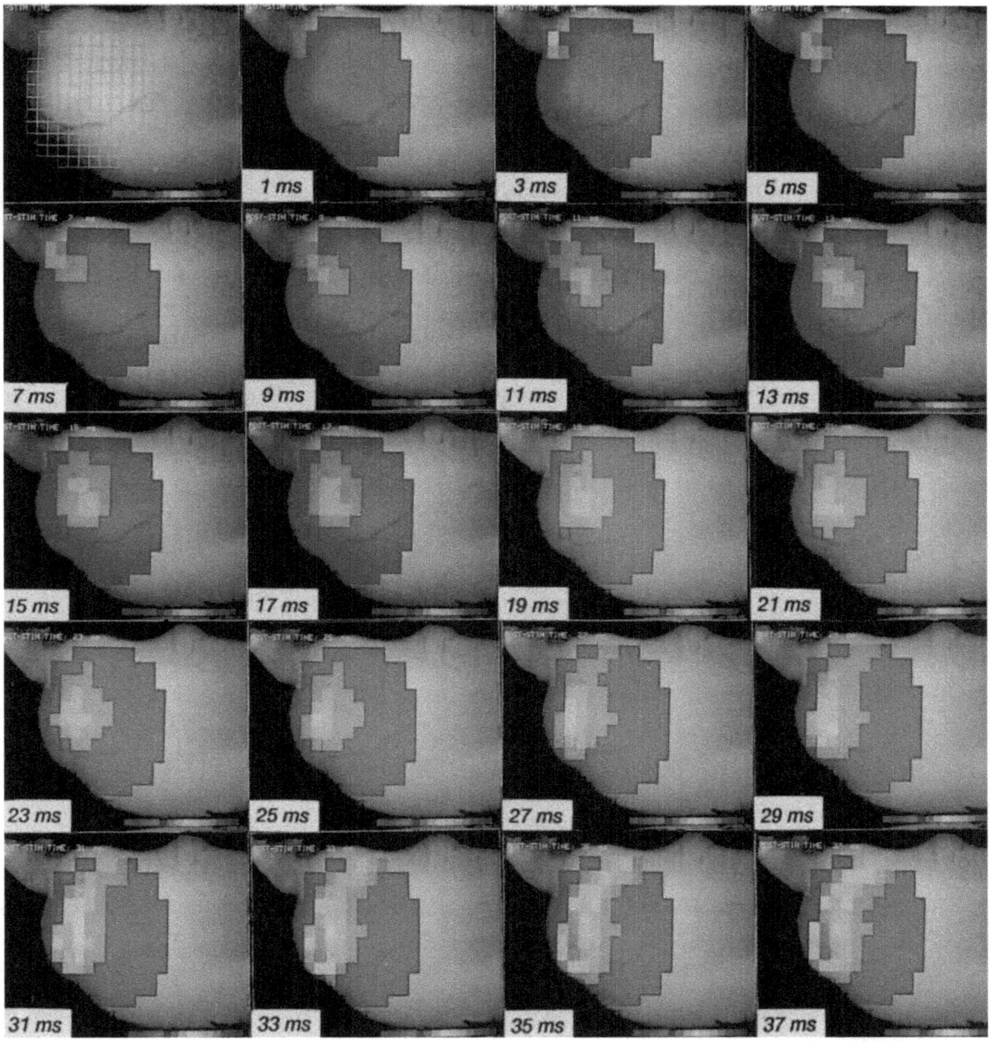

Fig. 3

nerve (0.7 ms, 50 v). The brain had been stained 45 min with the voltage-sensitive dye WW375 (0.8 mg/ml). Signals were recorded under brightfield illumination using a 720 ± 20 nm bandpass interefernce filter and a KG-1 heat filter. Dotted-lines indicate optical traces due that were omitted due to low signal-to-noise ratios. Signals used in Fig. 6 to calculate conduction velocity along the implantation cone are indicated by arrows which also indicate the point of stimulus onset in these traces.

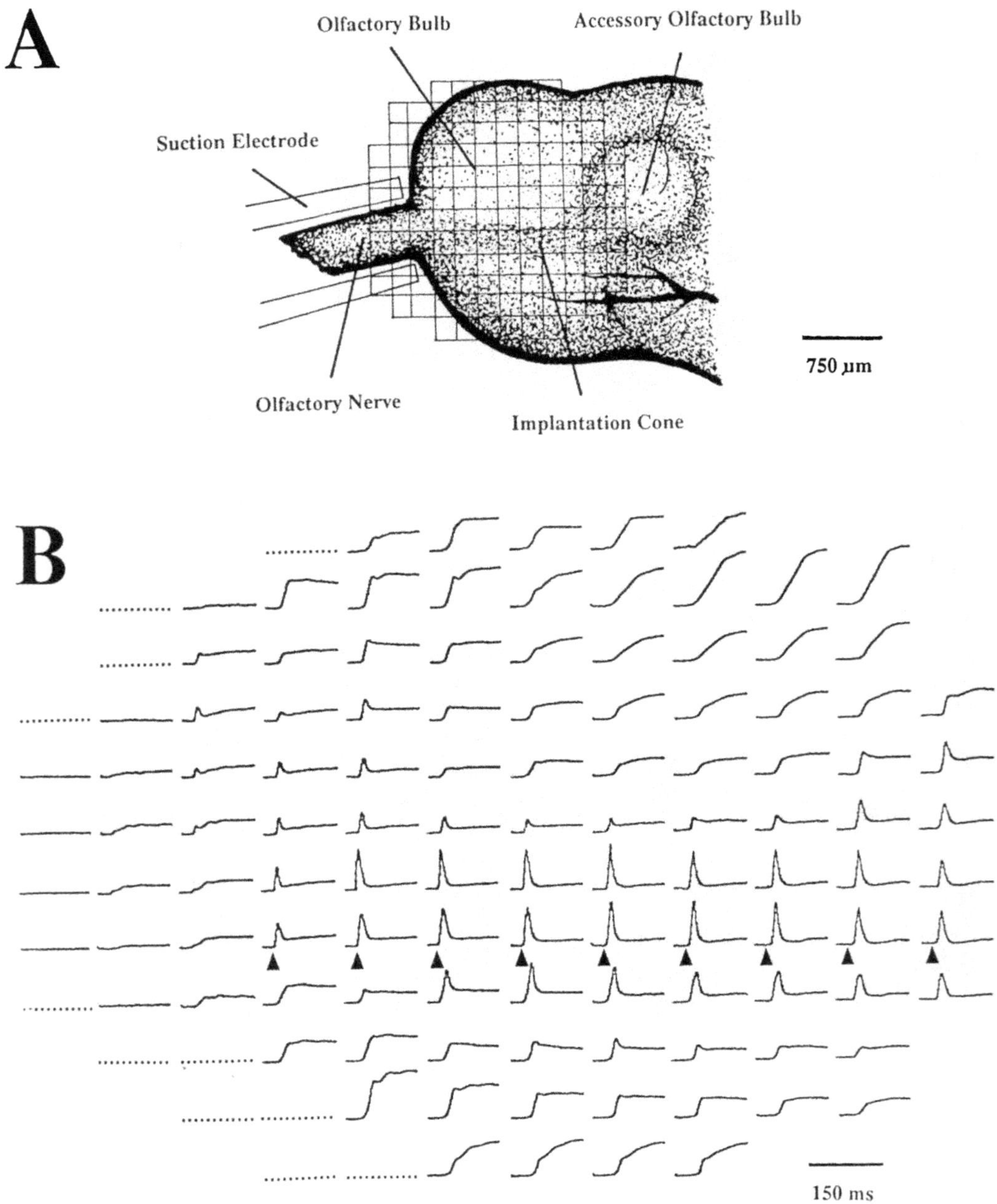
A
Olfactory Bulb
Accessory Olfactory Bulb
Suction Electrode
750 µm
Olfactory Nerve
Implantation Cone
B
150 ms

Fig. 4

When viewed as an animated PAM display, the dynamic nature of the evoked response in Figure 4 was striking. Figure 5 shows the each frame recorded during the first 18 msec following nerve stimulation. Two milliseconds post-stimulation, the leading edge of the afferent volley can be seen to enter the bulb at the junction of the olfactory nerve. During the next millisecond (time = 3 msec), the response rapidly spreads dorsally into the olfactory nerve layer, and caudally along the implantation cone. Between 4 and 5 msec post-stimulation, the intensity of the response continues to build (as indicated by the 'red-shift') especially in the rostral regions of the implantation cone.

Beyond 6 msec post-stimulation, activity can be observed to spread within three distinct pathways. First there is focus of activity that moves towards the dorsal bulbar surface. By 12 msec, the center of this focal activity has separated from activity occurring along the implantation cone. From 12 to 18 msec, the amplitude of this activity continues to increase with the most intense activity progressively shifting to more dorsal locations. Based on its time course and its anatomical location, we believe this particular activity spread corresponds to the activity observed from the dorsal perspective (Fig. 3) that seem to 'hover' from 15 to 25 msec post-stimulation. While the growth of the evoked response stabilized in its lateral extent, it is possible that it continued to move in a dorsal direction. When viewed from the dorsal bulbar surface, an upward (dorsal) movement of activity, along the optical axis, could not be easily detected with our current optical apparatus.

Figure 5. PAM display of the analog optical data presented in Figure 4. The first 18 msec of the response presented in Fig. 4 is shown as a 19-frame PAM sequence. Normalized optical signals were mapped to a 14-level pseudocolor scale and are displayed as colored polygons superimposed on a grey-scale video image of the preparation. Dark green area (polygons) indicate optical signals within ± 10% of baseline activity. Each color to the right of dark green on the pseudocolor scale ('warmer colors') indicates a 10% increase in signal amplitude in the direction of net membrane depolarization. The rightmost color value (dark red) indicates the instantaneous amplitude of an optical traces was within 80% to 100% of the maximum signal amplitude recorded during the trial. Each color to the left of dark green on the pseudocolor scale ('cooler colors') indicates a 10% increase in signal amplitude in the direction of net membrane hyperpolarization. (No brain areas with a net membrane hyperpolarization were observed in this trial). Frame at the extreme upper left shows registration of the diode array outline with the video image. Time of recording since stimulus onset is indicated at the lower left of each frame. See text for a description of the spatio-temporal spread of activity.

Figure 6. Graphical analysis of afferent propagation velocity. Plotted on the same time axis are the analog optical signals recorded from 9 adjacent photodiodes positioned over the implantation cone (see Fig. 4 diode locations). Amplitude calibration bar at left indictes a change of 1 part in 1,000 in light absorbance (Δ I) relative to resting light level. Inset at upper right shows a graph in which the distance from the entrance of the olfactory nerve of the brain area monitored by a diode element has been plotted as a function of time since stimulus onset for the diode's signal to reach its maximal amplitude (time to peak). Slope of the 'best fit' line determined by linear regression (r = 0.9) gives a conduction velocity of 0.1 m/sec for spike propagation within the implantation cone.

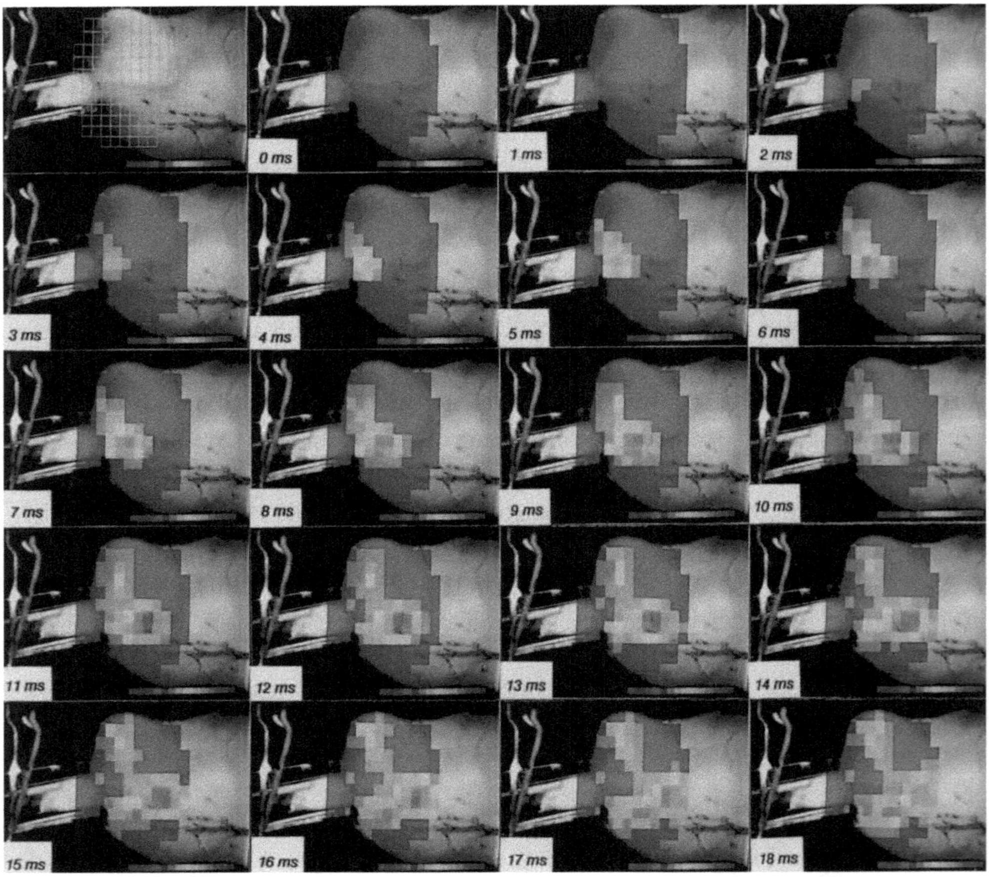
0 ms
1 ms
2 ms
3 ms
4 ms
5 ms
6 ms
7 ms
8 ms
9 ms
10 ms
11 ms
12 ms
13 ms
14 ms
15 ms
16 ms
17 ms
18 ms

Fig. 5

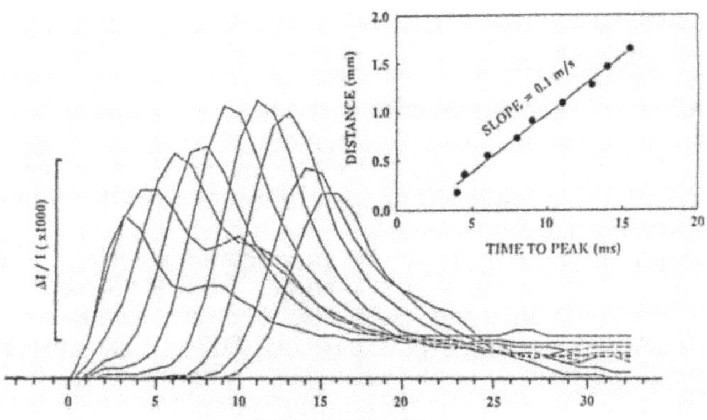
ΔI / I (x1000)
POST-STIMULUS TIME (ms)
0
5
10
15
20
25
30
DISTANCE (mm)
TIME TO PEAK (ms)
SLOPE = 0.1 m/s
0.0
0.5
1.0
1.5
2.0
0
5
10
15
20

Fig. 6

A second, but less distinct movement of activity begins at 13 msec following stimulation. This activity spreads first ventrally followed by extension into more caudal bulbar regions. Figure 5 shows only the initial stages of increased activity within the ventral olfactory bulb that continued to grow in both amplitude and anatomical extent later in the response (data not shown).

The third and most pronounced spread of activity occurred along the implantation cone. In the animated PAM sequence, this activity appeared as a relatively compact reddish 'block' that moved across the bulb in a caudal direction. By 17 msec post-stimulation, the center of this activity reached the level of the accessory bulb (see Fig. 4A) where it began a dorsal and to a lesser extent ventral expansion. Since the accessory bulb was beyond the field of view of the diode array, the extent to which activity in the implantation cone evoked activity in the accessory bulb could not be determined.

To calculate the conduction velocity along the implantation cone of the afferent volley, we have plotted in Figure 7 the optical signals generated by 9 adjacent diode array elements on the same time axis. Optical traces selected for analysis are indicated (▲) in Figure 4. Linear regression of the time-to-peaks (inset) gave a conduction velocity of 0.1 m/sec. This value is somewhat lower than 0.2 m/sec conduction velocity found by Orbach & Cohen for their earliest rapid peak (*P1*) but is identical to the conduction velocity (0.1 m/sec) cited by Kauer (1988) for afferent spike potentials in the salamander olfactory nerve.

DISCUSSION

Through the integration of diode array-based optical recording with image processing techniques, we have shown how animated Pseudocolor Activity Maps (PAM's) can be generated and used to visualize the spatio-temporal patterns of electrical activity evoked in the intact brain. Unlike the page display format, the animated PAM display allows information on brain structure (i.e. surface morphology) to be combined visually with brain function (i.e. optically recorded neural activity). In essence, the visualization techniques we have developed represents a logical extension of a new approach to neuroanatomy pioneered by Orbach & Cohen (1983). It is a kind of functional neuroanatomy in which fiber pathways and synaptic interactions can be directly visualized, not by histological staining, but by the dynamic spread of evoked electrical activity from one anatomical locus to another.

Because of the high temporal resolution afforded by the photodiode diode array detector (1 msec/frame), fast neural events such as propagated spike activity can be readily observed. Moreover, the large size of the optical signals (relative to the background noise) allow such evoked activity to be followed clearly, in a single trial, without the need to average combined data obtained in multiple trials.

Despite the advantages of the PAM display format, we have found that the animated PAM display does not replace the page display as the principal format for data analysis. Rather, our experience has been that PAM and the page displays provide distinctly different views of the optical signal data that are mutually reenforcing. As described in the Methods section, data analysis begins with a visual examination of the optical traces as a page display. Displayed in their analog form, optical traces corrupted by noise and/or channel malfunctioning can be easily detected and eliminated from further analysis. The elimination of traces with large amplitude noise is especially important since the pseudocoloration procedure used for the PAM displays requires optical traces to be normalized to the maximum signal amplitude recorded by the collective diode elements in the recording trial.

Conversely, the PAM display can be used to identify optical traces of particular interest that can be analyzed subsequently in their analog form (e.g. Fig. 6). For example, detailed analysis of the analog waveforms recorded from anatomically restricted regions, such as a individual glomeruli, might provide new insights into the role played by reciprocal dendrodentric synaptic connections between mitral and granular cells.

Examination of the STP Hypothesis

From the data presented above and from a large number of experiments that could not be presented, we have been impressed by the richness and complexity of the electrically-evoked response in the salamander olfactory bulb as revealed by the animated PAM display. Because of this complexity we need to redefine more precisely the relatively vague notion of 'spatio-temporal patterning' with respect to the following questions of what, where and when:

1. What do we mean by a 'pattern of activity'? How can we define such 'patterns' quantitatively so that we can statistically compared one pattern to another in a meaningful way?

2. Where in the brain should we look for these 'patterns'? Should we confine our search to simply the glomerular layer or expand the recording area to cover the whole bulb. What if the 'significant pattern' is not to be seen in the bulb but further 'downstream' in the anterior olfactory nucleus, pyriform cortex or perhaps even in the hippocampus? Maybe the 'significant pattern' is to be found in the reentrant (reciprocal) flow of activity between two (or more) of these nuclei?

3. When during the response should we look for the 'significant pattern' to appear and how long should we expect this pattern to remain visible? Bulbar responses evoked by odor stimulation can last for thousands of milliseconds, but are the most important 'patterns' formed during the first 20 millisconds of the response? How do we put the time dimension into a quantitative description of the pattern?

In addition to these questions related to the STP hypothesis, there remains the more general technical question as to the cellular basis of the optically recorded voltage-sensitive dye signals. To what extent can we attribute the amplitude and waveform of our optical signals to activty in the various cell types *i.e.* olfactory afferent fibers, mitral cells, tufted cells, granular cells, periglomerular cells and transitional cells to be found in the salamander olfactory bulb (Herrick, 1948)? Answers to these questions must await further experiments and perhaps even further development of the optical recording technique.

Future Technological Improvements

As mentioned above one way to improve the utility of diode array-based optical recording would be to increase its spatial resolution from the current 124 'pixel' (i.e. diode) elements while maintaining an acceptable temporal resolution (~ 1 msec/frame). To this end we have designed and constructed, in conjunction with Centronic Electro-Optics Inc. (Newbury Park, CA), a new silicon photodiode array with an increased number of active elements (Nash *et al.*, 1988). This new array is based on a 24 x 24 matrix grid with 24 elements removed from each corner to give a roughly circular detector area containing 464 independent silicon photodiodes. At present two detector arrays have been fabricated and are undergoing final performance testing.

Efforts are also underway to extend the current recording technology to support 3-dimensional localization of optical signals within intact brain tissue. Here the basic technological problem is the reduction of optical signal contamination by scattered light arriving from adjacent tissue above and below the plane of focus. As pointed out to us by Dr. Peter Saggou, it might be possible to reduce scattered light contamination by combining a photodiode array recording system with a laser illumination system. The laser system would be use to illuminate only a single restricted patch of brain tissue at any given instant. If the area of illumination closely matched the area monitored by a single array element, the amount of scattered light received by that array element would be greatly reduced compared to whole field illumination. By rapid scanning of the laser microbeam across the brain surface in registration with the diode array, optical signals with a relatively narrow plane of focus (optical sectioning) could be obtained (*e.g.* Liljeborg, 1988). Combining voltage-sensitive dye data recorded at different tissue depths with high resolution morphological data provided by conventional confocal microscopy, it should be possible to generate 3-D PAM's in which the spatio-temporal flow of evoked electrical activity in the brain could be visualized and studied in 3 dimensions.

ACKNOWLEDGEMENTS - This work was supported by grants (to DMS) from the San Antonio Area Foundation, The University of Texas Chancellor's Council, The University of Texas Higher Education Coordinating Board (ARP 2227), the United States Department of Defense (AFOSR-89-0118), the National Institutes of Health (R01-DE0-6904-05,RR-08194-10,GM-07717-10) and from the National Science Foundation (BNS-07594).

REFERENCES

Adrian ED (1950) Sensory discrimination with some recent evidence from the olfactory organ. Br Med Bull 6:330-333

Adrian ED (1951) Olfactory discrimination. Annee Psychol 50:107-113

Clarke E, Jacyna, LS (1987) Nineteenth-century origins of neuroscientific concepts. Univ of Calif Press, Berkekey, CA

Cohen LB, Lesher S (1986) Optical monitoring of membrane potential. In: DeWeer P, Salzberg BM (eds) Optical methods in cell physiology. New York, pp 71-99

Gesteland RC (1971) Neural coding in olfactory receptor cells. In: Beidler LM (ed) Handbook of sensory physiology, olfaction. Springer, New York, pp 132-150

Getchell TV (1974) Unitary responses in the frog olfactory epithelium to sterically related molecules at low concentrations. J Gen Physiol 64:241-261

Getchell TV, Shepherd GM (1978) Responses of olfactory receptor cells to step pulses of odour at different concentrations in the salamander. J Physiol (Lond) 282:521-540

Grinvald A, Frostig RD, Lieke E, Hildesheim R (1988) Optical imaging of of neuronal activity. Physiol Rev 68:1285-1366

Herrick CJ (1948) The brain of the tiger salamander. Univ of Chicago Press, Chicago, IL

Holley A, Duchamp A, Revial MF, Juge A, MacLeod P (1974) Qualitative and quantitative discrimination in the frog olfactory receptors: analysis of electrophysiological data. Ann NY Acad Sci 237:102-114

Kauer JS, Senseman DM, Cohen LB (1987) Odor-elicited activity monitored simultaneously from 124 regions of the salamander olfactory bulb using a voltage-sensitive dye. Brain Res 418:255-261

Kauer JS (1988) Real-time imaging of evoked activity in local circuits of the salamander olfactory bulb. Nature (Lond) 331:166-168

Liljeborg A (1988) Digital position encoding of galvanometer scanner in a laser microscope. Optical Eng 27:818-822

Moulton DG (1967) Spatio-temporal patterning of response in the olfactory system. In: Hayashi T (ed) Olfaction and Taste II, Pergamon, Oxford pp 109-116

Moulton DG (1976) Spatial patterning of responses to odors in the peripheral olfactory system. Physiol Rev 56:578-593

Moulton DG, Tucker D (1964) Electrophysiology of the olfactory system. Ann NY Acad Sci 116:380-428

Moulton DG, Beidler LM (1967) Structure and function in the peripheral olfactory system. Physiol Rev 47:1-52

Nash PL, Muljadi P, Wayner MJ, Senseman DM (1988) High-speed imaging of olfactory of electrical activity: watching the brain think on MTV/2. Neural Networks 1:268

O'Connell RJ, Mozell MM (1969) Quantitative stimulation of frog olfactory receptors. J Neurophysiol 32:51-63

Orbach HS, Cohen LB (1983) Optical monitoring of activity from many areas of the in vivo and in vitro salamander olfactory bulb: a new method for studying functional organization in the vertebrate central nervous system. J. Neurosci 3:2251-2262

Salzberg BM, Obaid AL, Senseman DM, Gainer H (1983) Optical recording of action potentials from vertebrate nerve terminals using potentiometric probes provides evidence for sodium and calcium components. Nature (Lond) 306:36-40

Senseman, DM, Horwitz IS, Salzberg BM (1987) MSORTV imaging of electronic conduction in an electrical syncitium: optical recording of polarization spread in a simple salivary gland. J Exp Zool 244:79-89

CHEMOSENSORS WITH PATTERN RECOGNITION

R. Müller

Techn. Universität München

1. INTRODUCTION

The idea of using the information present in the combination of sensorsignals for the identification of substances rather than to rely on the selectivity of single sensor elements /1, 2, 3, 4/ has lead to the application of the methods of pattern recognition (PARC) to chemo sensor signals. These methods are described e.g. in /5/. While the methods of PARC levels 1 and 2 serve the purpose of identification only, the methods of level 3 and 4 yield the identification and concentration at the same time. The latter two methods however are applicable only if the superposition law holds for the sensorsignals, which is rarely the case with chemosensors. Frequently however it is possible to find an artifical superposition plane by transformation of the sensor - signals /6/.

In the present paper the experimental results obtained with two kinds of sensorarrays (modified MOS - sensors /7/ and commercially available SnO_2 - Sensors /Figaro/) are presented in conjunction with three methods of pattern recognition.

2. MOS - SENSOR ARRAY (CORRELATION METHOD)

Fig. 1 shows an integrated array of four MOS - gassensors /8/. The sensitivities of the sensorelements are modified by the application of zeolite layers with different pore size /9/. The transfer - characteristics for one of them are shown in fig. 2 for the gases H_2, CH_4 and C_2H_2 /10/. The sensitivity is highest for hydrogen but as expected other gases yield signals too and it is not possible to distinguish for instance between CH_4 of high concentration and H_2 of low concentration. Fig. 3 shows the signal patterns for CH_4

NATO ASI Series, Vol. H 39
Chemosensory Information Processing. Edited by D. Schild

and C_2H_2 for different pressures as obtained from the sensor array. Already the visual inspection allows to distinguish between the two gases.

Fig. 1. Realization of the MOS gas sensor array.

Fig. 2. Signal S for three gases as a function of the concentration σ. Zeolite foil 3 Å.

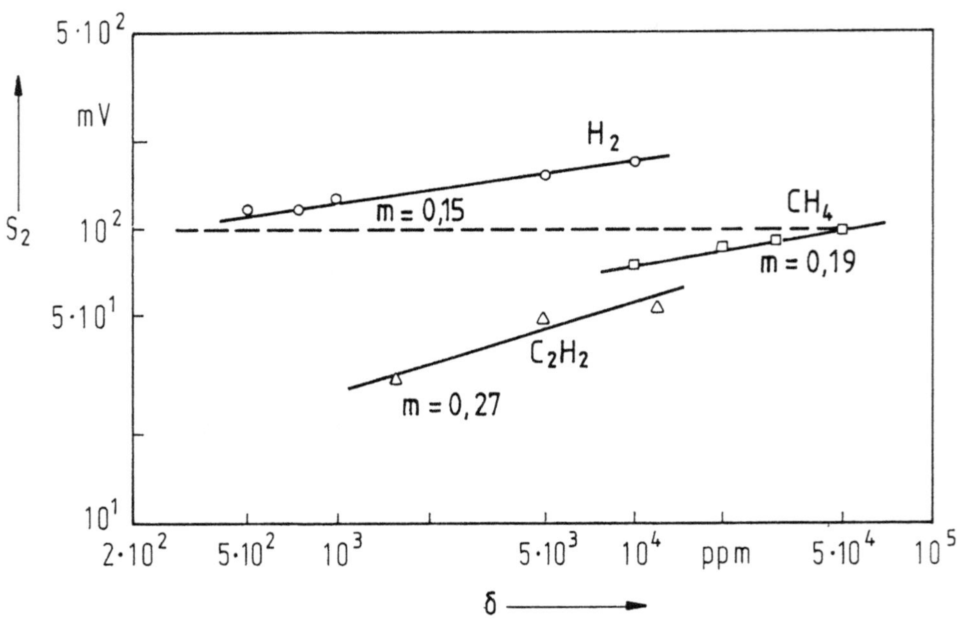

fig. 2

Fig. 3. Change of signal pattern with gas concentration.

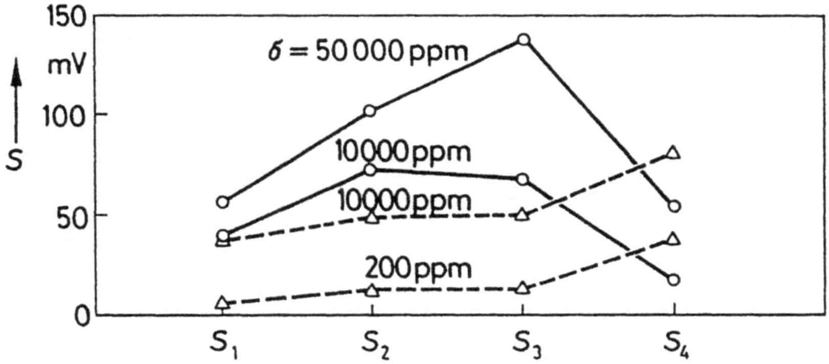

Proper quantitative expressions for the evaluation of the patterns are the <u>similarity coefficient</u> /3/ and the <u>correlation coefficient</u> ϱ_{jkp}:

$$\varrho_{jkp} = \frac{\sum_{i=1}^{n} (S_{ijp} - S_{jp})(N_{ik} - N_k)}{\sqrt{\sum_{i=1}^{n} (S_{ijp} - S_{jp})^2 \sum_{i=1}^{n} (N_{ik} - N_k)^2}}$$

with $N_{ik} = \frac{1}{m} \sum_{p=1}^{m} N_{ikp}$, $N_k = \sum_{i=1}^{n} N_{ik}$ and $S_{jp} = \frac{1}{n} \sum_{i=1}^{n} S_{ijp}$

The quantity N_{ikp} is the sensor signal of sensor i for the calibration gas k with pressure p and S_{ijp} is the corresponding measurement signal for gas j. The indices have the following meanings: i for sensor element i, j for measurement gas j with pressure index p and k for the calibration gas k. The quantity N_{ik} is the "mean pattern" of the calibration process (averaged over the pressure values) and N_k and S_{jp} are the average signals (averaged over the sensor elements) of the calibration and the measurement process respectively.

The correlation coefficient defined above may assume values between +1 and -1. For identical patterns it is +1. Fig. 4 shows the evaluations for the gases hydrogen, methane and acetylene. It will be noted that the type of gas is reliably identified in each case irrespective of the concentration /10/.

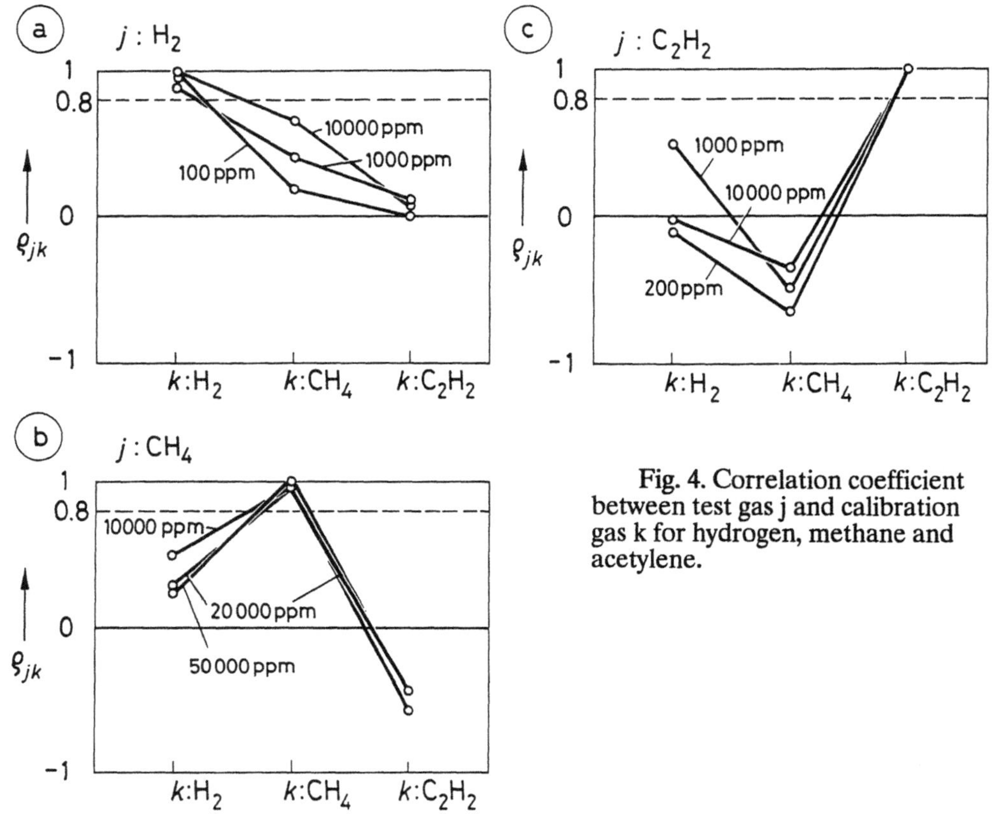

Fig. 4. Correlation coefficient between test gas j and calibration gas k for hydrogen, methane and acetylene.

3. SNO_2 - SENSORS (VECTOR METHOD)

Fig. 5 shows a test chamber with four different Figaro SnO_2 - sensors. Several transfer characteristics are shown in fig. 6 for two of the sensor elements. Again the sensitivities differ for different gases but there is no ideal selectivity. Fig. 7 shows the signal pattern of four sensors for the two gases CO and CH_4 /11/. Here a normalized quantity P_{ij} is derived from the actual sensor signal, its resistance R_{ij}.

$$\frac{R_{ij}}{R_{io}} = r_{ij} = (a_j\,[X_j] + 1)^{m_{ij}} ,$$

$$P_{ij} = \frac{\log r_{ij}}{\sum_{i=1}^{n} \log r_{ij}} = \frac{m_{ij}}{\sum_{i=1}^{n} m_{ij}} .$$

The quantity $[X_j]$ is the concentration of gas X_j and R_{io} is the resistance of sensorelement i in air. Due to the normalization above the vectors P_{ij} all end in a plane of order i-1, as shown in fig. 8 for there sensor elements. For each substance there is a special region, which can be used for identification.

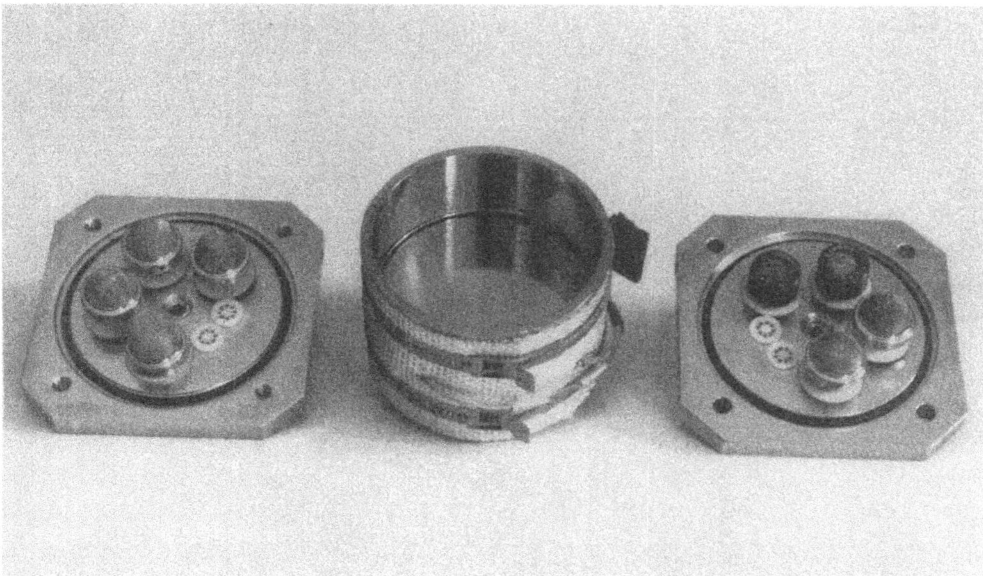

Fig. 5. Test chamber with 4 Figaro sensors.

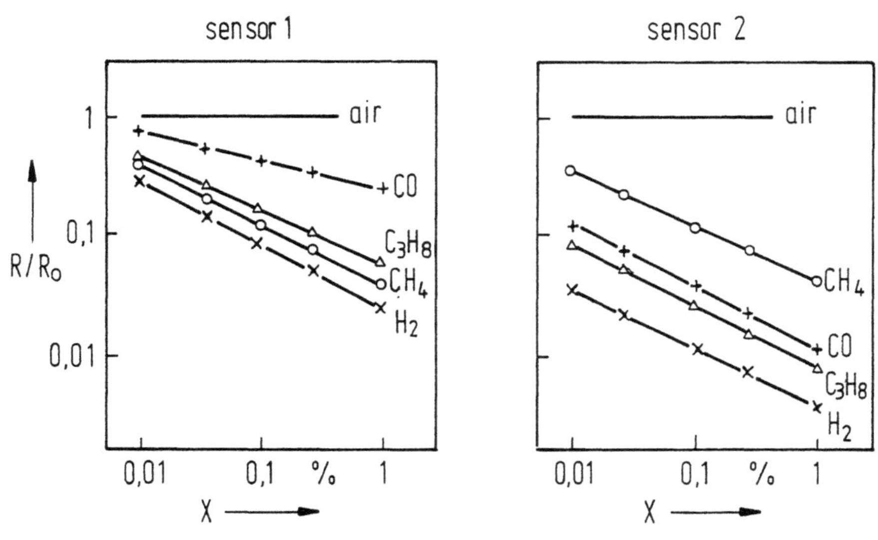

Fig. 6. Normalized resistance of SnO_2 - sensors as function of the gas concentration.

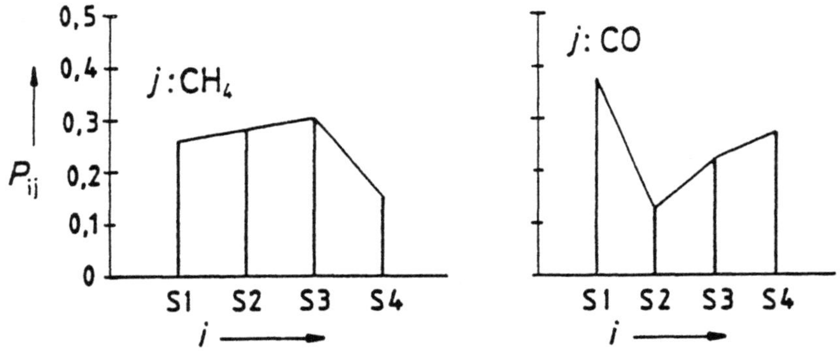

Fig. 7. Signal pattern for methane and carbon monoxide obtained by four different Figaro sensors

Fig. 8. Feature space for two substances A and B.

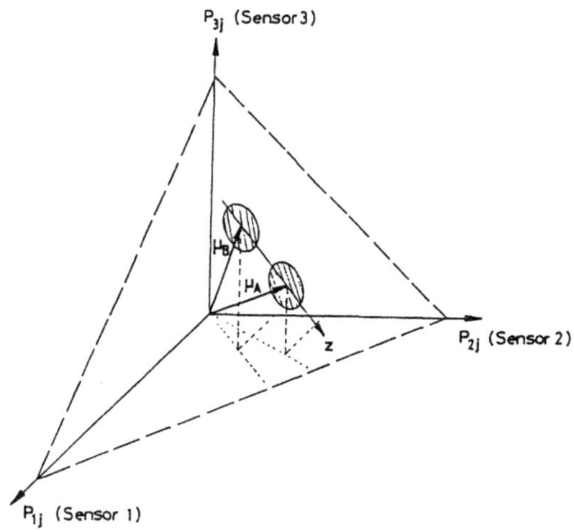

Fig. 9 shows (an assumed Gaussian) distribution for the two gases CO and CH_4. With this plot it is possible to define a measure $S_{A/B}$ for the selectivity of the array using the variance σ_A.

$$S_{A/B}\ (98\ \%) = \frac{|\mu_A - \mu_B|}{2\ \sigma_A} \quad ; \quad S_{B/A}\ (98\ \%) = \frac{|\mu_A - \mu_B|}{2\ \sigma_B}$$

A value α for this quantity means, that with a probability of 98 % the sensor systems indicates a ratio of the two concentration of the gas A to the gas B is greater than α /11/. With a similar definition for single sensor elements, the selectivity of the array can be compared with the selectivities of the single elements. Fig. 10 shows the comparison for two practically non - selective single elements (left side) and an array of 2, 3 and 4 similar elements. The advantage of the application of pattern recognition is evident.

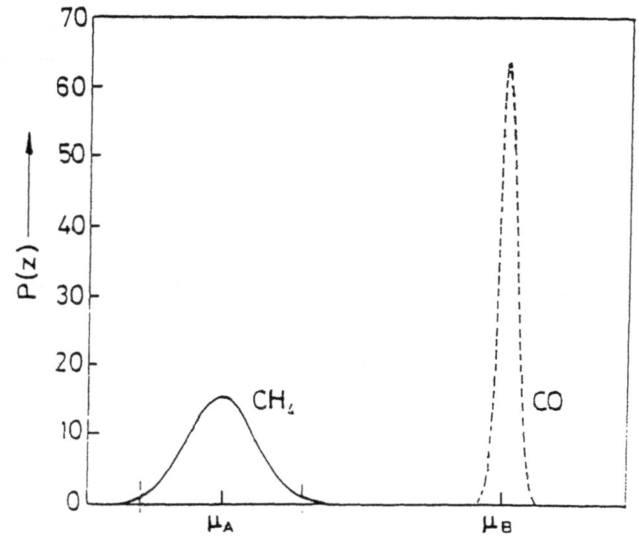

Fig. 9. Distribution in the feature space for methane and carbon monoxide.

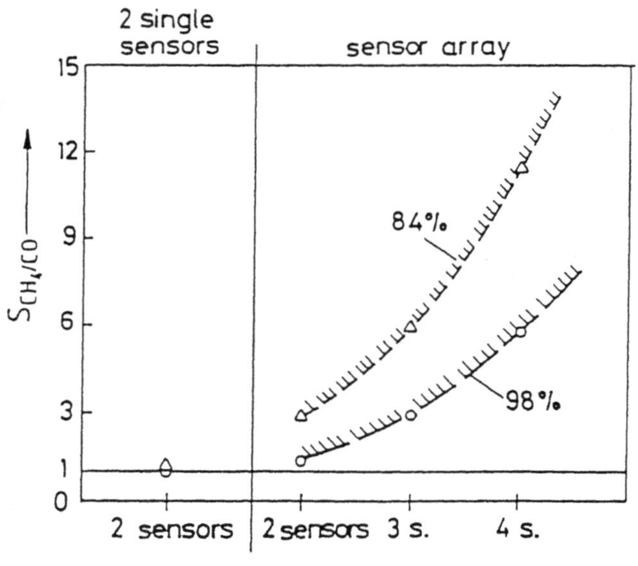

Fig. 10. Selectivity for methane versus carbon monoxide for single sensor elements (left side) and 2, 3 or 4 sensors with pattern recognition (right side).

4. SNO_2 - SENSORS (TLS METHOD)

The application of linear regression methods like for instance PLS (partial least square) /5/ rely on the validity of the superposition law, that is for two substances A and B the signals must superimpose to give the signal of the sum of the two gases:

$$s_{A+B} = s_A + s_B .$$

Frequently this law does not hold for the sensor signals, but a transformation is possible, so that the superposition law holds in an artifical superposition plane /6/, /12/:

$$f(S_{A+B}) = f(S_A) + f(S_B).$$

The application of this transformed least square method (TLS) usually lead to good results.

Fig. 11 shows the crossensitivity of a CH_4 sensor against the simultaneous application of CO. The indicated CH_4 concentration (after correction for the nonlinear transfer characteristic) is plotted versus the simultaneous CO concentration. Without cross - sensitivity the dashed line would result; the shaded area is an indication of the error (about 60% for equal concentration of CH_4 and CO). With the aid of TLS one obtaines fig. 12, which exhibits a reduction of the crossensitivity to about 6%.

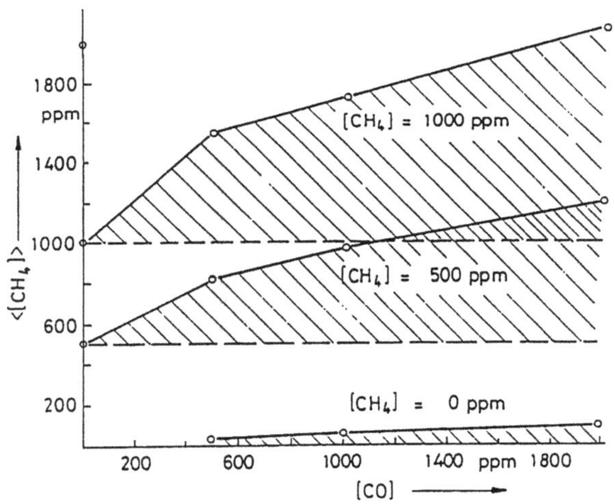

Fig. 11. Dependence of indicated concentrations of methane versus additionally applied carbon monoxide concentration for a single sensor element (Figaro TGS 812); the hatched area represents the crossensitivity.

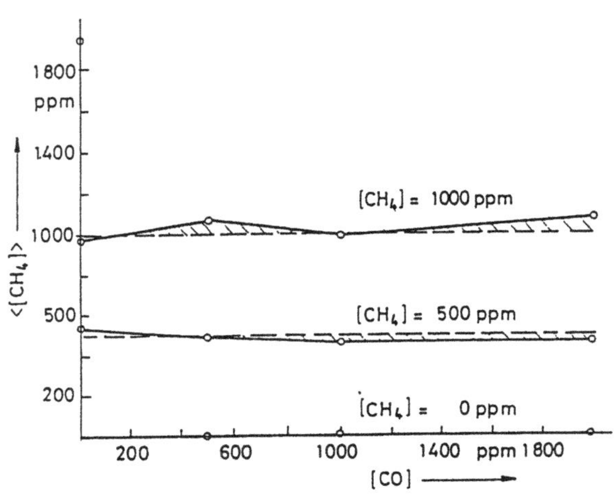

Fig. 12. Dependence of indicated concentration of methane versus additionally applied carbon monoxide concentration for a sensor array and TLS-evaluation.

5. DISCUSSION

Although the experimental results are not sufficient to yield a thorough comparison of the different methods of pattern recognition, a coarse comparison will be made partly on a heuristic basis in Table 1.

		KM	VM	PLS	TLS
TASK	IDENTIFIKATION	X	X	X	X
	MEASUREMENT OF CONCENTRATION	(X)	(X)	X	X
GAS MIXTURE	SINGLE COMPONENTS	X	X	X	X
	ARBITRARY MIXTURE			X	X
TRANSFER CHARACTERISTIC	LINEAR	X	X	X	X
	NON LINEAR	X	X		X
DIVERSITY OF GAS	LIMITED			X	X
	VERY LARGE	X	X		
FAILURE MESSAGE	LARGE SCALE			X	X
	SIMPLE	X	X		
CALIBRATION	LARGE SCALE			X	X
	SIMPLE	X	X		
MEASUREMENT	LARGE SCALE				
	SIMPLE	X	X	X	X

Table 1

REFERENCES

1. Hirschfeld T.: Microfabrication and Sensor Technology International Conference on solid state sensors and actuators. 1985. Digest of technical papers: New York, IEEE Press 1985
2. Stetter, J.R.; Zaromb, S.: Theoretical basis for identification and measurement of air contaminants using an array of sensors having partly overlapping selectivities. Sensors and Actuators 6 (1984), 225-243
3. Ikegami, A.; Kaneyasu, M.: Olfactory Detection Using Integrated Sensor. Proc. Int. conf. on Solid-State Sensors and Actuators (1985). New Yor: IEEE Press 1985, 136-139
4. Müller, R.; Lange, E.: Multidimensional sensors for gas analysis. Proc. Int. Conf. on Solid State Sensors and Actuators, 1985. New York: IEEE Press 1985, S. 81-84
5. Sharaf, M.A.: Illman, D.L.; Kowalski, B.,R.: Chemometrics. New York: John Wiley & Sons, 1986.
6. Hierold, Chr.; Müller, R.: Quantitative Analysis of Gas Mixtures with Non-Selective Gas Sensors.
7. Horner, G.; Lange, E.; Albertshofer, W.; Nuscheler, F.: MOS-Gassensoren mit Zeolith-Filterschichten, Sensoren - Technologie und Anwendung. NTG-Fachberichte 93, Bad Nauheim, 1986. 108-115.
8. Lundström, I.: Hydrogen sensitive MOS-structures, Part I: Principles and Applications. Sensors and Actuators 1, 1981.
9. Müller, R., Lange, E.: Multidimensional Sensor for Gas Analysis. Sensors and Actuators, 9, 1986. 39-48.
10. Müller, R., Horner, G.: Chemosensors with Pattern Recognition. Siemens Forsch.- u. Entwickl.-Ber. Bd. 15 (1986) Nr. 3, Springer-Verlag 1986.
11. Horner, G, Albertshofer, W.: Sensorarrays mit nicht-selektiven Chemosensoren - Analyse von Gasgemischen und Verbesserung der Selektivität. AEÜ Band 42, 1988 Heft 2.
12. Gall, M., Müller, R.: Investigation of Gasmixtures with different MOS-Gassensors with regard to Pattern Recognition.

MONTE CARLO GENERATION OF CHEMOSENSORY MAPS IN THE OLFACTORY BULB: GLOMERULAR ACTIVITY PATTERNS

D. Schild and H. Riedel

Department of Physiology, University of Göttingen
FRG

1. INTRODUCTION

Olfactory coding is probably not as well understood as the coding in other sensory systems. Most of the earlier theories dealing with olfactory coding consider only interactions between odorants and receptor cells (Amoore 1962, Amoore 1977, Davies 1965, Wright 1976, Beets 1957). A unified theory of these interactions may eventually evolve from biochemical and patch clamp studies on the olfactory transduction mechanisms (for ref., see Chapters 1 through 4 and Chap. 6 through 9 of this volume); especially promising is the use of olfactory receptor m-RNA to express membrane receptors and channels in oocytes (Getchell, this volume). However, understanding the transduction processes alone does not resolve the question of how olfactory coding is done, i.e. how receptor activities are processed by the following stages of the olfactory pathway. Consequently there have been only a very few attempts to treat rigorously the transmission of olfactory information along the chain from receptor cells to olfactory bulb and cortex (for ref., see the chapters by Freeman and Kauer).

Recently, a formal concept of olfactory coding has been proposed (Schild, 1988) which takes into account events at the level of the receptor cells and the olfactory bulb. In the present study we apply this concept to some simple, ontogenetically early situations in which lateral interactions within the system are only sparsely developed. In this way we get some ideas about pre- and perinatal odour learning. We have tried to base the general idea on morphological evidence as well as on results from single cell recordings and optical recordings (Kauer, and Sensemann, this volume). From this host

NATO ASI Series, Vol. H 39
Chemosensory Information Processing. Edited by D. Schild

of available data it was possible to derive a fundamental structure for the model olfactory system which represents a kind of synthesis of many different olfactory systems. A further motivation for simplifying its structure was to obtain a model which can be understood both in mathematical terms as well as in terms of visualization. It appeared to be a convenient starting point to begin with such a "simple" wild-type system, which will be extended later.

2. MAPS OF ODOUR STIMULI

2.1 Receptor cell activity patterns

Figure 1 shows the three stages of the periphery of the olfactory system: a stimulus S enters the system and binds to receptor cells, which are herein represented as an array of filter elements.

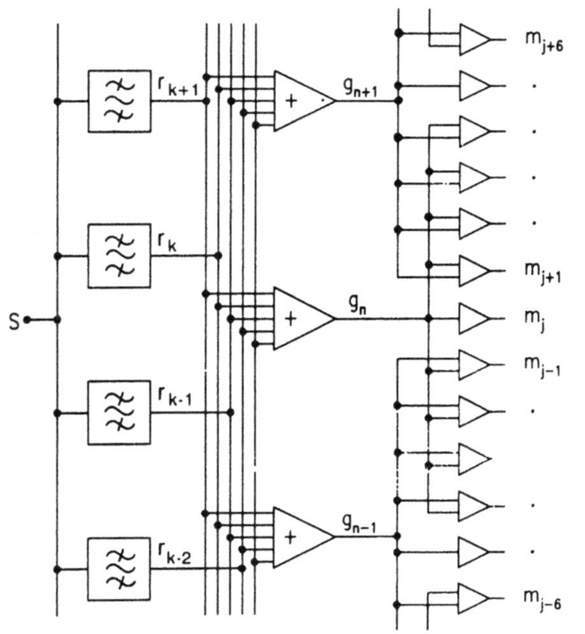

Fig. 1. Simplified one-dimensional block diagram showing the stages of information processing from stimuli S to mitral cells. Receptor cell classes are shown as filters. Glomerula and mitral cells form one-dimensional arrays g_1, g_2, ..., and m_1, m_2, ..., respectively. For further information and extension to two dimensions, see text.

Let us assume that there is a finite number K of receptor cell classes. A receptor cell class is defined as an assembly of receptor cells each of which would respond to <u>any</u> sequence of arbitrarily chosen stimuli with a sequence of activities which is identical for all cells of the receptor class (exept for stochastic variations). Two cells are thus in the same receptor cell class if they respond in the same way not just to one stimulus but to any other one. Though the assumption of receptor cell classes is suggested by biochemical evidence (Allen and Akeson 1985, Fujita et al. 1984), it must be regarded as a hypothesis given the present level of knowledge regarding stimulus binding to receptor neurones. It is however important to recognize that receptor cell responses to many stimuli, when analyzed by principal component analysis or factor analysis, they can be described by ten or less factors (Duchamps et al. 1974, Holley et al. 1974). This does not of course mean that there are ten or less receptor cell classes but, if there are more than ten classes their respective activities cannot be independent from each other. The output r_k of the k-th filter in figure 1 represents the ensemble activity of the receptor cells of the k-th class.

What is the difference between odours and how can it be measured, how are different odours mapped relatively to eachother, is their difference invariant or is resolution increased by the mapping, and how many different odours can be resolved? To treat these questions a measure of dissimilarity or similarity has to be used. As receptor cell class activities are vectors, e.g. $\underline{r}^a$ and $\underline{r}^b$ for odours a and b , it would be possible to take the Euclidean distance

$$D = |\underline{r}^a - \underline{r}^b| = \sqrt{\sum_{i=1}^{L} (r_i^a - r_i^b)^2} \quad .$$

Another reasonable measure would be the correlation coefficient C(a,b) between the components of both vectors:

$$C(\underline{r}^a, \underline{r}^b) = \frac{\sum_i (r_i^a - \langle \underline{r}^a \rangle)(r_i^b - \langle \underline{r}^b \rangle)}{\sqrt{\sum_i (r_i^a - \langle \underline{r}^a \rangle)^2 \sum_i (r_i^b - \langle \underline{r}^b \rangle)^2}}$$

where $\langle \underline{r} \rangle$ means the mean value of the vector components r_i of the vector $\underline{r}$.
$C(\underline{r}^a, \underline{r}^b)$ gives a value between -1 and 1.

2.2. Glomerular activity patterns

The signals from every receptor cell class can theoretically reach almost all glomerula whereby the *strength* of its influence on various glomerula will generally vary from glomerulum to glomerulum. In the language of computer hardware one would say that the receptor cell classes feed their outputs onto a data bus and that every glomerulum receives a linear combination of the data present on the different bus lines. Glomerula can be viewed as structures in which receptor cell class specificities are mixed. In mathematical terms this corresponds to a linear superposition

$$g(x,y) = \sum_k c_k(x,y)\, r_k, \tag{1a}$$

where g(x,y) is the overall input of the glomerulum at position x,y, with x and y being two coordinates which determine uniquely the position of the glomerulum. As far as small and more or less spherical olfactory bulbs are concerned x and y can be imagined to be the spherical angle coordinates. However, in more extended olfactory bulbs the exact geometrical structures become less important and x and y can for convenience be assumed to be cartesian coordinates. $c_k(x,y)$ in the above formula is the strength by which the k-th receptor cell class output r_k is coupled to the glomerulus at position x,y. The sum in eq.(1) is a scalar product and can also be written as such

$$g(x,y) = \underline{c}(x,y)\, \underline{r} \tag{1b}$$

or, when a particular stimulus, say S^a, is indicated

$$g^a(x,y) = \underline{c}(x,y)\, \underline{r}^a = |\underline{c}(x,y)||\underline{r}^a| \cos\beta \tag{1c}$$

with β being the angle between the two vectors $\underline{c}$ and $\underline{r}$. This formulation of the receptor-glomerular projection suggests a geometrical interpretation: It is possible to assign a convergence vector $\underline{c}(x,y)$ to every glomerulum and the more the direction of the receptor cell activity vector matches the glomerulum convergence vector $\underline{c}$ the larger is the overall glomerulum input to this stimulus. Fig. 2a and b show the responses of ten receptor cell classes to two different stimuli, S^a and S^b. These activity profiles look very much like spectra which describe an odour at receptor cell level. The only serious objection against the term "spectrum" is that the activity profile components are usually not linearly independent. So we prefer the term stimulus activity profile or simply profile. Assume now that a certain glomerulum at x,y has an input convergence pattern given by $\underline{c}(x,y)$. Just as $\underline{r}^a$, it can be visualized by its components (fig. 2 c). Because the convergence pattern $\underline{c}(x,y)$ has the same components as the stimulus profile $\underline{r}^a$, the angle between these vectors is zero and the scalar product, i.e. the overall glomerulum input is maximum. In this case it corresponds to a certain value, say A. When stimulus b or the stimulus profile $\underline{r}^b$ is the system input, the vector components do not match well and the corresponding glomerular input $g^b(x,y)$ is B < A. In this study we represent stimulus profiles and all glomerular input activities in the same plot by coding the amplitudes both of $\underline{r}$ and g in a grey code. Fig. 3 shows an example. The bars on the left side of the figure represent the stimulus profile beginning with r_1 (lower bar) and ending with r_{10} (upper bar). The connectivities c_k from the receptor cell classes (bars) to the glomerula are chosen randomly in this example. The image of glomerula input activities (large grey spots in fig. 3) is thus random, too.

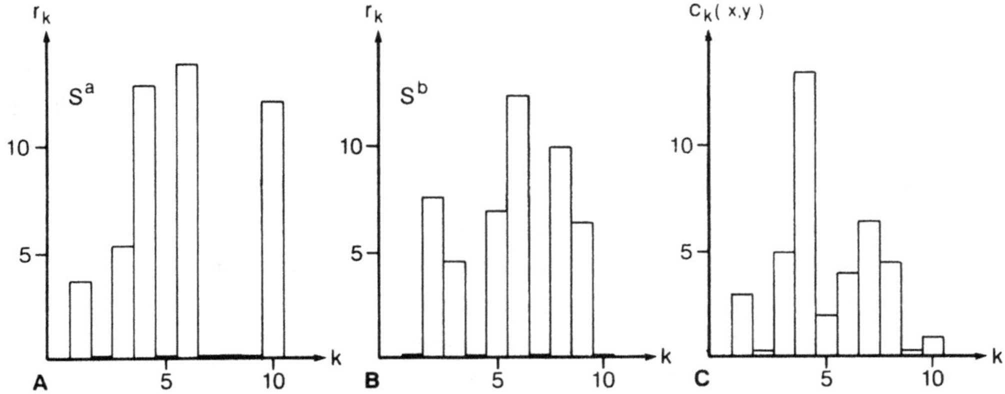

Fig. 2. Response activities of receptor classes and connectivities to glomerula. Response activity to two different stimuli: (a) S^a activates five classes, (b) six classes are responsive to S^b. (c) Connectivity pattern of a particular glomerulum represented in the same form as the receptor class activities.

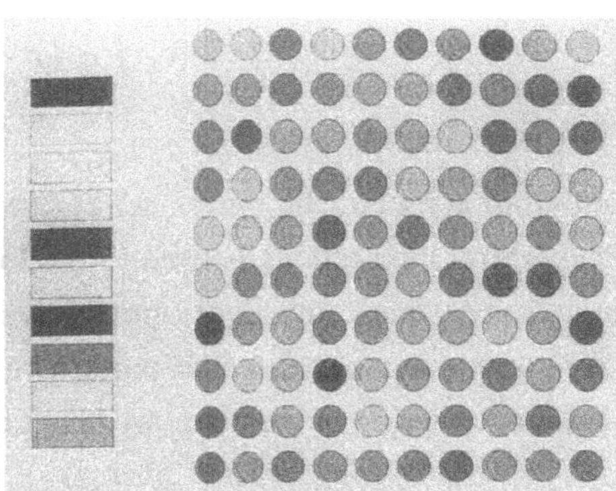

Fig. 3. Receptor cell class and glomerular activities. The receptor activity profile is identical to that in fig. 2 a and in fig. 4 a at the highest concentration. Glomerula are represented as dots in a 10*10 array. The activity of a glomerulum is coded by a grey scale such that the lowest and highest activity correspond to white and black, respectively.

2.3 Mitral cell activity patterns

Though this paper does not explicitly deal with mitral cell images, we outline briefly the connections between glomerula and mitral cells in order to assess the meaning of glomerula images for the information processing in the olfactory bulb. One line of the glomerula matrix in fig. 3 corresponds to the array of glomerula in fig. 1. In this one-dimensional simplification it can be seen that every glomerulum feeds the mitral cells in its neighbourhood whereby the adjacent ranges of glomerula influence overlap. For the present model it is not relevant whether mitral cells get their inputs through several dendrites from the neighbouring glomerula (as in many fish, e.g., Kosaka and Hama 1983, Oka 1983) or whether they get their input from one primary dendrite from "their" glomerulum and also through excitatory connections from adjacent glomerula (as in higher vertebrates). The principle of overlapping signal divergence from glomerula to mitral cells thus seems to be realized in more than one way. In the two-dimensional representation, the activity image of mitral cells can be added by simply superimposing 25-times more mitral cells over the glomerular image.

The formal description of the glomerula-mitral cell projection is

$$m(v,w) = \sum_{x,y} p_{x,y}(v,w)\, g(x,y) \tag{2}$$

whereby m(v,w) is the input of the mitral cell at the coordinates v,w. $p_{x,y}(v,w)$ is greater than zero in the vicinity of the mitral cell at x,y, i.e.,

$$p_{x,y}(v,w) \begin{cases} = 0 & \text{for } |x-v| > \delta_{x,v} \text{ and } |y-w| > \delta_{y,w} \\ > 0 & \text{otherwise} \end{cases} \tag{3}$$

The output a(v,w) of the mitral cell at (v,w) is determined not only by its input but also by the lateral interactions between mitral cells. These are essential for the system as a whole since they introduce nonlinearities which influence the coding process considerably (they enhance the pattern resolution, lead to attraction basins, and allow efferent modulation of the olfactory bulb's overall transfer function).

2.4 Regeneration of receptor neurones and plasticity

The coding concept in its general form has another important feature. The connectivities between receptor cells and glomerula as well as those between glomerula and mitral cells are generally time-dependent, i.e. the system can show plasticity. One type of plasticity is based upon receptor cell regeneration whereas the second one is a short term plasticity as found in the olfactory bulb (Schild and Zippel, 1986). In this paper we deal only with the first type of plasticity. Regeneration of olfactory receptor neurones, which has been well described, means that the connectivity pattern between receptor neurones and glomerula is variable. As the regeneration process needs days or weeks, the connectivity pattern changes only very slowly (slowly in comparison to all typical neural time scales). In the model we describe the connection strength between the k-th receptor cell class and the a glomerulum at the coordinates x,y by a number $c_k(x,y)$, which has the following dynamics:

a) Degeneration process

We assume that the degeneration or death of receptor cells is homogenously distributed over all receptor classes. In mathematical terms we can state, that the decrease dc/dt of a certain value c is proportional to itself.

$$\frac{\partial c_k(x,y)}{\partial t} = -\frac{1}{\tau_c}\, c_k(x,y)$$

b) Learning and growth of connections

A Hebb-like learning rule is assumed in our model: If the activity of the k-th receptor class multiplied with the glomerular activity exceeds a certain threshold, the corresponding strength is increased:

$$r_k\, g_k(x,y) > \text{threshold} \Rightarrow \frac{\partial c_k(x,y)}{\partial t} > 0$$

Of course, this process cannot continue to infinity, because the synaptic input to a glomerulum has to be limited. The glomerulum is then saturated. A renormalisation of the connection strengths is therefore performed when the saturation limit has been exceeded. In addition, we determine an upper limit for the total connection strength of a receptor class. In this way the suppression by one fast growing receptor class of all other classes is avoided.

2.5 Interpretation of the maps

It is tempting to interpret fig. 1 and eqs.(1) and (2) from a systems theory point of view: because the receptor cell class activities are the output of an array of filters, they may be considered to form a spectrum, and therefore the question arises how olfactory "spectra" are processed in the olfactory bulb. Basically, there would be two possibilities, either they retain their spectral character and are somehow modified or they are transformed from the spectral domain into the spatial domain to give real images of the odours.

For the following reason the second alternative seems more attractive at first glance: assume that the coefficients $p_{x,y}(v,w)$ in eq. (2) do not depend on the absolute coordinates v and w but only on the differences v-x and w-y, eq.(2) would become

$$m(v,w) = \sum_{x,y} p(v-x, w-y)\, g(x,y)$$

This is a convolution in space and, together with eq.3, corresponds to a spatial low-pass filter. Under this hypothesis, the $\underline{c}(x,y)$-map transforms the spectrum $\underline{r}^a$ into an image $g^a(x,y)$ which is then low pass filtered by the p(v-x, w-y)-map (Fig. 5a). This concept, though attractive, has some serious drawbacks:

(i) there is no experimental evidence that the divergence from glomerula to mitral cells is identical for all mitral cells. This even seems to be highly improbable so that the assumption $p_{x,y}(v,w) = p(v-x,w-y)$ cannot be sufficiently justified;

(ii) the receptor cell class activities r_i probably do not correspond to a one-dimensional spectrum in the usual sense, since the existence of more receptor classes than factors (see p.3) means that the spectral components are not linearly independent; furthermore it is likely that there is more than one parameter according to which the receptor cell class activities must be ordered. Ordering relations among such parameters are currently unknown. One has to assume therefore that activity profiles at the receptor level consist of partly linearly dependent components of a multi-parameter "spectrum";

(iii) the transform $\underline{c}(x,y)$ is not an orthogonal transform such as an inverse Fourier- or Hadamard transform, which could transform a spectrum into an image. Such transforms must have negative elements whereas all $c_k(x,y)$ are positive or zero. However, at a later stage of the model, inhibitory periglomerular cells will be considered which might contribute negative coefficients to the transform from receptor cells to glomerula.

For these reasons we suggest that the idea of the recomposition of glomerular image from a receptor cell activity spectrum is not appropriate. An alternative interpretation of the first steps in olfactory coding is therefore proposed here:

1. $\underline{r}$ is a (possibly multiparameter and spectrum-like) activity profile the components of which are not all linearly independent.
2. $c_k(x,y)$ is a transform which maps the high-dimensional receptor cell activity space to a function of two variables; regions of the receptor cell class activity space with high information content are spread over relatively large areas on the glomerular layer while other regions with little information are only scarcely projected to the glomerula; similar stimuli (in the receptor cell activity space) are mapped to similar glomerular activity patterns thereby introducing a spatial order among odours on the glomerula layer. The map $\underline{c}$ is self-organized in a way which is similar to the concepts proposed by Hebb (1949) and Kohonen (1987).
3. $p_{x,y}(v,w)$ together with mitral cells work as matched filters i.e., every constellation of glomerular activation stimulates a certain set of mitral cells maximally. This set, by suppressing their neighbouring cells, can be distinguished very well from slightly different sets representing similar stimuli. In this way, the odour resolution is considerably enhanced at the third layer of the olfactory network.
4. The parameters of the spatial map between receptor cell classes and mitral cell activities are controlled by several efferent systems (above all ACh, 5-HT, and norepinephrine).

In this study we treat the generation of receptor cell class activities and glomerular activity patterns.

3. RECEPTOR CELL CLASS ACTIVITIES

Recording from olfactory receptor neurones in vivo has proved extremely difficult. Intracellular recordings usually show reduced membrane potentials indicating a too low seal (leak) resistance between pipet and cell membrane. Under such circumstances intracellular recording yields results which do not reflect normal physiological conditions since all membrane currents are partly short-circuited by the leak. Extracellular recordings (for review, see Getchell 1986) have however shown some important features of the cells' responses to natural stimuli: (i) the response activity is of the phasic-tonic type, (ii) a single (vertebrate) cell usually responds to many odours though with different spiking rates, (iii) responses show a high degree of reproducibility (i.e.,there is little inter-trial adaptation), (iv) when subjected to factor analysis, the responses of many cells to many different odours can be described by less than 10 factors, (v) concentration coding over a wide concentration range is done by more than one cell since single olfactory neurones have dose- response ranges of about one logarithmic unit with a variability between half a log-unit to three log-units, (vi) the statistics of single unit discharges resembles a Poisson process, i.e., spikes occur fairly randomly and a spike train (after the dynamic response peak) is well described by its mean firing rate, (vii) spontaneous firng rates are well below 1 spike/s, the exact spontaneous rate is not known for two methodological reasons: first the recording electrode might influence the spontaneous rate and, second, cells firing at a very low rate are very likely not to be detected, and (viii) maximum firing rates vary between 5 to 15 spikes/s.

All published data on olfactory neuron activities do probably not provide sufficient material for modelling a representative input to a model olfactory bulb. It is however possible to create numerically receptor cell responses which could represent real odour responses in the sense that these responses incorporate the above listed physiological properties.

The programme which calculated the odour responses at receptor cell class level was developed and run on a VME-computer (Motorola MVME 147). It was written in "C" and worked as follows:

10 receptor cell classes were assumed and the model system was considered to be responsive over eight logarithmic units whereby the receptor cell class activities were calculated in half-logarithmic steps. For every odour stimulus, we had thus to determine 170 values, i.e. activity values of 10 cell classes, each at 17 concentration steps. (Fig.4).

In order to find these 170 values per odour, random variables were drawn from a random number generator (drand48, UNIX V.3). For each odour it was first determined how many, and which out of the ten receptor cell classes were active. In fig. 4 a there are, for instance, five active classes. In addition, we added a constraint to this choice: if class k was already active the probability p_c of classes $k \pm 3n$, $n \in N$ to be active as well was set to a certain higher value. This means that if certain classes are active there is an elevated probability for other classes to be active, too, i.e., a correlation beween receptor cell class activities is introduced this way (e.g. classes 2, 5 and 8 as well as 3, 6 and 9 in fig. 4 b). Without this constraint the data must be described in a ten-dimensional space, whereas with this constraint, the dimensionality at receptor cell level was reduced by correlations between receptor cell class activities. By setting the joint-probability for receptor classes to be active, it is thus possible to gradually vary the dimensionality (or number of sufficient factors). In order to approximate the experimental data, we generated simulated receptor cell data in seven dimensions.

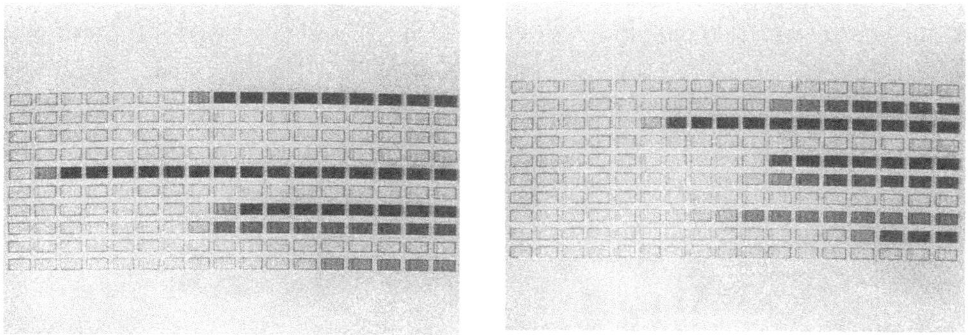

Fig. 4. Representation of two odours as activities of receptor cell classes in 17 different concentrations. The different classes are drawn from bottom to top, so that the lowest row represents the activities of the first class. Seventeen increasing concentrations, c_0 through c_{16}, are drawn from left to right, i.e., the highest field in the rightmost row corresponds to the highest concentration of the tenth receptor cell class.

After having established the indices of active receptor cell classes (in fig. 4 a, e.g., receptor cell classes 1,3,4,6 and 10), three more random variables were selected for each receptor cell class: first it was determined at which concentration the respective class had its threshold, i.e., at which concentration the response exceeded spontaneous activity. In fig. 4 a this was, e.g., c_{11} for receptor cell class 1 and c_1 for receptor cell class 6, etc.. Second, the dynamic range for every receptor cell class had to be chosen. In fig. 4 b receptor cell class 9 has a dynamic range of three log units while receptor cell class 6 has a dynamic range of only one log unit. Finally for every receptor cell class, a random variable was drawn to set its maximum firing rate. Spontaneous rate was in any case assumed to be between 0.01 and 0.1 s^{-1}. From fig. 4 a it is clear that receptor cell class 1 has a much lower saturation firing rate than the other receptor cell classes. In this way, activities were calculated for 100 odour stimuli each at 17 different concentrations and saved in a file.

4. GLOMERULAR IMAGES

In our model we used an array of 10*10 glomerula, the activites of which were calculated according to eq.(1a) and visualized as circles. The amplitudes g(x,y) were coded by a grey scale. In fig. 5 four different glomerular patterns corresponding to four different odours are shown.

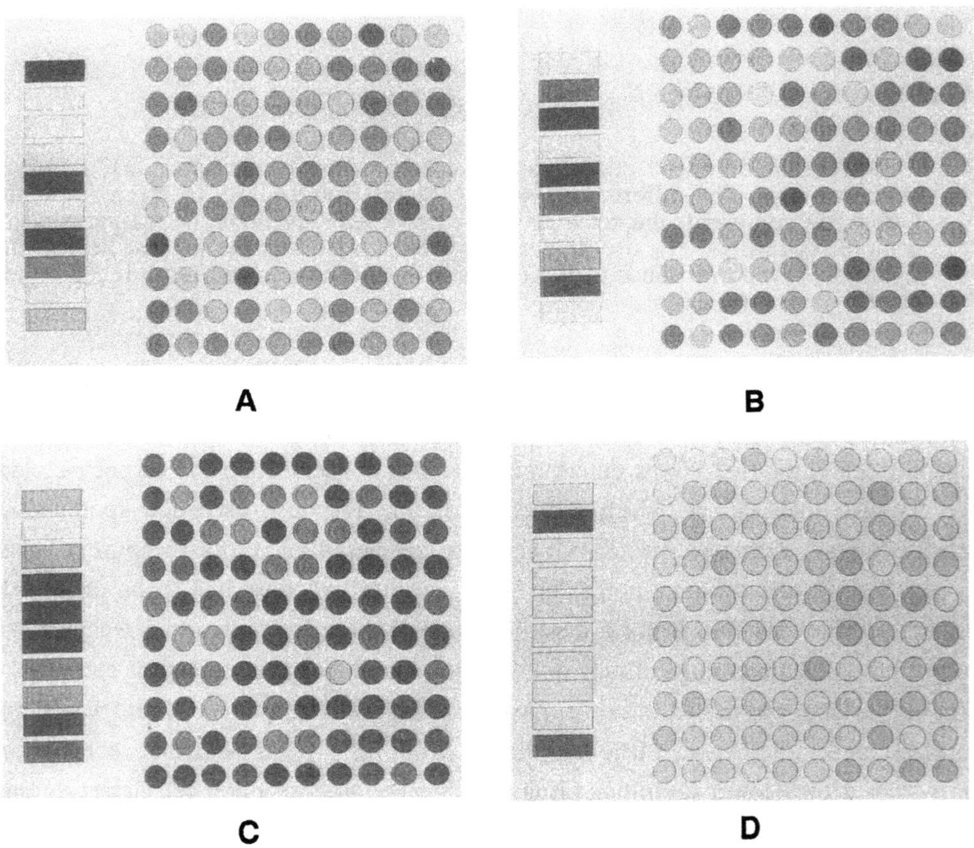

Fig. 5. Receptor cell class activities and their glomerular patterns for four different odours. With respect to the overlap between the patterns, this sample of four glomerular patterns can be regarded as typical.

Obviously many glomerula take part in the coding of any one of the odours, and the patterns show thus a high degree of overlap. Given the present stimulus distribution there is no glomerulum (or small group of glomerula) assigned exclusively to one odour. If each glomerulum code just one odour, only 100 odours could be stored in the model, while, by allowing overlapping glomerular activities, the number of possible patterns (storable odours) is increased by many orders of magnitude. The differences between complex glomerular patterns are, of course, more difficult to visualize. The differences between glomerular images can be assessed by a variety of mathematical tools. Similar to the distance measure at the receptor cell level, one can calculate for example Euclidean distances or correlation coefficients. Although the global performance can only be derived mathematically, some interesting features of the maps can be easily visualized. Similar odours, for example odour stimuli of the same quality but in different concentrations, elicit similar patterns (fig. 6) reflecting the linearity of the map.

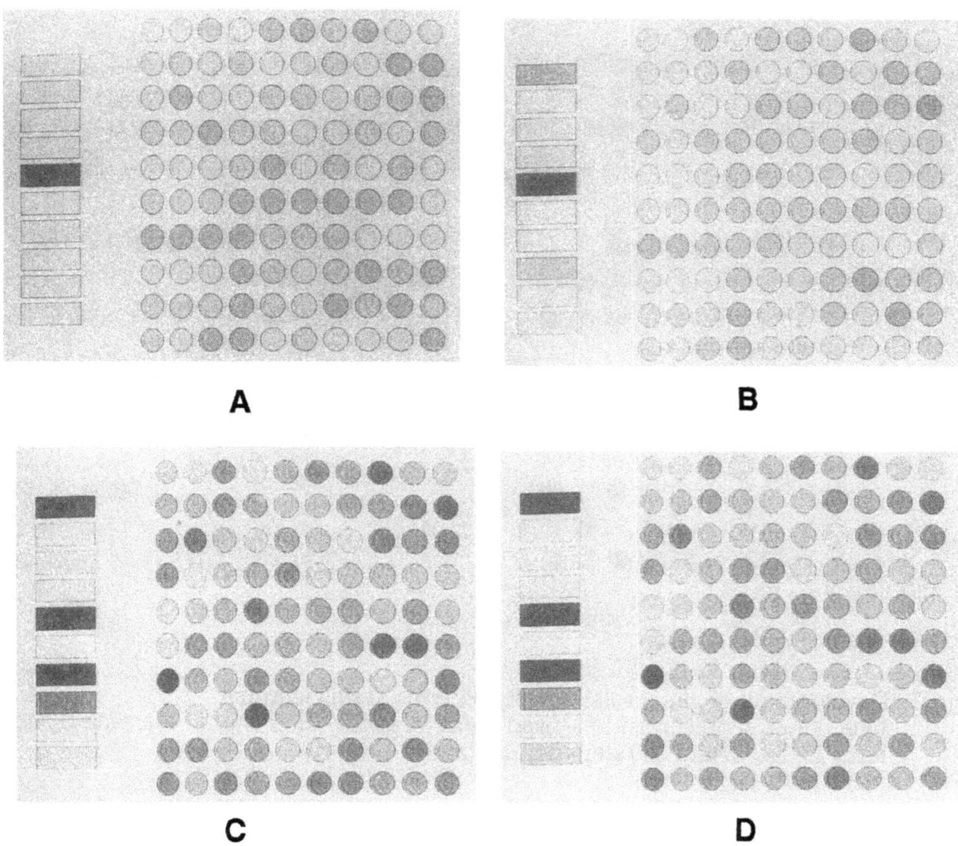

Fig. 6. Glomerular pattern for the same odour quality in four different concentrations. The distance between every two stimuli is two log units.

The stimulus image corresponding to the higher concentrations appears somewhat darker whereby the intensity increase is not homogenous. If only one odour is presented during a learning process, it can be observed that the activation of 'dark' glomerula increases monotonically while other glomerular activities are gradually supressed over time. This demonstrates the increase of connection strengths if glomerular input signals are highly correlated and the decrease of connection strengths if the correlation is small (fig. 7).

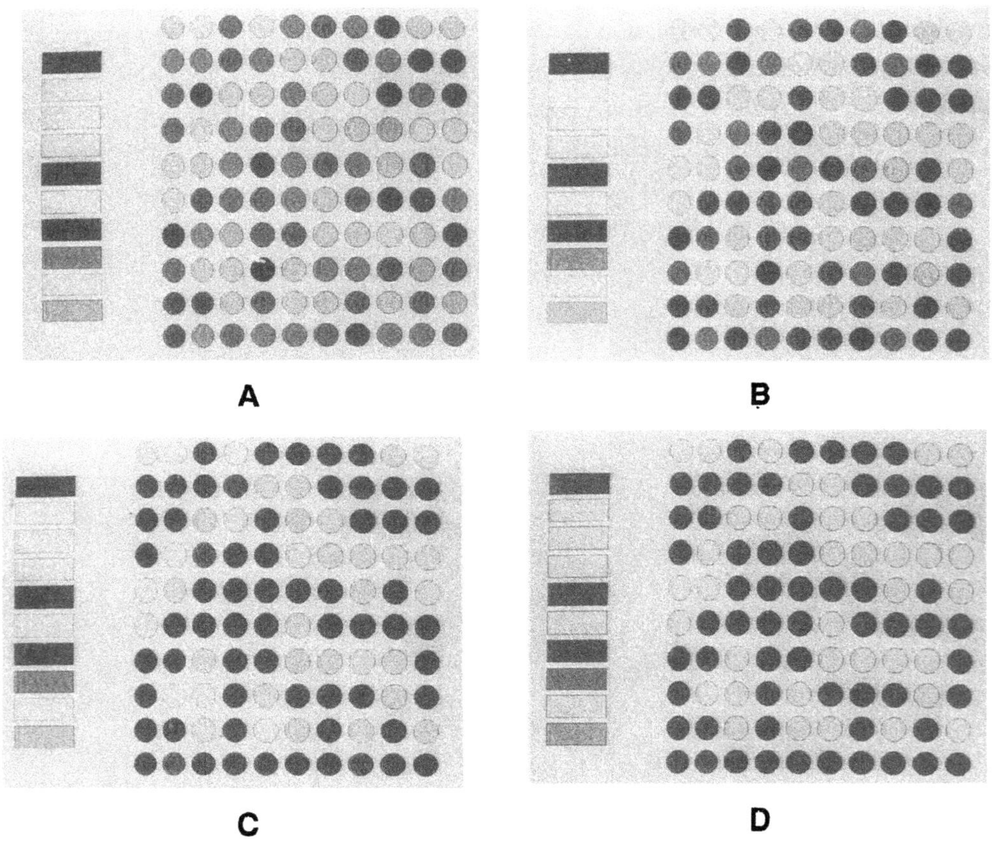

Fig. 7. Development of glomerular activity patterns during learning. One odour is presented in a fixed concentration. The result of learning is shown after 25, 50, 75 and 100 steps of adaptation.

5. DISCUSSION AND CONCLUSIONS

Though the present model neglects important features of the olfactory bulb and deals mainly with the plasticity between receptor cells and glomerula, some interesting features of the information processing in the olfactory system can be shown. First of all, it is seen that the map from the high-dimensional receptor activity space to the glomerular image leads necessarily to a considerable overlap of glomerular activities if many different stimuli are offered. Second, it can easily be visualized that odours which are similar in the high-dimensional receptor activity space give similar glomerular images. Third, it can be observed how a Hebbian learning rule leads to increasing or decreasing of single glomerular activities according to the correlation of the input signals.

One important question refers to the function of the glomerula. Considering the two mappings from receptor cells to glomerula and from glomerula to mitral cells, it can be seen (eqs. 1a and 2) that both are linear, which means that they could be taken together as one linear map. But this mathematical step, which is consistent with the present model, would correspond to cancelling the glomerula from the model. In fact, in lower vertebrates such as goldfish there exists a glomerular plexus rather than spherically shaped glomerula. It might be hypothesized that the glomerula arose phylogenetically together with interneurones such as periglomerular cells or, in the case of lower vertebrates, with those granule cells that mediate signals between the tiny dendritic or axonal processes in the glomerular layer. In this way, a third stage of information processing in the olfactory pathway could be established to improve performance by dividing the information channel between receptor neurones and mitral cells in two subchannels (or maps) which cannot be summed up in one single map because of the nonlinear interaction of interneurones. In addition, there is a second reason for assuming two differnt maps. The connections between receptor cells and glomerula develop on a very slow time scale involving the regeneration of the receptor cells, while the projections between glomerula and mitral cells can presumably change on a much faster time scale. In this way, the second map (the p-map) is adiabatically decoupled from the slower c-map.

A principle drawback of the present model might be that sets of mean activities (i.e. constant values) are considered to be the response to an odour. From experiments from all stages of the system we known that every single recorded trace is subject to a certain amount of "noise". By neglecting this "noise" we possibly lose an interesting interpretation suggested by Freeman (1987): the "noise" of every dimension could actually make up a strange attractor in an appropriate multidimensional space. This attractor could characterize a perceived odour. Freeman's and our point of view of olfactory coding are thus not as dissimilar as they appear at the first glance. By introducing the dynamics of interactive elements, the present model could be developed to describe odours as state space structures (Schild, 1984), which can be fixed points, limit cycles, or strange attractors.

We hope that the interaction between experiments and further theoretical investigations including the mitral cell layer, lateral interactions, but also concepts from information theory, will provide us with deeper insights in the information processing in the olfactory pathways.

REFERENCES

Allen WK, Akeson R (1985) Identification of a cell surface glycoprotein family of olfactory r eceptor neurons with a monoclonal antibody. J.Neurosci. 5:284-296

Amoore JE (1962) The stereochemical theory of olfaction. Identification of the seven primary odours. Special Supplement to No.37. Sci.Sect. Toilet Goods Assn. 37:13

Amoore JE (1977) Specific anosmia and the concept of primary odors. Chem.Sens. Flav. 2:267-281

Beets MGJ (1957) Structure and odour. In: Molecular Structure and Organoleptic Quality. Soc. of Chemical Industry, London 54-90

Davies JT (1965) A theory of the quality of odours. J.Theor.Biol. 8:1-7

Duchamp A, Revial MF, Holley A, MacLeod P (1974) Odor discrimination by frog olfactory receptors. Chem.Sens.Flavor 1:213-233

Freeman WJ (1987) How brains make chaos in order to make sense of the world. Behavioral and Brain Sciences 10:161-195

Fujita SC, Mori K, Imamura K, Obata K (1984) Subclasses of olfactory receptor cells and their segregated central projections demonstrated by a monoclonal antibody. Brain Res. 326:192-196

Getchell TV (1986) Functional properties of vertebrate olfactory receptor neurons. Physiol.Rev. 66:772-817

Hebb DO (1949) Organisation of Behaviour. Wiley, New York

Holley A, Duchamps A, Revial M-F, Juge A (1974) Qualitative and quantitative discrimination in the frog olfactory receptors: Analysis from electrophysiological data. Ann.N.Y. Acad.Sci. 237:102-114

Kohonen T (1988) Self-Organization and Associative Memory. Springer-Verlag, Berlin,Heidelberg,NewYork,London

Kosaka T, Hama K (1983) Synaptic organization in the teleost olfactory bulb. J.Physiol., Paris 78:707-719

Oka Y (1983) Golgi, electron-microscopic and combined golgi-electron-microscopic studies of the mitral cells in the goldfish olfactory bulb. Neuroscience 8:723-742

Schild D (1984) Coordination of neuronal signals as structurs in state space. Intern.J.Neuroscience 22:283-298

Schild D (1988) Principles of odor coding and a neural network for odor discrimination. Biophys.J. 54:1001-1011

Schild D, Zippel HP (1986) The influence of repeated natural stimulation upon discharge patterns of mitral cells of the goldfish olfactory bulb. J.Comp.Physiol.A 158:563-571

Wright RH (1976) Odour and molecular vibration: A possible membrane interaction mechanism. Chem.Sens.Flav. 2:203-206

OLFACTORY EEG CHANGES UNDER SERIAL DISCRIMINATION OF ODORANTS BY RABBITS[1]

Walter J. Freeman
G. Wesley Davis[2]

Department of Physiology-Anatomy
University of California
Berkeley, CA 94720

I. INTRODUCTION

The basic hypothesis in this study is that odor-specific information in the olfactory bulb is carried in the spatial patterns of cooperative activity of masses of bulbar neurons. We predict that the bulbar response to an odorant in any given time frame (on the order of 0.1 sec) is manifested in a spatial pattern of electrical activity over the neurons comprising the bulb (on the order of 100 mm^2 in surface area) at a surface grain corresponding to the mean distance between glomeruli (0.25 mm center to center) that spatially coarse-grain the receptor input. Axons from receptors project directly into the bulb and excite second order neurons (the mitral cells) that project in turn to the olfactory cortex. Unit studies of olfactory receptors in several species have shown that cells differ markedly in their sensitivities to odors. Anatomical and electro-physiological studies have shown that there is a degree of topographic order in the axonal connections from the receptors to the bulb (Freeman, 1975). Adrian (1950) predicted that different odorants would cause different spatial activity patterns of receptors in the mucosa, and that their axons would establish different patterns of activity in the spatial array of mitral cells.

In previous studies we attempted to show a relationship between spatial patterns of mitral cell unit activity and odorants presented to the nose. This was a different

1) Supported by a grant from NIMH (MH06686)
2) Present address: 1201 South Second Street
Milwaukee WI 53204

NATO ASI Series, Vol. H 39
Chemosensory Information Processing. Edited by D. Schild

question from asking whether single neurons responded differentially to various odors; it was based in part on existing evidence that they did so differ. We made simultaneous recordings from 10 microelectrodes in waking minimally restrained rabbits; we found no consistent relation between odorant and the spatial pattern of unit activity from up to 10 microelectrodes (Freeman, 1975). We concluded that the size of the spatial sample was too small for the test of the hypothesis; yet technical limitations precluded adding more channels for recording. We did find close temporal and spatial relationships between unit activity of mitral cells and the dendritic field potential activity of interneurons (the granule cells) that largely determines the EEG (Freeman, 1975). We repeated the attempt by presenting odorants to waking rabbits with 64 electrodes implanted epidurally over the surface of the bulb, but found no relationship between odorant presented and the spatial patterns of the EEG (Freeman, 1978).

In these as well as preceding studies we found that the presence of significant unit and EEG activity required some degree of motivation irrespective of kind. We provided that in the above attempts by imposing a standard period of food deprivation prior to testing. In ensuing work we used classical aversive conditioning by giving a brief electric shock as the US to rabbits 2 seconds after onset of each odorant presentation as the CS (Freeman et al., 1983). During the acquisition of a CR to each new odorant we observed a change in EEG spatial pattern; a new pattern emerged in the first session and stabilized over the next two or three sessions, so that it was present whether or not the odorant was present (Freeman and Schneider, 1982). These changes did not take place if the odorants were unreinforced or if the CS was visual or auditory. A new pattern appeared with each conditioning to a new odor. We concluded that the EEG patterns manifested a change in attentiveness, specifically a change in expectancy relating to odorant stimulation (Freeman, 1983).

We extended the work to classical appetitive conditioning in which water-deprived rabbits were induced to lick in response to an odorant CS+ and not to CS-. In a pilot study to develop our techniques we introduced one CS+ odorant at a time and no CS-, finding the same results as with aversive conditioning (Viana Di Prisco and Freeman, 1985). Thereafter, we introduced both CSs to a new group of rabbits in the same starting session on randomly interspersed trials, and we reversed the contingencies in later sessions. We found that the rabbits responded with licking to the CS+ and not the CS-, but that they responded to both odorants with sniffing, which occurred at shorter latencies than for licking and at response rates to the CS previously found for pseudo-conditioning to odorants (Freeman et al., 1983). We observed EEG spatial pattern changes between sessions as before, but we also found pattern differences within trials between control ("air") and test ("odor") EEGs on the CS+ trials. We did not find differences between odorant EEG patterns on adjacent CS+ and CS- trials as previously found (Viana Di Prisco and Freeman, 1985). We used this same discrimination procedure but with aversive conditioning of rabbits in a pharmacological study of the bulb (Gray et al., 1986). Again the rabbits sniffed reliably (> 80%) on CS+ trials and at pseudoconditioning rates (50-60%) on CS- trials; EEG patterns changed under conditioning; a transient difference appeared between control and test EEGs; but no reliable difference emerged between CS+ and CS- EEG patterns.

In the study reported here under classical aversive conditioning we introduced each new CS+ or CS- after training to the previous CS for at least 3 sessions. We undertook the analysis of EEG patterns particularly in those sessions wherein a discriminative response was being generated as measured by sniffing to only one odor. Our specific aim was to compare EEG patterns under conditions of responding only to one odorant with EEG patterns under conditions of responding to either of two odors.

2. METHODS

Arrays of 64 electrodes (0.25 mm diameter stainless steel wire) were prefabricated (Eastman, 1975) in 8 x 8 format (0.5 mm center to center) and surgically placed epidurally over the lateral aspect of the left olfactory bulb of 6 adult rabbits. A 64 pin connector was cemented to the skull. Ground and reference electrodes for monopolar recording were fixed in the ipsilateral orbit. After recovery each rabbit was placed in a shielded restraining box and the connectors were mated. A cone was fitted over the muzzle for delivery of odorants from a dilution olfactometer (Freeman and Schneider, 1982) as conditioned stimuli (CS). Two odorants were used throughout odorant A was amyl acetate (1:10,000) and odorant B was butyric acid (1:2,500). Skin clips were attached to the cheek for delivery of current pulses as the unconditioned stimulus (US), 3-5 pulses at 100/sec with sufficient intensity to elicit a twitch of the eyelid as an unconditioned response (UR). A pneumograph was fitted around the chest to measure respiration (Freeman et al., 1983); an abrupt and short-lasting increase in respiratory rate (sniff) was used as the indicator of a conditioned response (CR) (Davis and Freeman, 1982). The level for "significant response rate" was 50% of 10 or more trials (Freeman et al., 1983). The rabbits were familiarized with this apparatus in 4 one-hour sessions at weekly intervals without US prior to further training.

Trials were conducted at intervals of 50-120 sec with 20-30 trials/session. Each trial lasted 6 sec with 3 sec control and test periods. The CS was given at the start of the test period and the US at the end. In a single start-up Session O a CS was given on 10 trials without reinforcement (CS_B-). In Session 1-3 there were 10 control trials with no odorant ("air") and 10 trials with odorant A reinforced with the US. In Sessions 4-6 odorant CS_B- was introduced on 10 trials randomly interspersed between 10 control trials and 10 trials with CS_A+. In Sessions 7-9 odorant A was omitted and odorant B became CS_B+. In Session 10-12 odorant A was re-introduced as CS_A-. At least 3 sessions at weekly intervals were given with each odorant combination. Additional sessions were given to some rabbits if the EEG spatial pattern appeared not to have stabilized by the third session. In sessions 11-12 from 15 to 40 additional trials were given with CS_A- in an effort to extinguish the sniff response.

The EEG was amplified with 64 preamplifiers, band pass filtered (10Hz/160Hz), digitized (8 bits, 400/sec) and stored with measurements of CS, CR, US and UR. Off-line the EEG traces were edited for artifacts with an oscilloscope. Segments 75 millisec in duration (30 time points) containing an oscillatory burst at 40-80 Hz with inhalations that immediately preceded and followed odorant arrival were measured and displayed visually. Three bursts were taken from each period labeled T1, T2, T3 starting with the first burst after odorant arrival and at about 0.5 sec intervals thereafter, and one to three bursts from the control period labeled C1, C2, C3. Then each burst was reduced to a 64x2 vector for the 64 channels (x,y), by computing its root mean square (rms) amplitude V(x,y) and the phase P(x,y) of the oscillatory burst at the common frequency with respect to the ensemble average of the 64 traces. The phase and frequency were determined with the Fast Fourier Transform (Childers, 1978) applied to each of the 64 traces and to the ensemble time average. Each was weighted by a Hamming window and padded with zeroes for a total of 256 time points. The peak frequency was found within the range of 40 - 80 Hz to a precision of ± 0.78 Hz, and the gain and phase were calculated from the real and imaginary parts at that frequency. The most effective EEG measure for detecting significant differences between 10 air bursts and 10 odorant bursts was the product $U(x,y) = V(x,y)\cos P(x,y)$. This was proposed as the best local measure of the level of bulbar activity that was received by the central structures to which the bulb sent its output (Freeman and Schneider, 1982).

Differences in patterns of U(x,y) between bursts with or without an odorant were sought over the sets of 10 trials each yielding from 4 to 6 bursts. The small sample pooled t-test was applied to the difference between the means of U(x,y) for each of the 64 channels (x,y) in each pair of bursts. A total of 2304 t-values (36 x 64) from 6 sets of 10 control burst comparisons in each of 6 animals was found to conform to the theoretical t-distribution (Figure 1). The same conformance was found for CS- odorant burst pairs but not for pairs of control and CS+ odorant bursts from selected sessions in which EEG control-odorant pattern differences were expected. The chi square (X^2) test was applied to the difference between the distribution of the 64 experimental t-values from each set of 10 burst-burst differences and the distribution of theoretical t-values. The distribution of X^2 values with no odorant was assembled over 72 air-air burst differences (6 rabbits and 12 sessions). A level of $X^2 = 100$ ($\log_{10} X^2 = 2.0$) was determined empirically as the 95% upper confidence limit for background variations (Figure 2). This is, any 2 sets of 10 bursts that yielded a value of $\log_{10} X^2 > 2.0$ were

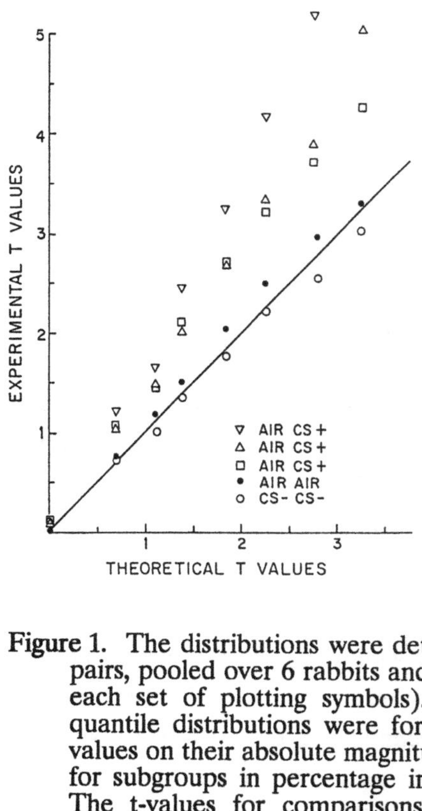

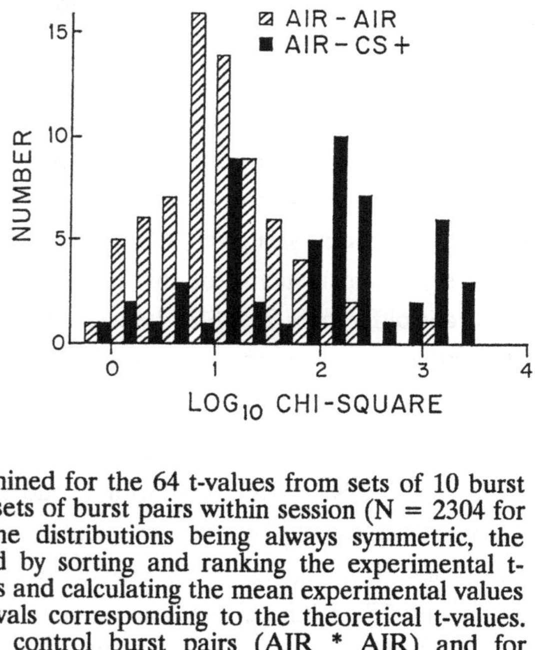

Figure 1. The distributions were determined for the 64 t-values from sets of 10 burst pairs, pooled over 6 rabbits and 6 sets of burst pairs within session (N = 2304 for each set of plotting symbols). The distributions being always symmetric, the quantile distributions were formed by sorting and ranking the experimental t-values on their absolute magnitudes and calculating the mean experimental values for subgroups in percentage intervals corresponding to the theoretical t-values. The t-values for comparisons of control burst pairs (AIR * AIR) and for comparisons of CS- burst pairs conformed to the theoretical t-distribution. Control-odorant CS+ burst pairs from sessions in which 2 new CS+ odorant was introduced and in which EEG pattern differences occurred did not conform. Three examples are shown. The statistical significance of the non-conformance was evaluated by use of the X^2 test.

Figure 2. Distributions are shown of X^2 values for air.air and air.CS+ EEG rms spatial pattern differences. The degrees of freedom in the use of X^2 test were not determined; an empirical distribution was formed from control burst pairs (N = 72), w2hich indicated that the 95% upper confidence level was $X^2 = 100$ $\log_{10} X^2 = 2$). This was consistent with previous findings in a larger sample (Freeman and Schneider, 1982). The distribution of 54 X^2 values from control-odorant CS+ burst pairs showed that 52% lay above this criterion.

inferred to differ significantly. As previously found (Freeman and Schneider, 1982) elevated X^2 values for some subjects were found not in the initial session following introduction of a new CS odorant but on the next succeeding. In combining results across subjects the X^2 values were taken from whichever of those two sessions yielded the highest values. Similarly some subjects showed elevations in X^2 values on presentation of a CS for three or more sessions. Up to three additional sessions were interpolated until a failure of increase in X^2 was observed, and these results were combined into the total.

3. RESULTS

3.1 Normalized amplitude pattern changes

The measurement of sniff responses and EEG differences are summarized in Figure 3 for the six rabbits. In Session 0 the presentation of an odorant without reinforcement (CS_B-) on 10 trials failed to elicit sniffing above the background rate, and there were no significant differences between control and CS_B- bursts on the same trials. In Session 1, wherein an odorant was given with reinforcement (CS_A+) for each of 6 rabbits there was a significant response of sniffing to the odor, and for all 6 rabbits there were significant differences in EEGs between control and CS_A+ bursts. By Session 3 the EEG differences were no longer significant, but the high response rates for the CS_A+ odorant persisted.

In Session 4 the unreinforced odorant (CS_B-) was reintroduced on randomly interspersed trials between those with no odorant or with CS_A+. There were significant sniff responses to CS_B- by all the rabbits. There were no significant differences between control and CS_B- bursts, but strong difference reappeared for all rabbits between control and CS_A+ bursts. By Session 6 these differences also became insignificant, and the behavioral responses continued to CS_A+ but not to CS_B-.

In Session 7 odorant A was deleted and odorant B was presented with reinforcement (CS_B+). The sniff response to CS_B+ appeared immediately and persisted through the session for all subjects, and for 4 of the 6 there was a significant difference between control and CS_B+ bursts. By Session 9 these EEG differences decreased below the significance level, but the behavioral response persisted to CS_B+.

In Session 10 the previously reinforced odorant A was reintroduced without reinforcement (CS_A-) while CS_B+ was continued on randomly interspersed trials; the

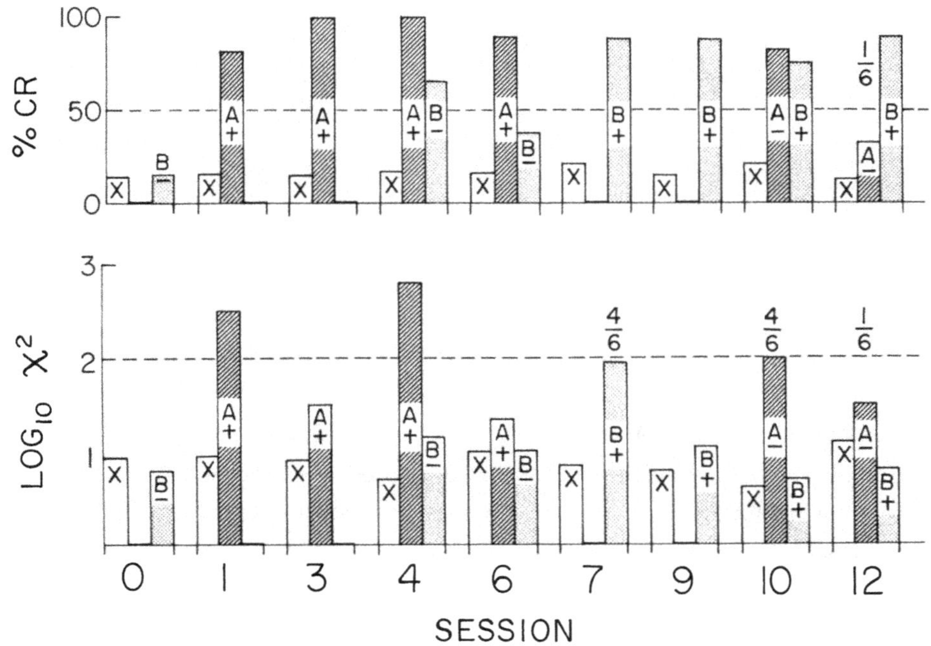

Figure 3. **Upper frame:** Mean response rates (relative frequency of sniffing) are pooled over 6 rabbits, each being the mean % of 10 trials on which the sniffing CR occurred. X denotes trials with no odorant CS; A and B denote the two odorant CSs. (+) denotes shock as reinforcement and (-) denotes no shock. The dashed lines indicate the levels of significant deviation from chance. The fractions show the proportion of subjects above chance levels when less than all or more than none. **Lower frame:** the mean $\log_{10}X^2$ values pooled for 6 subjects are shown for no odor, odorant A, and odorant B.

sniff response occurred for both CS_A- and CS_B+. There was no significant difference between control and CS_B+ bursts for any rabbit. For 4 of 6 rabbits there was a significant difference between control and CS_A- bursts. The difference did not occur for the first CS_A- bursts on odorant arrival, that is, with the first inspiration of air containing the odorant (average $\log_{10}X^2 = .85$) but with the second burst as shown in Figure 3 average ($\log_{10}X^2 = 1.98$) which took place between 0.2 and 0.7 sec after the odorant had arrived and had been inhaled. In Sessions 11 and 12 from 15 to 40 additional CS_A- trials were given to each animal prior to recording EEG data in the effort to extinguish the sniff response to CS_A- without impairing the responses to CS_B+. This was only partly successful. By the latter part of session 12 the EEG spatial pattern difference no longer occurred in 5 of the 6 rabbits, but it persisted in rabbit along with a significant sniff response rate to CS_A-.

Density plots were printed for the ensemble average rms patterns for each subject and session over the 10 control bursts and 3 sets of 10 test bursts for each CS odor. The method normalized the patterns by dividing the range into a 7-level graphic display. As shown by two examples in Figure 4 the rms pattern varied over sessions for each rabbit. Despite the variability across sessions for both control and odorant patterns the pattern characteristic for each subject (its "signature") persisted. A quantitative test of the degrees of similarity stability was applied across sessions; it was based on the product moment correlation coefficient, the chi square test being too sensitive for this purpose (Freeman and Schneider, 1982). A standard was first formulated by cross-correlating the 64 rms values of pairs of control * control bursts on adjacent trials from the same sessions for each animal in Sessions 6, 9, and 12 when the EEG patterns were apparently stabilized (Table 1). The average values by (z-transform) for the 10 pairs from each subject and session ranged from .79 to .96 with grand mean of r = .84. The correlation between animals averaged .19 with a range of -.27 to .42. Within Sessions 6 or 12 (at maximal stability of patterns) the average values for control * CS_A+ and control * $CS_{JB}+$ burst pairs were slightly but significantly higher, whereas those for

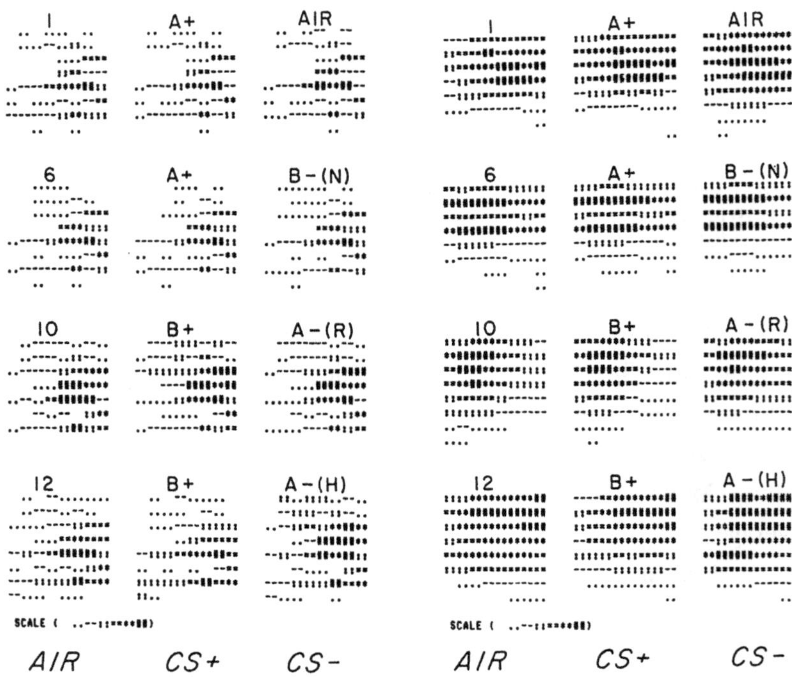

Figure 4. Two sets of density plots are shown from the designated sessions for two rabbits, each showing the mean for rms amplitude of 10 bursts in a 7-level intensity scale. Each rabbit had its characteristic spatial pattern (its "signature") that was distinct from any other. Differences were observed within and between sessions for each animal to the extent shown here. There were no persistences of spatial pattern for any odorant and no geometric regularities in the amounts or directions of change.

control * CS_A- and control * CS_B- burst pairs were not. The average values for CS_A+ * CS_A+ pairs and for CS_B+ * CS_B+ pairs both were higher, whereas those for CS_A- * CS_A- pairs were significantly lower. That is, under each CS+ the rms pattern tended to greater stability or reproducibility, despite the lower rms amplitude with burst depression, whereas under the CS_A-, which had previously been a CS+, there was a tendency to diminished correlation, implying decreased reproducibility.

Across sessions the mean values for control * control burst pairs ranged over subjects from .49 to .72 with a mean of .65, still well above the mean level of comparisons between subjects (r = .19). Of particular interest was the comparison between bursts with the same odorant in Sessions 6 and 10. The mean values both for

Table 1

burst pairs	$\bar{r}$	Δr	t	p
same session				
air.air	.844	-	-	
air.CS_A+	.891	+.047	3.48	.001
air.CS_B-	.847	+.003	- .22	ns
air.CS_B+	.873	+.029	2.15	.05
air.CS_A-	.823	-.021	-1.56	ns
CS+A.CS+A	.909	+.065	4.52	.001
CS_A-.CS_A-	.810	-.034	-2.52	.015
CS_B+.CS_B+	.883	+.039	2.89	.005
CS_B-.CS_B-	.840	-.004	- .30	ns
CS_A+.CS_B-	.802	-.042	-3.32	.001
CS_B+.CS_A-	.788	-.056	-3.91	.001
across sessions				
air.air	.653	-	-	
CS_A+.CS_A-	.547	-.106	-12.83	.001
CS_B+.CS_B-	.557	-.096	-11.62	.001

Table 1. These are the means r (by Fisher's z-transform) of the product moment correlations of the 64 rms amplitudes pooled over 30 burst pairs for each of 6 subjects (d.f. = 10, 980); the changes in correlation r from the air.air standard (.844); the t-values of the differences; and the probability p of occurrence by chance. The control burst pairs (air . air) served as a reference standard. The main findings were the relatively high within-session correlations (reflecting the characteristic spatial pattern "signature" of each rabbit); the lower correlation across sessions (reflecting changes in spatial pattern with learning); the still lower correlations between pairs of bursts with the same odorants but with reversed contingencies; and the tendencies for CS+ odorant bursts to have slightly higher coherence and CS- bursts slightly lower coherence than the standard.

$CS_A+ * CS_A-$ and for $CS_B+ * CS_B-$ comparisons were significantly lower, averaging .55. This result supported the conclusion from inspection of density plots that with neither odorant A nor odorant B was a spatial EEG pattern conserved over changes in odorant discrimination.

3.2 Changes in amplitude, frequency and phase with conditioning.

The rms amplitude of the 64 channel x 30 time point traces was used routinely for estimating burst amplitude. The pooled means for the 6 rabbits are shown in Figure 5. Over Sessions 4 to 6 with CS_A+ and CS_B- there was a decline in control amplitude for 5 of the 6 rabbits, but on the introduction of the new CS_B+ the decline was reversed. Particularly during Session 4 the amplitude decreased further on presentation of the CS_A+ and the CS_B-. The average decrease in amplitude with CS_B+ was not statistically significant in Sessions 10 to 12. The frequency of oscillation in bursts in the control periods declined in all animals over the 12 sessions on the average by about 10%. In those sessions in which the arrival of the CS resulted in a decrease in amplitude, a transient decrease in burst frequency also took place. Despite this obvious relation to the CS there was near zero correlation between amplitude and frequency among sets of bursts within control or within test periods for all of the animals.

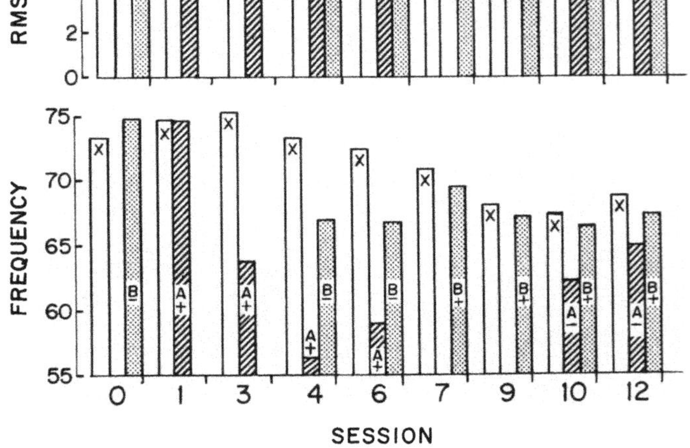

Figure 5. The rms amplitude of burst in control periods (open bars) decreased across Sessions 4 to 6 and across Sessions 10 to 12 in which two CS odorants were given alternately on randomly interspersed trials. The burst frequency in Hz of control bursts on the average declined over the sessions. Large, transient decreases in burst amplitude and frequency occurred with CS_A+ but not with CS_B+. The data are the pooled means of 6 rabbits.

Further study was focussed on three sets of measurements of rms amplitude and frequency from sessions in which some degree of olfactory discrimination was displayed by the rabbits in respect to CS+ and CS- odors. These included from the present study Sessions 5 and 6 (denoted A+,B-), Sessions 11 and 12 (B+,A-), and sessions 16 to 18 of the previously reported study (Freeman and Viana Di Prisco, 1985) on appetitive conditioning (C+,A-). These sets differeds from each other in respect to the CR as well as the CSs and USs. In the first set the rabbits had never learned to respond to CS_B-. In the second set the CR to CS_A- was attenuated or extinguished. In the third set the rabbits responded reliably to CS_C+ and less frequently to CS_A- (at the level of pseudoconditioning).

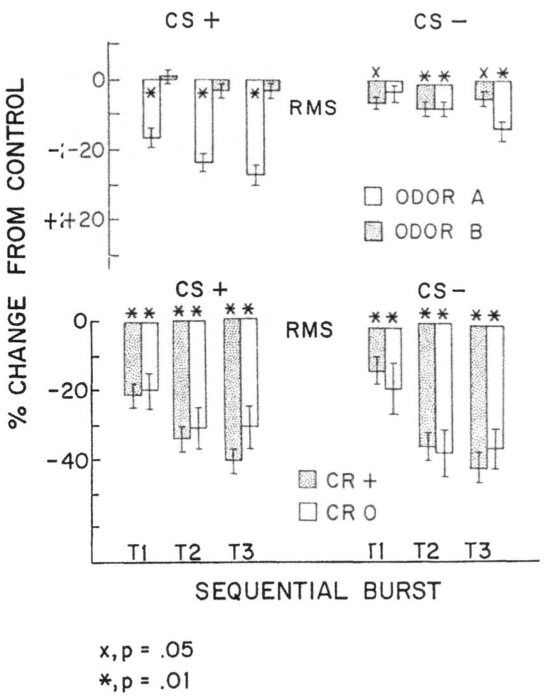

Figure 6. The changes in peak frequency (mean and S.E. of % change from control values on the same trials) are shown for the three successive odorant bursts in test trials (T1, T2, T3) pooled over 6 rabbits. X, p= .05; *, p = .01.

Upper graphs: Data were from Sessions 5-6 and 11-12 of the present study, with CS+ on the left and CS- on the right.

Lower graphs: Data were taken from the last 3 Sessions (16-18) of the previous study (Viana Di Prisco and Freeman, 1985) pooled over 5 rabbits. The values were segregated by whether a licking CR occurred to the CS+ (CR+) or not (CR 0) and whether a sniffing CR occurred to the CS- (CR+) or not (CR0).

The search for burst peak frequency was restricted to the range greater than 40 Hz in order to concentrate on the dominant characteristic frequency and to avoid other peaks in the multimodal distribution of frequencies in the bulbar EEG. The burst peak frequency was measured from the spectrum of the burst ensemble average and also from the average of the 64 peak frequencies from the spectra of the ensemble. These two estimates differed by as much as ± 8% and on the average by ± 3%. The grand mean coefficient of variation of channel frequency (SD divided by mean channel frequency) was about 10%. Inspection of sets of the EEG traces showed that each channel had the same number of peaks and zero crossings as all others within each burst; by theory, as well, the ensemble had a common instantaneous frequency (Freeman, 1979c). Therefore, the frequency SD was treated as an estimate of error of measurement of channel frequency, and the ensemble average peak frequency was adopted as the best estimator of burst peak frequency. Analysis of variance over the 6 rabbits under aversive conditioning showed that they differed in two respects, first that one rabbit had a lower mean peak frequency than the others, and second that the amount of frequency decrease with CSs for this rabbit was lower than for the others. Unlike the others it was not by itself statistically significant. The directions of change being the same, the data were pooled. In set (A+,B-) the mean peak frequency in all rabbits decreased with CS_A+ (Figure 6, upper graphs) but did not change significantly with CS_B-. In set (B+,A-) the mean frequency decreased with CS_B+ to a lesser extent and significantly at p = .01 only on T3,. whereas it continued to decrease strongly with CS_A-. In set (C+,A-) the mean peak frequency decreased for both CS_C+ and CS_AA- (Figure 6, lower graphs), although the change was not statistically significant for the first test burst T1, and for CS_C+ it was significant only on trials in which the licking CR occurred.

As noted in Section 3.1 the rms amplitude in set (A+,B-) decreased significantly with CS_A+ and also slightly with CS_B-. The decrease was observed as early as the first test burst T1 (Figure 7, upper graph). In set (B+,A-) the decrease was significant for CS_A- prior to extinction in T2 and T3 but not for CS_B+ nor for CS_A- following extinction of the CR in those 5 of the 6 rabbits in which the procedure succeeded (not shown in Figure 7). In set (C+,A-) the reduction occurred with both CSs (Figure 7, lower graphs) whether or not a CR occurred, either sniffing or licking.

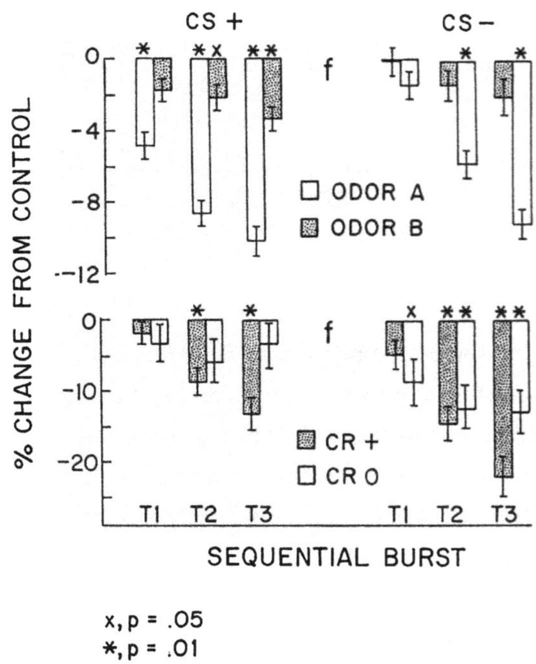

Figure 7. The changes in rms burst amplitude are shown as % change from control values on the same trails (mean and S.E.). See legend to Figure 6.

Analysis of variance again showed that the rabbits differed in respect to mean burst amplitude, and that the amount of decrease was significantly less in those with lower mean amplitude; the directions of change being the same, the date were pooled. In brief, reduction of rms amplitude and burst frequency occurred together in trials with either CS+ or CS- in circumstances in which a CR was likely to take place whether or not it did so.

The mean value for phase at the peak frequency in each burst of course tended to zero, the phase reference being the phase of the peak frequency of ensemble average of each burst. The measure of interest was the dispersion of phase values about the mean and how this changed with presentation of odors. As shown by example in Figure 8 the phase values appeared to be normally distributed about zero mean, so the standard deviation (SD) was used to estimate phase dispersion. The grand average over rabbits from the pooled variance was 26° with a range over rabbits from 19° to 44°. Those with lower mean rms burst amplitudes had higher phase SDs. The only significant effect of odorant presentation of phase SD was an increase, and then only in the presence of an rms amplitude decrease as in the example in Figure 8. This increase in SD was traced directly to an increase in the error of measurement of phase with decrease in the gain at the peak frequency relative to the average gain over the dispersion with CS+ presentation as previously predicted (Freeman, 1979c).

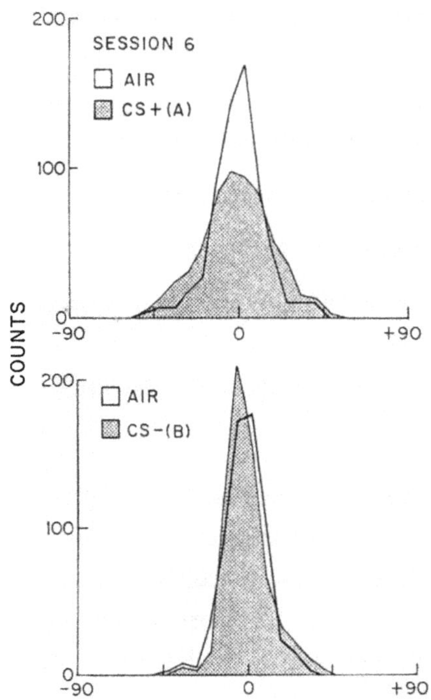

Figure 8. Representative distributions are shown for 640 phase values from 10 bursts of a rabbit. They appeared to approximate the normal density function about zero mean. The increase in the scatter of phase values in odorant bursts (T3) was due to an increase in the errors of measurement of phase at lower burst amplitudes and frequencies. No other changes in phase or phase patterns were found associated with presentations of CSs.

Three other aspects of the data that were also explored are noteworthy. First, the channel amplitudes at the peak frequency were examined in the same way as the channel rms amplitudes in regard to evaluation of burst amplitude and pattern changes, yielding the same results. Second, the nonlinear mapping algorithm of Sammon (1969) used in the preceding study f(Freeman and Viana Di Prisco, 1985) for classification of bursts in respect to odorant condition here also confirmed the results using the X^2 test; further applications to the present data similarly confirmed the outcomes of the X^2 test, including failure of burst patterns to show reliable and significant differences between CS+ and CS- conditions within Sessions. Third, the amplitudes of peaks in the frequency range from 10 to 40 Hz were examined in relation to odorant condition. Briefly, when the burst rms amplitude decreased as described for various CS conditions, the reduction on the average was concentrated in the upper frequency range; again on the average there were small relative and absolute increases in the power contained in the lower frequency range, but with no distinction between CS+ and CS-.

4. DISCUSSION

The main finding of this study confirmed previous results that the spatial patterns of the bulbar EEG did not reflect direct and immediate dependence on odorant input. This was most clearly shown in 5 instances:

(a) the lack of differency by X^2 test for B- in Sessions 0, 4 and 6

(b) the difference by X^2 for A+ in Session 1 and not in Session 3;

(c) the difference by X^2 for A+ in Session 4 when B- was reintroduced;

(d) the difference in X^2 for B+ in Session 7 when B was changed from CS- to CS+; and

(e) the differency in X^2 for A- that occurred several hundred millisec after odorant A had arrived in Session 10.

The characteristic spatial pattern or "signature" for each rabbit was preserved under these changes in detailed shape. The short-term and long-term patterns were neither regular nor progressive. There was no evidence for the preservation or retention of a particular spatial pattern with a recurring odorant CS. Correlations among spatial patterns of rms amplitude revealed a slight increase in spatial coherence with CS+ odorants and a slight decrease with CS- odors. We did not find distinctive patterns of rms amplitude between CS+ and CS- odorant bursts on adjacent trials, although the pair-wise correlation between control bursts on adjacent trials. This difference in correlation could be attributed to the fact that odorant bursts on the average had lower amplitude and might be "noisier" than control bursts, except that the CS+ odorant bursts (also lower in amplitude) had a higher average correlation within the group.

The characteristic EEG change with odorant onset consisted of decreases in burst amplitude and frequency and an increase in phase SD, the latter being attributed to increased error of measurement at lower amplitudes and frequencies of oscillation. (Preliminary results suggesting that phase might converge (Freeman, 1980) were traced to an artifact of data processing. The phase was previously computed from the correlation of the wave form (which was then measured with a 6-bit instead of an 8-bit ADChere) on each channel with the ensemble average. When the amplitude decreased with burst suppression, the coarse-graining effect of the ADC increased; this inflated the correlation co-efficients and caused an illusory decrease in phase dispersion with the CS+). This configuration occurred on initial presentations of novel odorants f(Gray et al., 1986) but abated over 2-3 presentations and typically was no longer detectable by the 4th presentation, provided that no reinforcement was added or changed in conjunction with the novel odor. Extinction of the sniff response to odorants previously reinforced required from 20 to 40 presentations repeatedly on 3 or more sessions, under conditions where an occasional CS+ odorant trial was given with shock in order to maintain responding. The procedure was not always successful, but if the sniff response was extinguished, the transient EEG change also disappeared.

The main difference in the design of this experiment from two preceding studies (Viana Di Prisco and Freeman, 1985; Gray et al., 1986) lay in the introduction of one new odorant or US contingency in the first of several sessions, as compared with the previous introduction fo both CS+ and CS- in the first of several sessions. The main difference in results were two-fold. Whereas previously the rabbits sniffed in response to both odors, though less reliably to the CS- than to the CS+, here they sniffed only to the CS+ after familiarization or extinction with the CS-. And whereas previously the EEG rms spatial pattern showed a statistically significant difference between control and test bursts persisting for 6 sessions, here that difference became statically insignificant by the third session with a CS-. Differences persisted between control and odorant bursts in the present study, but they were no greater than those among sets of control bursts.

Our present interpretation of these combined results is that the bulb serves as an adaptive filter for receptor input that can be modified to correspond to the expectation of input. The first requirement is to detect difference from the background odorant complex. This implies that a mechanism exists for the identification for the control input, such as that departures of any kind can be detected, including novel odors. The fact that regular, coherent bursts occur with most inhalations in waking, motivated animals indicated that the majority of control bursts manifest a signal from the bulb to the brain that the status quo holds. The sudden decrease in EEG frequency and amplitude that accompanies a novel odorant (On the first few trials) may suffice to signal the failure of the status quo. Sniffing invariably results. Without reinforcement these responses appear to habituate rapidly; under reinforcement or prior to extinction we see continued orienting and the development of EEG pattern changes.

In Conditions where 2 or more odorants are to be discriminated, this mechanism cannot suffice. In conditions where the identification of a single odorant is paramount (as in Sessions 1-9 and in Session 12 after extinction), it may suffice. That is, detection may not be followed by convergence to a discrimination. This may account for the disappearance of control-odorant pattern differences several sessions after the introduction of a new CS+ (or CS-), though the need for repeated sessions (up to 5 or 6 total) indicates that reliance on this short-cut may require practice.

We conclude that the bulb maintains a mechanism for detecting and discriminating the background odorant complex as the basis for detecting novel odors. Rapid habituation to CS- odorants may give the appearance of odorant discrimination of a CS+ odor, whereas if only one CS+ is used to elicit one type of CR, identification may be based on "deviation from the background" rather than on selective responding to the CS+. Therefore, the search for the neural mechanisms of odorant discrimination must be done in conditions of differential responding to two or more odors. The findings in this study have led us to new techniques for analyzing the EEG (Freeman, 1985) and a new dynamic basis for comprehending brain function (Freeman and Skarda, 1985). Present evidence shows that EEG pattern difference do distinguish bursts with CS+ and CS- odorants (Freeman and Viana di Prisco, 1986) but only in discriminative learning.

REFERENCES

Adrian, E.D. (1950) The electrical activity of the mammalian olfactory bulb. Electroencephalog. clin. Neurophysiol. 2:377-388

Childers, D.G. (1978) Modern Spectrum Analysis. New York, IEEE Press

Davis, G.W. and Freeman, W.J. (1982) On-line detection of respiratory events applied to behavioral conditioning of rabbits. IEEE Trans. Biomed. Engineering BNE 29 (6):453-456

Freeman, W.J. (1975) Mass Action in the Nervous System. New York, Academic

Freeman, W.J. (1978) Spatial properties of an EEG event in the olfactory bulb and cortex. Electroencephalog. clin. Neurophysiol. 44:586-605

Freeman, W.J. (1979) EEG analysis gives model of neuronal template-matching mechanism for sensory search with olfactory bulb. Biological Cybernetics 35:221-234

Freeman, W.J. (1983) The physiological basis of mental images. Academic Address. Biological Psychiatry 18:1107-1125

Freeman, W.J. (1987) Analytic techniques used in the search for the physiological basis of the EEG. Gevins, A. and Remond, A. (eds) Handbook of Electroencephalography and clinical Neurophysiology, Vol. 3A, Part 2, Ch. 18. Amsterdam, Elsevier. pp. 583-664

Freeman, W.J. and Schneider, W.S. (1982) Changes in spatial patterns of rabbitolfactory EEG with conditioning to odors. Psychophysiol. 19:44-56

Freeman, W.J. and Skarda, C.A. (1985) Spatial EEG patterns, nonlinear dynamics and perception. Brain Research Reviews 10:147-175

Freeman, W.J. and Viana Di Prisco, G. (1986) EEG spatial pattern differences with discriminated odorants manifest chaotic and limit cycle attractors in olfactory bulb of rabbits. Proceedings, Conference on Brain Theory, Trieste 1984. Berlin, Springer-Verlag. pp. 97-119

Freeman, W.J., Viana Di Prisco, G., Davis, G.W. and Whitney, T.M. (1983) Conditioning of relative frequency of sniffing by rabbits to odors. J. Comp. Psychol. 97:12-23

Gray, C.M., Freeman, W.J. and Skinner, J.E. (1986) Chemical dependencies oflearning in the rabbit olfactory bulb: Acquisition of the transientspatial pattern change depends on norepinephrine. Behavioral Neuroscience 100:585-596

Sammon, J.W. (1969) A nonlinear mapping for data structure analysis. IEEE Trans. on Computers 18:401-409

Viana Di Prisco, G. and Freeman, W.J. (1985) Odor-related bulbar EEG spatial pattern analysis during appetitive conditioning in rabbits. Behavioral Neuroscience 99:964-978

Index

NATO ASI Series H

Vol. 1: **Biology and Molecular Biology of Plant-Pathogen Interactions.** Edited by J.A. Bailey. 415 pages. 1986.

Vol. 2: **Glial-Neuronal Communication in Development and Regeneration.** Edited by H.H. Althaus and W. Seifert. 865 pages. 1987.

Vol. 3: **Nicotinic Acetylcholine Receptor: Structure and Function.** Edited by A. Maelicke. 489 pages. 1986.

Vol. 4: **Recognition in Microbe-Plant Symbiotic and Pathogenic Interactions.** Edited by B. Lugtenberg. 449 pages. 1986.

Vol. 5: **Mesenchymal-Epithelial Interactions in Neural Development.** Edited by J.R. Wolff, J. Sievers, and M. Berry. 428 pages. 1987.

Vol. 6: **Molecular Mechanisms of Desensitization to Signal Molecules.** Edited by T.M. Konijn, P.J.M. Van Haastert, H. Van der Starre, H. Van der Wel, and M.D. Houslay. 336 pages. 1987.

Vol. 7: **Gangliosides and Modulation of Neuronal Functions.** Edited by H. Rahmann. 647 pages. 1987.

Vol. 8: **Molecular and Cellular Aspects of Erythropoietin and Erythropoiesis.** Edited by I.N. Rich. 460 pages. 1987.

Vol. 9: **Modification of Cell to Cell Signals During Normal and Pathological Aging.** Edited by S. Govoni and F. Battaini. 297 pages. 1987.

Vol. 10: **Plant Hormone Receptors.** Edited by D. Klämbt. 319 pages. 1987.

Vol. 11: **Host-Parasite Cellular and Molecular Interactions in Protozoal Infections.** Edited by K.-P. Chang and D. Snary. 425 pages. 1987.

Vol. 12: **The Cell Surface in Signal Transduction.** Edited by E. Wagner, H. Greppin, and B. Millet. 243 pages. 1987.

Vol. 13: **Toxicology of Pesticides: Experimental, Clinical and Regulatory Perspectives.** Edited by L.G. Costa, C.L. Galli, and S.D. Murphy. 320 pages. 1987.

Vol. 14: **Genetics of Translation. New Approaches.** Edited by M.F. Tuite, M. Picard, and M. Bolotin-Fukuhara. 524 pages. 1988.

Vol. 15: **Photosensitisation. Molecular, Cellular and Medical Aspects.** Edited by G. Moreno, R.H. Pottier, and T.G. Truscott. 521 pages. 1988.

Vol. 16: **Membrane Biogenesis.** Edited by J.A.F. Op den Kamp. 477 pages. 1988.

Vol. 17: **Cell to Cell Signals in Plant, Animal and Microbial Symbiosis.** Edited by S. Scannerini, D. Smith, P. Bonfante-Fasolo, and V. Gianinazzi-Pearson. 414 pages. 1988.

Vol. 18: **Plant Cell Biotechnology.** Edited by M.S.S. Pais, F. Mavituna, and J.M. Novais. 500 pages. 1988.

Vol. 19: **Modulation of Synaptic Transmission and Plasticity in Nervous Systems.** Edited by G. Hertting and H.-C. Spatz. 457 pages. 1988.

Vol. 20: **Amino Acid Availability and Brain Function in Health and Disease.** Edited by G. Huether. 487 pages. 1988.

NATO ASI Series H

Vol. 21: **Cellular and Molecular Basis of Synaptic Transmission.**
Edited by H. Zimmermann. 547 pages. 1988.

Vol. 22: **Neural Development and Regeneration. Cellular and Molecular Aspects.**
Edited by A. Gorio, J.R. Perez-Polo, J. de Vellis, and B. Haber. 711 pages. 1988.

Vol. 23: **The Semiotics of Cellular Communication in the Immune System.**
Edited by E.E. Sercarz, F. Celada, N.A. Mitchison, and T. Tada. 326 pages. 1988.

Vol. 24: **Bacteria, Complement and the Phagocytic Cell.**
Edited by F.C. Cabello und C. Pruzzo. 372 pages. 1988.

Vol. 25: **Nicotinic Acetylcholine Receptors in the Nervous System.**
Edited by F. Clementi, C. Gotti, and E. Sher. 424 pages. 1988.

Vol. 26: **Cell to Cell Signals in Mammalian Development.**
Edited by S.W. de Laat, J.G. Bluemink, and C.L. Mummery. 322 pages. 1989.

Vol. 27: **Phytotoxins and Plant Pathogenesis.**
Edited by A. Graniti, R.D. Durbin, and A. Ballio. 508 pages. 1989.

Vol. 28: **Vascular Wilt Diseases of Plants. Basic Studies and Control.**
Edited by E.C. Tjamos and C.H. Beckman. 590 pages. 1989.

Vol. 29: **Receptors, Membrane Transport and Signal Transduction.**
Edited by A.E. Evangelopoulos, J.P. Changeux, L. Packer. T.G. Sotiroudis, and K.W.A. Wirtz. 387 pages. 1989.

Vol. 30: **Effects of Mineral Dusts on Cells.**
Edited by B.T. Mossman and R.O. Bégin. 470 pages. 1989.

Vol. 31: **Neurobiology of the Inner Retina.**
Edited by R. Weiler and N.N. Osborne. 529 pages. 1989.

Vol. 32: **Molecular Biology of Neuroreceptors and Ion Channels.**
Edited by A. Maelicke. 675 pages. 1989.

Vol. 33: **Regulatory Mechanisms of Neuron to Vessel Communication in Brain.**
Edited by F. Battaini, S. Govoni, M.S. Magnoni, and M. Trabucchi. 416 pages. 1989.

Vol. 34: **Vectors as Tools for the Study of Normal and Abnormal Growth and Differentiation.**
Edited by H. Lother, R. Dernick, and W. Ostertag. 477 pages. 1989.

Vol. 35: **Cell Separation in Plants: Physiology, Biochemistry and Molecular Biology.**
Edited by D.J. Osborne and M.B. Jackson. 449 pages. 1989.

Vol. 36: **Signal Molecules in Plants and Plant-Microbe Interactions.**
Edited by B.J.J. Lugtenberg. 425 pages. 1989.

Vol. 37: **Tin-Based Antitumour Drugs.** Edited by M. Gielen. 226 pages. 1990.

Vol. 38: **The Molecular Biology of Autoimmune Disease.**
Edited by A.G. Demaine, J-P. Banga, and A.M. McGregor. 404 pages. 1990.

Vol. 39: **Chemosensory Information Processing.** Edited by D. Schild. 403 pages. 1990.

www.ingramcontent.com/pod-product-compliance
Lightning Source LLC
Chambersburg PA
CBHW061154240926
51846CB00006B/1014
* 9 7 8 3 6 4 2 7 5 1 2 9 5 *